Current Topics in Developmental Biology

Volume 30

Series Editors

Roger A. Pedersen   and   Gerald P. Schatten
Laboratory of Radiobiology   Department of Zoology
and Environmental Health   University of Wisconsin, Madison
University of California   Madison, WI 53706
San Francisco, CA 94143

Editorial Board

John C. Gerhart
University of California, Berkeley

Peter Grüss
Max Planck Institute of Biophysical Chemistry,
Göttingen-Nikolausberg, Germany

Philip Ingham
Imperial Cancer Research Fund, Oxford, United Kingdom

Story C. Landis
Case Western Reserve University, Cleveland, Ohio

David R. McClay
Duke University, Durham, North Carolina

Yoshitaka Nagahama
National Institute for Basic Biology, Okazaki, Japan

Susan Strome
Indiana University, Bloomington

Virginia Walbot
Stanford University

Mitsuki Yoneda
Kyoto University

Founding Editors

A.A. Moscona
Alberto Monroy

Current Topics in Developmental Biology
Volume 30

Edited by

Roger A. Pedersen
Laboratory of Radiobiology
and Environmental Health
University of California, San Francisco
San Francisco, California

Gerald P. Schatten
Department of Zoology
University of Wisconsin, Madison
Madison, Wisconsin

Academic Press
San Diego New York Boston London Sydney Tokyo Toronto

Front cover photograph: Photomicrograph of mouse sperm bound to the zona pellucida of an unfertilized mouse egg *in vitro*. Photo courtesy of Wassarman and Litscher. (For more details see Chapter 1, Figure 1.)

This book is printed on acid-free paper. ∞

Copyright © 1995 by ACADEMIC PRESS, INC.

All Rights Reserved.
No part of this publication may be reproduced or transmitted in any form or by any means, electronic or mechanical, including photocopy, recording, or any information storage and retrieval system, without permission in writing from the publisher.

Academic Press, Inc.
A Division of Harcourt Brace & Company
525 B Street, Suite 1900, San Diego, California 92101-4495

United Kingdom Edition published by
Academic Press Limited
24-28 Oval Road, London NW1 7DX

International Standard Serial Number: 0070-2153

International Standard Book Number: 0-12-153130-9

PRINTED IN THE UNITED STATES OF AMERICA
95 96 97 98 99 00 BC 9 8 7 6 5 4 3 2 1

Contents

Contributors ix
Notice xi
Preface xiii

1

Sperm–Egg Recognition Mechanisms in Mammals
Paul M. Wassarman and Eveline S. Litscher

    I. Introduction   1
   II. Characteristics of the Zona Pellucida   4
  III. ZP3 Functions during Fertilization   4
  IV. ZP3 Genes from Different Mammalian Species   4
   V. ZP3 Polypeptides from Different Mammalian Species   6
  VI. Oligosaccharides at the ZP3 Combining Site for Sperm   8
 VII. Structure of Oligosaccharides at the ZP3 Combining Site for Sperm   10
VIII. Location of the ZP3 Combining Site for Sperm   11
  IX. Does the ZP3 Polypeptide Consist of Two Domains?   13
   X. Final Comments   15
      References   16

2

Molecular Basis of Mammalian Egg Activation
Richard M. Schultz and Gregory S. Kopf

    I. Introduction   21
   II. Description of Fertilization of Mammalian Eggs   23
  III. Intracellular Second Messengers That Regulate Egg Activation   26
  IV. Maturation-Associated Changes in CG Exocytosis in Response to Second Messengers   34
   V. Signal Transduction of Pathways Leading to Egg Activation   37
  VI. Receptors on the Egg That Mediate Sperm-Induced Egg Activation   47
 VII. Sperm Surface Ligands Involved in Egg Activation   51
VIII. Is There a Need for an Egg-Specific Receptor and a Sperm-Specific Ligand in Egg Activation?   52
      References   53

3

Mechanisms of Calcium Regulation in Sea Urchin Eggs and Their Activities during Fertilization
Sheldon S. Shen

    I. Changes in $[Ca^{2+}]_i$ during Fertilization   65
    II. Regulation of Intracellular Ca^{2+} Release   71
    III. How do Sperm Activate the Rise in $[Ca^{2+}]_i$?   82
    IV. Conclusions   87
        References   89

4

Regulation of Oocyte Growth and Maturation in Fish
Yoshitaka Nagahama, Michiyasu Yoshikuni, Masakane Yamashita, Toshinobu Tokumoto, and Yoshinao Katsu

    I. Introduction   103
    II. Gonadotropin   105
    III. Structure of Follicles   105
    IV. Oocyte Growth   108
    V. Oocyte Maturation   117
    VI. Summary   137
        References   138

5

Nuclear Transplantation in Mammalian Eggs and Embryos
Fang Zhen Sun and Robert M. Moor

    I. Introduction   147
    II. Historical Background   149
    III. Nuclear Transplantation Procedures   151
    IV. Early Events in Cells Reconstituted by Nuclear Transplantation   154
    V. Biological Factors Influencing Development of Cells Reconstituted by Nuclear Transplantation   156
    VI. Serial Nuclear Transplantation   166
    VII. Epigenetic Modifications of Genomic Totipotency   167
    VIII. Summary and Prospects   168
        References   170

Contents

6

Transgenic Fish in Aquaculture and Developmental Biology
Zhiyuan Gong and Choy L. Hew

    I. Introduction   177
    II. Fish as a Transgenic System   179
    III. Production of Transgenic Fish: Methodology   180
    IV. Transgenesis in Fish: Integration, Inheritance, and Expression   187
    V. Application of Transgenic Fish in Aquaculture   192
    VI. Application of Transgenic Fish in Developmental Biology   197
    VII. Concluding Remarks   205
         References   206

7

**Axis Formation during Amphibian Oogenesis:
Re-evaluating the Role of the Cytoskeleton**
David L. Gard

    I. Introduction   215
    II. Stage VI *Xenopus* Oocytes Are Structurally and Functionally Polarized along the A–V Axis   218
    III. Cytoskeletal Organization and Axis Specification during Early Oogenesis   226
    IV. Cytoplasmic and Cytoskeletal Polarity Are Established during Stage IV of Oogenesis   235
    V. Ectopic Spindle Assembly during Maturation of *Xenopus* Oocytes: Evidence for Functional Polarization of the Oocyte Cortex   241
    VI. Concluding Remarks   244
         References   246

8

**Specifying the Dorsoanterior Axis in Frogs:
70 Years since Spemann and Mangold**
Richard P. Elinson and Tamara Holowacz

    I. Introduction   253
    II. How Does the Organizer Form?   262
    III. How Does the Organizer Act?   274
    IV. Concluding Remarks   277
         References   278

Index   287

Contributors

Numbers in parentheses indicate the pages on which the authors' contributions begin.

Richard P. Elinson Department of Zoology, University of Toronto, Toronto, Ontario, Canada M5S 1A1 (253)

David L. Gard Department of Biology, University of Utah, Salt Lake City, Utah 84112 (215)

Zhiyuan Gong Research Institute, Hospital for Sick Children, and Departments of Clinical Biochemistry and Biochemistry, University of Toronto, Toronto, Ontario, Canada M5G 1L5 (177)

Choy L. Hew Research Institute, Hospital for Sick Children, and Departments of Clinical Biochemistry and Biochemistry, University of Toronto, Toronto, Ontario, Canada M5G 1L5 (177)

Tamara Holowacz Department of Zoology, University of Toronto, Toronto, Ontario, Canada M5S 1A1 (253)

Yoshinao Katsu Laboratory of Reproductive Biology, Department of Developmental Biology, National Institute for Basic Biology, Okazaki 444, Japan (103)

Gregory S. Kopf Division of Reproductive Biology, Department of Obstetrics and Gynecology, University of Pennsylvania, Philadelphia, Pennsylvania 19104 (21)

Eveline S. Litscher Roche Institute of Molecular Biology, Roche Research Center, Nutley, New Jersey 07710 (1)

Robert M. Moor Development and Differentiation Laboratory, Babraham Institute, Babraham, Cambridge CB2 4AT, England (147)

Yoshitaka Nagahama Laboratory of Reproductive Biology, Department of Developmental Biology, National Institute for Basic Biology, Okazaki 444, Japan (103)

Richard M. Schultz Department of Biology, University of Pennsylvania, Philadelphia, Pennsylvania 19104 (21)

Sheldon S. Shen Signal Transduction Training Group, Department of Zoology and Genetics, Iowa State University, Ames, Iowa 50011 (63)

Fang Zhen Sun Development and Differentiation Laboratory, Babraham Institute, Babraham, Cambridge CB2 4AT, England (147)

Toshinobu Tokumoto Laboratory of Reproductive Biology, Department of Developmental Biology, National Institute for Basic Biology, Okazaki 444, Japan (103)

Paul M. Wassarman Roche Institute of Molecular Biology, Roche Research Center, Nutley, New Jersey 07710 (1)

Masakane Yamashita Laboratory of Reproductive Biology, Department of Developmental Biology, National Institute for Basic Biology, Okazaki 444, Japan (103)

Michiyasu Yoshikuni Laboratory of Reproductive Biology, Department of Developmental Biology, National Institute for Basic Biology, Okazaki 444, Japan (103)

Notice

The Editor and Editorial Board would like to encourage authors of topical reviews in any aspect of developmental biology to submit them for consideration for publication in *Current Topics in Developmental Biology*. Such submissions will be peer-reviewed by members of the Editorial Board or external reviewers, at the Editor's discretion. Authors with questions about this process may wish to contact the Editor(s) directly in writing or by facsimile [(415) 476-6951, Dr. Pedersen; or (608) 262-7319, Dr. Schatten]. If possible, please include an abstract of the proposed manuscript in this initial correspondence, including information about manuscript length and number and type of illustrations.

Preface

The field of developmental biology is burgeoning as a result of the fantastic progress derived from molecular, cellular, genetic, and biophysical approaches, and the implications of these discoveries affect all of us. The study of development is of importance not only for the scientists and students interested in the fundamental mechanisms that transform a fertilized egg into a highly differentiated adult, but also for the patients and physicians in clinical settings, farmers striving for improved crops or breeding stock, and even aquaculturalists seeking swiftly growing, freeze-resistant salmon.

This volume explores the manner in which the quiescent, unfertilized egg is activated during fertilization, and how it is transformed into a differentiated organism with an intricate body plan. The chapter by Wassarman and Litscher reviews the mechanisms of sperm–egg recognition. Egg activation is considered by Schultz and Kopf in mammals, and by Shen in sea urchin eggs; these chapters focus on signal transduction and calcium regulation, respectively. The field of cell cycle regulation largely owes its origins to investigations on oocytes and eggs, and Nagahama, Yamashita, Tokumoto, and Katsu consider oocyte growth and maturation in fish. The transfer of nuclei from various-staged blastomeres into unfertilized oocytes has led to the discovery of genomic imprinting in mammals, as well as various commercial applications in domestic species, and Sun and Moor review the implications of these powerful approaches. A refinement to bulk nuclear transfer is the introduction of specific genes into oocytes, and Gong and Hew consider transgenic fish; this chapter discusses the methods and molecular expression of transgenic fish as well as their implications for aquaculture and developmental biology. A central theme in development is the transformation of the seemingly homogeneous oocyte into the differentiated heterogeneous embryo: Axis specification is considered by Gard, who explores the role of the oocyte cytoskeleton, and by Elinson and Holowacz who consider the specification of the dorsoanterior axis in amphibians.

We are delighted to dedicate this volume to the memory of Professor Alberto Monroy and to Professor Aaron Moscona, the founding editors of this Series. This thirtieth volume of *Current Topics in Developmental Biology* underscores the cogency of the vision they shared when they began this series. We are also pleased to continue in their tradition by forming a new editorial partnership; we hope that our complementary perspectives on developmental biology will provide our current and future audience a balanced selection of the thrilling discoveries in our field. We wish to express our gratitude to our editorial board members: John Gerhart, Peter Gruss, Philip Ingham, Story Landis, David McClay, Yoshi Nagahama, Susan Strome, Virginia Walbot, and Mitsuki Yoneda. We are also grateful to Heather Aronson for help in editing, and to Liana Hartanto and Diana Myers for administrative assistance. We thank the scientists who prepared articles for this volume and to their funding agencies for supporting their research.

Gerald P. Schatten
Roger A. Pedersen

1
Sperm–Egg Recognition Mechanisms in Mammals

Paul M. Wassarman and Eveline S. Litscher
Roche Institute of Molecular Biology
Roche Research Center
Nutley, New Jersey 07710

    I. Introduction
   II. Characteristics of the Zona Pellucida
  III. ZP3 Functions during Fertilization
  IV. ZP3 Genes from Different Mammalian Species
   V. ZP3 Polypeptides from Different Mammalian Species
  VI. Oligosaccharides at the ZP3 Combining Site for Sperm
 VII. Structure of Oligosaccharides at the ZP3 Combining Site for Sperm
VIII. Location of the ZP3 Combining Site for Sperm
  IX. Does the ZP3 Polypeptide Consist of Two Domains?
   X. Final Comments
      References

I. Introduction

Several different mechanisms are employed by animals to insure maintenance of speciation during reproduction. For example, binding of sperm to unfertilized eggs during fertilization is a relatively species-specific event (Rothschild, 1956; Gwatkin, 1977; Yanagimachi, 1994). The basis of such specificity has been investigated vigorously in both vertebrates and invertebrates. As Lillie (1919) suggested 75 years ago, "No theory of fertilization which fails to include the factor of specificity as one of the prime elements can be true."

    Free-swimming mammalian sperm must bind tightly to the ovulated egg extracellular coat, or zona pellucida, for fertilization to occur (Fig. 1). As in invertebrates, such as mollusks (Vacquier and Lee, 1993) and echinoderms (Palumbi and Metz, 1991), binding of mammalian sperm to unfertilized eggs can exhibit species specificity (Yanagimachi, 1994). Once the egg has been fertilized, the zygote zona pellucida becomes refractory to the binding of free-swimming sperm (Fig. 2). Researchers can conclude that mammalian sperm recognize and bind to a specific component of the unfertilized egg zona pellucida. The component is modified shortly after fertilization, presumably by egg cortical granule exudate, so that it is rendered unrecognizable to free-swimming sperm. Historically, in

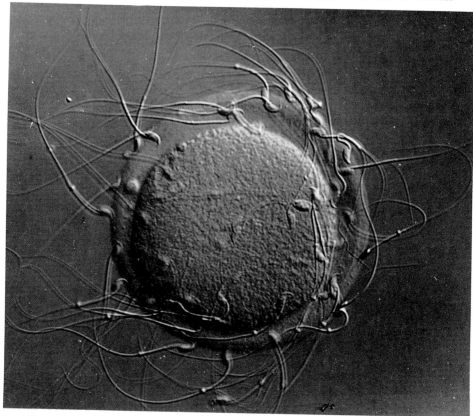

Fig. 1 Photomicrograph of mouse sperm bound to the zona pellucida of an unfertilized mouse egg *in vitro*. The micrograph was taken using Nomarski differential interference contrast microscopy.

fertilization literature dating back more than 50 years, this zona pellucida component, like the vitelline coat component of marine eggs, has been termed a "sperm receptor;" whether or not this nomenclature is appropriate remains somewhat controversial (Wassarman, 1993a,b).

During the past 15 years or so, gamete surface components have been described that support tight species-specific binding of free-swimming mammalian sperm to unfertilized eggs. Interestingly, good reasons suggest that such binding is mediated in part by carbohydrate. In this chapter, the discussion focuses on the zona pellucida glycoprotein named ZP3 that serves as sperm receptor during fertilization in mammals (Bleil and Wassarman, 1980a,b; Wassarman, 1990). Close inspection of the characteristics of this glycoprotein, coupled with some speculation, may shed light on sperm–egg recognition mechanisms in mammals. Overall, we can conclude that species-specific binding of sperm to eggs in mammals is a carbohydrate-mediated event that is attributable to a zona pellucida glycoprotein, called ZP3, and a sperm-surface protein that recognizes and binds to ZP3 oligosaccharides. We suggest that ZP3 consists of at least two different functional domains, perhaps separated from each other by a hinge region of the polypeptide. One of the domains is thought to include the C-terminal third of the ZP3 polypeptide, which possesses the O-linked oligosaccharides that constitute, at least in part, the combining site for sperm.

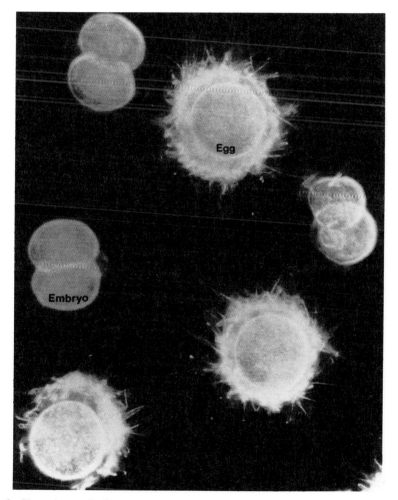

Fig. 2 Photomicrograph of mouse sperm bound to unfertilized mouse eggs in the presence of two-cell mouse embryos *in vitro*. Sperm bind to the zona pellucida of unfertilized mouse eggs, but not to the zona pellucida of two-cell embryos because of inactivation of sperm receptors (ZP3) following fertilization. The micrograph was taken using dark-field optics.

II. Characteristics of the Zona Pellucida

All mammalian eggs are surrounded by a relatively thick extracellular coat called the zona pellucida (Dietl, 1989). The coat is composed of long interconnected filaments that create a loose extracellular matrix that is permeable to large macromolecules and even to small viruses (Greve and Wassarman, 1985; Wassarman and Mortillo, 1991; Yanagimachi, 1994). The mouse zona pellucida is constructed of three glycoproteins called ZP1 (200 kDa), ZP2 (120 kDa), and ZP3 (83 kDa), all of which have relatively low isoelectric points and exhibit considerable heterogeneity on electrophoretic gels (Bleil and Wassarman, 1980a; Wassarman, 1988). The latter characteristics are attributable primarily to the Asn- (N-) linked and Ser/Thr- (O-)linked oligosaccharides on these glycoproteins. Zona pellucida filaments are polymers of [ZP2–ZP3] dimers, present every 150 Å or so along the filaments, and are cross-linked by ZP1; all interactions between these glycoproteins are noncovalent in nature (Greve and Wassarman, 1985; Wassarman, 1988; Wassarman and Mortillo, 1991).

III. ZP3 Functions during Fertilization

Several lines of evidence suggest that when free-swimming mouse sperm interact with the egg zona pellucida, they recognize and bind to ZP3. Paramount among this evidence is the finding that sperm are inhibited from binding to ovulated eggs *in vitro* as a result of pre-incubation in the presence of nanomolar concentrations of purified egg ZP3, but not ZP1 or ZP2, nor embryo ZP3 (Bleil and Wassarman, 1980b; Wassarman *et al.*, 1985; Wassarman, 1990). Thus, as expected for a bona fide sperm receptor, ZP3 is inactivated as a result of fertilization. Purified egg ZP3 binds to tens of thousands of sites on sperm-head plasma membrane (i.e., to acrosome-intact sperm) and, in this manner, prevents sperm from binding to ovulated eggs (Bleil and Wassarman, 1986; Mortillo and Wassarman, 1991). Furthermore, sperm bind to silica beads to which purified egg ZP3 is linked covalently (Fig. 3; Vazquez *et al.*, 1989) and sperm pre-incubated in the presence of recombinant mouse ZP3 are inhibited from binding to ovulated eggs *in vitro* (Kinloch *et al.*, 1991; Beebe *et al.*, 1992).

IV. ZP3 Genes from Different Mammalian Species

Mouse (Kinloch *et al.*, 1988; Chamberlin and Dean, 1989), hamster (Kinloch *et al.*, 1990), and human (Chamberlin and Dean, 1990; van Duin *et al.*, 1992) ZP3 genes have been cloned and characterized. All three genes contain eight exons, and their coding regions are 75% (mouse vs. human) or more identical (Fig. 4). The ZP3 transcription unit is not significantly similar to any other entries in DNA

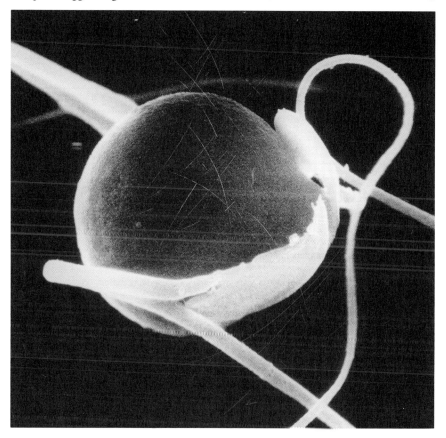

Fig. 3 Scanning electron micrograph of mouse sperm bound to a silica bead possessing covalently linked mouse ZP3. (For details, see Vazquez *et al.*, 1989.)

sequence databases, encodes a 1.5-kb polyadenylated messenger RNA, and has unusually short 5' and 3' noncoding regions. The ZP3 gene is expressed exclusively in growing oocytes, during a 2- to 3-wk period of oogenesis, and is regulated by *cis*-acting sequences located very close (within ~140 bp) to the gene's transcription start site, as well as by *trans*-acting factors present in growing oocytes (Kinloch *et al.*, 1993; Wassarman, 1993b). A fully grown mouse oocyte has $2.5-3 \times 10^5$ copies of ZP3 messenger RNA, but by the time it is converted to an unfertilized egg during ovulation (~10–12 hr) only 2% of the messenger RNA remains; a fertilized egg has undetectable levels of ZP3 messenger RNA (Kinloch and Wassarman, 1989a; Roller *et al.*, 1989; Kinloch *et al.*, 1993). Therefore, the pattern and extent of ZP3 synthesis and secretion during oogenesis are consistent with levels of ZP3 messenger RNA present in mouse

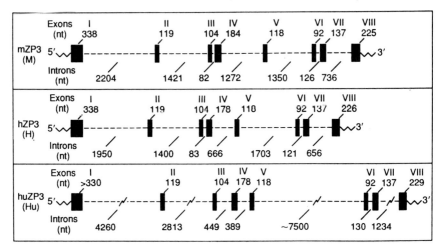

Fig. 4 Diagrammatic representation of the mouse (M), hamster (H), and human (Hu) ZP3 genomic loci. Shown are the positions and sizes (numbers of nucleotides) of the 8 exons (black boxes) and 7 introns (dashed lines) that, in conjunction with the 5' and 3' noncoding regions, constitute the ZP3 transcription unit. (For details, see Kinloch et al., 1988,1990; Chamberlin and Dean, 1989,1990; Kinloch and Wassarman, 1989a,b; Wassarman, 1993b.)

oocytes, eggs, and early embryos (Bleil and Wassarman, 1980c; Salzmann et al., 1983; Shimizu et al., 1983; Philpott et al., 1987; Kinloch and Wassarman, 1989a; Roller et al., 1989; Kinloch et al., 1993).

V. ZP3 Polypeptides from Different Mammalian Species

As a result of molecular cloning, the polypeptide primary structures of ZP3 from four mammalian species—mouse (Kinloch et al., 1988; Chamberlin and Dean, 1989), hamster (Kinloch et al., 1990), marmoset (Thillai-Koothan et al., 1993), and human (Chamberlin and Dean, 1990; van Duin et al., 1992)—have been reported. The sequences of these polypeptides are shown in Fig. 5 using the

Fig. 5 Comparison of primary structures of mouse (M), hamster (H), human (U), and marmoset (R) ZP3 polypeptides using the single letter amino acid code. For hamster, human, and marmoset ZP3, only those positions at which the amino acid sequence differs from that of mouse ZP3 are shown. A 2-amino-acid deletion in hamster ZP3 is indicated by dashes at positions 223 and 224 of mouse ZP3. A 2-amino-acid deletion in human ZP3 is indicated by dashes at positions 220 and 221 of mouse ZP3. A 2-amino-acid deletion in marmoset ZP3 is indicated by dashes at positions 223 and 224 of mouse ZP3. Two added amino acids, E and V in human ZP3 and E and L in marmoset ZP3, are indicated just below the sequences (U and R, respectively). The predicted potential signal sequences are

1. Sperm-Egg Recognition Mechanisms in Mammals

```
M   MASSYFLFLCLLLCGGPELCNS QTLWLLPGGTPTPVGSSS    40
H   .GL..Q.L........AKQ.C. .P........GKLT.       40
U   .EL..R..I....W.ST...YP .P....Q..ASH.ET.VQ    40
R   .EL..R..I....W.ST.L.YP .P.R..Q...SH.ETALQ    40

M   PVKVECLEAELVVTVSRDLFGTGKLVQPGDLTLGSEGCQP      80
H   S.E...┴................I..E.......N┴R.       80
U   .....┴Q..T.M.M..K........IRAA.....P.A┴E.     80
R   ..V..┴Q..T......K.....R..IRAV.....P..┴E.     80

M   RVSVDTDVVRFNAQLHECSSRVQMTKDALVYSTFLLHDPR     120
H   L...A......K......┴.N...V.E.......V...Q..   120
U   L..M.......EVG...┴GNSM.V.D...............   121
                     +E
R   L..T.......EVG...┴GNSM.V.D...............   121
                     +E

M   PVSGLSILRTNRVEVPIECRYPRQGNVSSHPIEPTWVPFR     160
H   ..P.............AD............A.R.......S   160
U   ..GN...V....A.I...┴.....┴....QA.L...L...    161
R   ..GN...V....A.I...┴....R.┴...QA.L...L...    161

M   ATVSSEEKLAFSLRLMEENWNTEKSAPTFHLGEVAHLQAE     200
H   T........V...............LS........Y....    200
U   T..F.....T..............A..RS......DA....   201
R   T..F.....T.............S...RT......D.....   201

M   VQTGSHLPLQLFVDHCVATPSPLPDPNSSPYHFIVDFHGC     240
H   .........L....R┴......--.QTA....V......┴    238
U   IH....V..R....┴...--.T..Q┴A....T.......┴    239
R   IH....V..R....┴.....T.--.Q┴A....T.......   239

M   LVDGLSESFSAFEVPRPRPETLQFTVDVFHFANSSRNTLY     280
H   ...........Q.............┴.............I.   278
U   ....TDAS...K....G.D..........┴D...MI.       279
R   ....TDAS...QA......D.........┴D...MI.       279

M   ITCHLKVAPANQIPDKLNKACSFNKTSQSWLPVEGDADIC     320
H   ..┴....T..┴.T..E.....┴..RS.K..S......EV┴   318
U   ..┴....TL.E.D..E......┴..S.P.N..F....┴...  319
R   ..┴....TL.E.D..E......┴..S.A.N..F....P..┴  319

M   DCCSHGNCSNSSSSQFQIHGPRQWSKLVSRNRRHVTDEAD     360
H   G.┴.S.D┴GS..R.RY.A..VS..P.SA..R.....R...    358
U   Q.┴.NK.D┴GTP.H.RR.P.VMS...ASA........E...   359
R   Q.┴..K.D┴GTP.HARR.P.VVSLG.GSPA.D.....E...   359

M   VTVGPLIFLGKANDQTVEGWTASAQTSVALGLGLATVAFL     400
H   ............S..A...AS.....L........A....   398
U   .........DA.G.HE..Q.ALPSD...L..V...V.VS.   400
                                   +V
R   ........DRTG.HEM.Q.ALP.D..LL..T...V,.L.    400
                                   +L

M   TLAAIVLAVTRKCHSSSYLVSLPQ   424
H   .......G...S┴.TP.HV...S.   422
U   ..T.VI.VL..R┴RTA.HP..ASE   424
R   ..T.VI.VL..R┴RTA.LP..ASE   424
```

italicized (residues 1–22). Locations of 13 cysteine residues (C) and potential N-linked glycosylation sites (N) are underlined. (For details, see Kinloch *et al.*, 1988,1990; Chamberlin and Dean, 1989,1990; Thillai-Koothan *et al.*, 1993.)

single letter amino acid code. Several points can be made in connection with the primary structures of mouse, hamster, marmoset, and human ZP3:

1. The four ZP3 polypeptides are very similar to one another (e.g., mouse and hamster ZP3, 82% identical; mouse and human ZP3, 67% identical; mouse and marmoset ZP3, 65% identical; marmoset and human ZP3, 91% identical).
2. Each ZP3 polypeptide has a 22-amino-acid hydrophobic putative "signal sequence" at the N terminus that is not present on the secreted glycoprotein. The secreted forms of mouse, marmoset, and human ZP3 consist of 402 amino acids; hamster ZP3 consists of 400 amino acids.
3. Each ZP3 polypeptide has an N-terminal glutamine that is blocked, probably due to cyclization of glutamine to pyroglutamic acid.
4. Each ZP3 polypeptide is unusually rich in proline and serine plus threonine residues (e.g., mouse ZP3 has 7% proline and 17% serine plus threonine residues).
5. Each ZP3 polypeptide has several potential N-linked glycosylation sites (i.e., the consensus sequence Asn–Xaa–Ser/Thr) of which two are conserved in the four ZP3 polypeptides. These sites are located in relatively hydrophilic regions of polypeptide that are predicted to form reverse turns or coils.
6. Each ZP3 polypeptide has 13 cysteine residues, all in identical positions, suggesting that intramolecular disulfide bonds are important in determining the glycoprotein's three-dimensional structure. Note that ZP3 must have at least one free cysteine residue; a relatively unusual feature of a secreted glycoprotein.
7. The hydrophobicity profiles of the four ZP3 polypeptides are very similar to one another. Each polypeptide has a quite hydrophobic domain very close to the C terminus.

From these characteristics we can assume that the overall three-dimensional structures of these ZP3 polypeptides, representing four different mammalian species, are very similar to one another. This feature has important implications in the context of species specificity of sperm–egg interaction in mammals (discussed subsequently). In addition, it probably reflects the fact that ZP3 also functions as a structural component of zona pellucida filaments (Greve and Wassarman, 1985; Wassarman and Mortillo, 1991), which could place severe restraints on changes in polypeptide structure during evolution.

VI. Oligosaccharides at the ZP3 Combining Site for Sperm

Considerable experimental evidence supports the conclusion that sperm recognize and bind to oligosaccharides located at the ZP3 combining site. Furthermore, evidence supports the conclusion that, at least in the case of mouse ZP3, these oligosaccharides are covalently linked to serine or threonine residues (i.e., O-linked oligosaccharides) rather than to asparagine residues (i.e., N-linked oligosaccharides). This evidence comes from (1) experiments with inhibitors of *in vitro* fertilization, (2) experiments with purified ZP3, and (3) experiments with sperm proteins that recognize ZP3.

1. Inhibition of binding of sperm to eggs, and consequently of *in vitro* fertilization, by incubation of gametes with lectins, with monosaccharides, oligosaccharides, or glycoconjugates, or with glycosidases is consistent with the idea that sperm–egg interaction is carbohydrate mediated (Yanagimachi, 1977,1994; Wassarman, 1992; Litscher and Wassarman, 1993). However, several problems are inherent in this indirect experimental approach and the results, although consistent with a role for carbohydrates in gamete adhesion, could be viewed with some skepticism.

2. The first direct evidence for the role of carbohydrates in mammalian gamete adhesion was obtained by Florman and Wassarman (1985), who selectively removed N-linked oligosaccharides from purified mouse ZP3 with N-glycanase and found that the glycoprotein retained its sperm receptor activity *in vitro*. On the other hand, when they selectively removed O-linked oligosaccharides from purified mouse ZP3 with mild alkali (β elimination), the glycoprotein lost its sperm receptor activity. When the β elimination reaction was carried out in the presence of sodium borohydride, so O-linked oligosaccharides released from ZP3 could be recovered, the oligosaccharides themselves exhibited sperm receptor activity *in vitro*.

The direct demonstration that ZP3 oligosaccharides are responsible for the glycoprotein's sperm receptor activity is consistent with several other observations. For example, sperm receptor activity was retained even after boiling or exposure of ZP3 to detergents, denaturants, or reducing agents (Wassarman, 1988). Sperm receptor activity was also retained by glycopeptides derived from ZP3 after extensive digestion by Pronase (Florman *et al.*, 1984; Leyton and Saling, 1989a).

Although intact ZP3, ZP3 glycopeptides, and ZP3 O-linked oligosaccharides all exhibit sperm receptor activity *in vitro*, they are not equally effective at equivalent concentrations. Intact ZP3 is more effective than glycopeptides, which are more effective than O-linked oligosaccharides. At least two explanations for this disparate behavior are plausible. Either the ZP3 polypeptide influences the "presentation" of the O-linked oligosaccharides to sperm, so that binding is significantly enhanced (i.e., an indirect role for polypeptide), or sperm bind to ZP3 polypeptide as well as to oligosaccharides (i.e., a direct role for polypeptide). Since researchers have demonstrated that polypeptide can have a dramatic effect on oligosaccharide conformation and clustering (Lee, 1992), perhaps the former explanation should be favored.

At this point, note that whereas intact ZP3 binds to sperm and induces exocytosis in the form of the acrosome reaction, small ZP3 glycopeptides and ZP3 O-linked oligosaccharides bind to sperm but fail to induce the acrosome reaction (Bleil and Wassarman, 1983; Florman et al., 1984; Florman and Wassarman, 1985; Leyton and Saling, 1989a). These findings suggest that induction of the acrosome reaction may be due to multivalent interactions between ZP3 oligosaccharides and a binding protein located in the sperm head plasma membrane (Wassarman et al., 1985; Leyton and Saling, 1989a). Such interactions could lead to a redistribution of the binding protein, thereby converting this portion of sperm plasma membrane to a state capable of fusing with the outer acrosomal membrane.

3. Further support for the presence of essential O-linked oligosaccharides at the mouse ZP3 combining site comes from work on the potential roles of sperm β-1,4-galactosyltransferase (Shur and Hall, 1982; Miller et al., 1992) and another sperm protein called sp56 (Bleil and Wassarman, 1990; Cheng et al., 1994) in sperm–egg binding.[1] In each case, the protein is associated with plasma membrane overlying the sperm head and recognizes and binds specifically to O-linked oligosaccharides on ZP3, not to any other zona pellucida glycoprotein. Sperm galactosyltransferase and sp56 bind to oligosaccharides possessing different sugars at their reducing termini—GlcNAc and Gal, respectively. Therefore, it is unlikely that the two sperm proteins bind to the same ZP3 O-linked oligosaccharides. Rather, the two proteins either act in concert during binding of sperm to eggs or, alternatively, only one of the proteins is the authentic ZP3 binding protein. Definitive experiments must be carried out to resolve this issue.

Note that some evidence suggests that essential N-linked oligosaccharides may be located at the pig ZP3 combining site for sperm (Noguchi et al., 1992). Although this result conflicts with other reports that O-linked oligosaccharides are involved (Yurewicz et al., 1991), the interesting possibility remains that, although binding of sperm to eggs in mammals is always carbohydrate mediated, some species may utilize O-linked oligosaccharides whereas others may utilize N-linked oligosaccharides. Here again, definitive experiments must be carried out to resolve this intriguing issue.

VII. Structure of Oligosaccharides at the ZP3 Combining Site for Sperm

Unfortunately, the composition and sequence of essential O-linked oligosaccharides at the ZP3 combining site have not been established yet. On the other

---

[1]Note that sp56 and β-1,4-galactosyltransferase are not the sole candidates for the ZP3 binding protein present on sperm (Litscher and Wassarman, 1993; Ward and Kopf, 1993). For example, one candidate is a 95-kDa mouse sperm protein (p95) that is a tyrosine kinase substrate and reportedly binds exclusively to ZP3 on Western blots (Leyton and Saling, 1989b). Recently, p95 has been shown to be a unique tyrosine-phosphorylated form of hexokinase present in sperm (Kalab et al., 1994).

1. Sperm–Egg Recognition Mechanisms in Mammals

hand, certain features of the oligosaccharides have been reported, although they are the subject of some disagreement. One example is the identity of the sugar at the nonreducing terminus of these O-linked oligosaccharides on mouse ZP3. In one case the terminal sugar has been identified as Gal in α linkage with the penultimate sugar (Bleil and Wassarman, 1988), whereas in another it has been identified as GlcNAc (Miller et al., 1992). In the former case, either removal of the terminal Gal with α-galactosidase or conversion of its C-6 hydroxyl to an aldehyde with galactose oxidase was found to inactivate ZP3 as a receptor. Further support for the presence of a terminal Gal in α linkage comes from experiments utilizing mouse ZP3 and a monoclonal antibody called LA4 (Shalgi et al., 1990). The epitope recognized by this antibody contains a terminal Gal in α linkage with a penultimate Gal (Dodd and Jessell, 1985); LA4 binds to ZP3, but not to other zona pellucida glycoproteins, on immunoblots. Finally, results of recent experiments employing oligosaccharide constructs possessing defined structures strongly suggest that terminal Gal residues in either α or β linkage, but not β-linked GlcNAc residues, may be important in binding of sperm to mouse ZP3 (Litscher et al., 1993).

Although a terminal sugar such as Gal or GlcNAc may be essential for ZP3 receptor activity, the entire epitope recognized by sperm is very likely to consist of several sugars, not just a single residue. This feature could account for the binding of sperm from one species to eggs from another. The features of essential oligosaccharides of ZP3 from one species may resemble those of ZP3 from another species closely enough and, consequently, may support nonhomologous gamete adhesion. An example of this event is the binding of hamster sperm to mouse eggs and mouse sperm to hamster eggs *in vitro* (Moller et al., 1990; M. Weetall, E. Litscher, and P. Wassarman, unpublished results). Interestingly, the step after binding of heterologous sperm to eggs—induction of the acrosome reaction by ZP3 (Bleil and Wassarman, 1983; Kopf and Gerton, 1991)—may or may not take place. For example, hamster ZP3 induces mouse sperm to acrosome-react *in vitro*, but mouse ZP3 does not induce hamster sperm (Moller et al., 1990). A similar situation exists for several other heterologous crosses (Yanagimachi, 1994). Therefore, species specificity may apply not only to binding of sperm to eggs, but also to induction of the acrosome reaction.

VIII. Location of the ZP3 Combining Site for Sperm

Four different lines of experimental evidence suggest that the mouse ZP3 combining site for sperm is located on the C-terminal third of the polypeptide and includes amino acids 328–343. This evidence is summarized here:

1. Digestion of mouse ZP3 by either papain or V8 protease yields a heavily glycosylated peptide (~55 kDa) derived from the C-terminal half of the polypeptide that exhibits sperm receptor activity *in vitro* (Rosiere and Was-

sarman, 1992). [Note: Like intact ZP3 (Bleil and Wassarman, 1983), this ZP3 glycopeptide also induces sperm to undergo the acrosome reaction *in vitro*; E. Litscher and P. Wassarman, unpublished results.] When the region encompassing amino acids 328–343 of mouse ZP3 is removed from the 55-kDa glycopeptide by more extensive proteolysis, sperm receptor activity is lost, implicating this region in biological activity.

2. The presence of antibodies directed specifically against a synthetic peptide consisting of amino acids 328–343 in mouse ZP3 prevents binding of sperm to eggs *in vitro* (Rosiere and Wassarman, 1992; S. Mortillo and P. Wassarman, unpublished results). Furthermore, female mice immunized with a comparable synthetic peptide produce antibodies against ZP3 and exhibit long-term infertility (Millar *et al.*, 1989).

3. Embryonal carcinoma (EC) cells transfected with the mouse ZP3 gene synthesize and secrete an active form of the receptor, whereas EC cells transfected with the hamster ZP3 gene synthesize and secrete an inactive form of the receptor (Kinloch *et al.*, 1991). In the later case, presumably the C-terminal half of the ZP3 polypeptide is improperly glycosylated and does not possess the O-linked oligosaccharides essential for receptor activity. Support for this conclusion comes from exon-swapping experiments in which hamster ZP3 exons 6–8 were fused to mouse ZP3 exons 1–5 and mouse ZP3 exons 6–8 were fused to hamster ZP3 exons 1–5 (Kinloch *et al.*, 1994). EC cells stably transfected with the former recombinant construct produced an inactive receptor, whereas those transfected with the latter construct produced an active receptor. These results provide further evidence that the receptor activity of ZP3 is dependent on the C-terminal third of the molecule.

4. EC cells transfected with the mouse ZP3 gene synthesize and secrete an active form of the receptor (Kinloch *et al.*, 1991). However, when five serine residues (there are no threonine residues) located in the region encompassing amino acids 328–343 (encoded by exon 7) are converted to glycine, alanine, or valine by site-directed mutagenesis, and EC cells are transfected with the mutant mouse ZP3 gene, the cells synthesize and secrete an inactive form of the receptor (Kinloch *et al.*, 1994). Thus, elimination of the potential O-linked glycosylation sites in this region of mouse ZP3 correlates with production of an inactive receptor.

Collectively, these and other results strongly suggest that the region of the mouse ZP3 polypeptide including amino acids 328–343 possesses the O-linked oligosaccharides that constitute at least part of the combining site for sperm. Examination of the primary structure of this region for mouse, hamster, marmoset, and human ZP3 reveals that it has undergone considerable change during evolution compared with other regions of the polypeptide (Figs. 5 and 6). For example, a comparison of the primary structures of mouse and hamster

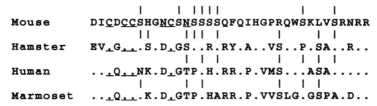

Fig. 6 Comparison of the putative combining site region of mouse, hamster, human, and marmoset ZP3 polypeptides. Using the single letter amino acid code, this diagram shows the region of mouse ZP3 encompassing residues 318–353, as well as the corresponding regions of hamster, human, and marmoset ZP3. Only when these sequences differ from that of mouse ZP3 is the amino acid shown. The positions of potential O-linked glycosylation sites within this region (i.e., serine and threonine residues) are indicated by a vertical line directly above the residue. Site-directed mutagenesis of Ser-329, -331, -332, -333, and -334 of mouse ZP3 (converted to glycine, valine, alanine) resulted in production of an inactive sperm receptor by transfected EC cells (R. Kinloch and P. Wassarman, unpublished results). Note that in the region encompassing residues 328–340, mouse, hamster, human, and marmoset ZP3 have 5, 4, 3, and 2 potential O-linked glycosylation sites, respectively. Ser-332 is conserved in all four species and Ser-334 is conserved in mouse, hamster, and human ZP3.

ZP3 polypeptides (82% identical), carried out in 40-residue increments from Pro-35 to Leu-394, reveals that the region encompassing residues 315–354 has undergone twice as many changes (18 changes; 55% identical) as any other region (Fig. 7). Perhaps this characteristic reflects the possibility that differences in ZP3 oligosaccharide structure among species accounts for the frequent failure of sperm from one mammalian species to bind to eggs of another. Although the rules that govern placement and structure of O-linked oligosaccharides on glycoproteins remain unclear (Sadler, 1984; Wilson et al., 1991), changes in polypeptide primary structure in and around the ZP3 combining site are likely to influence the location and nature of oligosaccharides added to nascent ZP3 and, in this manner, to determine species specificity of gamete interactions.

IX. Does the ZP3 Polypeptide Consist of Two Domains?

ZP3 performs two very different roles as a structural component of zona pellucida filaments and as a sperm receptor. Therefore, it might not be very surprising to find that the ZP3 polypeptide consists of at least two domains. In fact, evidence suggests that this is the case.

As mentioned earlier, a specific region of mouse ZP3 polypeptide is particularly susceptible to cutting by several proteases. This feature results in production of a large C-terminal fragment that possesses sperm receptor activity (Rosiere and Wassarman, 1992; similar results have been obtained with hamster

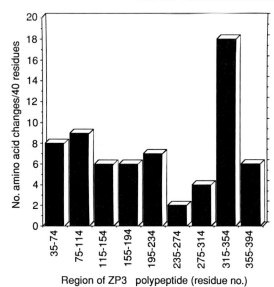

Fig. 7 Comparison of mouse and hamster ZP3 polypeptides. This histogram indicates the number of amino acid differences between mouse and hamster ZP3 in 40-amino-acid increments along the polypeptide chain, from residue 35 to residue 394. The signal sequence and extreme C-terminal region of the polypeptide are not included in the analysis. Note that although mouse and hamster ZP3 are about 82% identical, the region including amino acids 315–354 (ZP3 combining site for sperm) is only 55% identical (22 of 40 amino acids are the same).

ZP3—Weetall *et al.*, 1993; M. Weetall, E. Litscher, and P. Wassarman, unpublished results). Examination of the primary structure of mouse ZP3 polypeptide, in the region susceptible to digestion by proteases, suggests the interesting possibility that it represents a so-called "hinge." As in immunoglobulins (Branden and Tooze, 1991), such a region would serve as a flexible link between two different ZP3 polypeptide domains: one domain (C-terminal) that possesses the combining site for sperm and a second domain (N-terminal) that perhaps participates in zona pellucida filament assembly during oogenesis.

The sequence of the putative hinge is shown in Fig. 8 for mouse, hamster, marmoset, and human ZP3. The sequence consists of two relatively proline-rich stretches separated from each other by 24 amino acid residues that are highly conserved from mouse to human. The proline-rich regions presumably form the rigid parts of the hinge since proline residues provide less conformational freedom than other amino acids (Branden and Tooze, 1991). In this context, note that a comparison of the mouse ZP3 polypeptide with hamster, marmoset, and human ZP3 polypeptides reveals that the latter have all undergone a two amino acid deletion in the first proline-rich stretch. This change may be related to the

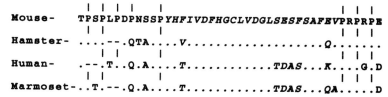

Fig. 8 Comparison of the putative hinge region of mouse, hamster, human, and marmoset ZP3 polypeptides. Using the single letter amino acid code, this diagram shows the region of mouse ZP3 including residues 219–260, as well as the corresponding regions of hamster, human, and marmoset ZP3. Only when these sequences differ from that of mouse ZP3 is the amino acid shown. The positions of proline residues are indicated by a vertical line directly above the residue. Note that two proline-rich regions [residues 219–230 (region I) and residues 255–260 (region II)] are separated by a 24-amino-acid region (italicized) that lacks proline residues and is highly conserved. All three proline residues in region II are conserved; two of five proline residues in region I are conserved.

possibility that the first proline-rich stretch serves as a border between two domains that have combined by fusion of genes during the course of evolution.

X. Final Comments

In 1977 Yanagimachi presented a working model for the possible mechanism of sperm–zona pellucida interaction during fertilization in mammals (Yanagimachi, 1977). To date, certain aspects of this model have received strong experimental support. In particular, carbohydrate–protein interactions appear to provide the molecular basis for species-specific binding of free-swimming sperm to the egg zona pellucida. These types of interaction are now thought to be employed in a number of instances of cellular adhesion, for example, in neuronal development (Hynes et al., 1989; Schachner, 1989) as well as in the well-publicized case of lymphocyte homing (Springer, 1991; Lasky, 1993). Of course, the potential for enormous variation in oligosaccharide structure, which permits extreme fine-tuning of recognition determinants, is an appealing feature of carbohydrate-mediated cellular adhesion in general. In the case of ZP3, the essential role of oligosaccharide at the sperm receptor combining site is also appealing in view of the fact that the glycoprotein is a structural component of a specialized extra-cellular matrix that has some "mucin-like" character (e.g., hydrated and viscous).

Acknowledgments

We thank past and present members of our laboratory who carried out research on ZP3, and are grateful to the National Institutes of Health, the National Science Foundation, the Rockefeller Foundation, and Hoffmann-La Roche, Inc., who provided support for the research. Finally, we hope that Alberto Monroy, who loved the subject of animal fertilization, would have enjoyed some of the speculation in this article.

References

Beebe, S., Leyton, L., Burks, D., Ishikawa, M., Fuerst, T., Dean, J., and Saling, P. (1992). Recombinant mouse ZP3 inhibits sperm binding and induces the acrosome reaction. *Dev. Biol.* **151**, 48–54.
Bleil, J., and Wassarman, P. (1980a). Structure and function of the zona pellucida: Identification and characterization of the mouse oocyte's zona pellucida. *Dev. Biol.* **76**, 185–203.
Bleil, J., and Wassarman, P. (1980b). Mammalian sperm–egg interaction: Identification of a glycoprotein in mouse egg zonae pellucidae possessing receptor activity for sperm. *Cell* **20**, 873–882.
Bleil, J., and Wassarman, P. (1980c). Synthesis of zona pellucida proteins by denuded and follicle-enclosed mouse oocytes during culture *in vitro*. *Proc. Natl. Acad. Sci. USA* **77**, 1029–1033.
Bleil, J., and Wassarman, P. (1983). Sperm–egg interactions in the mouse: Sequence of events and induction of the acrosome reaction by a zona pellucida glycoprotein. *Dev. Biol.* **95**, 317–324.
Bleil, J., and Wassarman, P. (1986). Autoradiographic visualization of the mouse egg's sperm receptor bound to sperm. *J. Cell Biol.* **102**, 1363–1371.
Bleil, J., and Wassarman, P. (1988). Galactose at the nonreducing terminus of O-linked oligosaccharides of mouse egg zona pellucida glycoprotein ZP3 is essential for the glycoprotein's sperm receptor activity. *Proc. Natl. Acad. Sci. USA* **85**, 6778–6782.
Bleil, J., and Wassarman, P. (1990). Identification of a ZP3-binding protein on acrosome-intact mouse sperm by photoaffinity crosslinking. *Proc. Natl. Acad. Sci. USA* **87**, 5563–5567.
Branden, C., and Tooze, J. (1991). "Introduction to Protein Structure." Garland Publishing, New York.
Chamberlin, M., and Dean, J. (1989). Genomic organization of a sex specific gene: The primary sperm receptor of the mouse zona pellucida. *Dev. Biol.* **131**, 207–214.
Chamberlin, M., and Dean, J. (1990). Human homolog of the mouse sperm receptor. *Proc. Natl. Acad. Sci. USA* **87**, 6014–6018.
Cheng, A., Le, T., Palacios, M., Bookbinder, L., Wassarman, P., Suzuki, F., and Bleil, J. (1994). Sperm-egg recognition in the mouse: Characterization of sp56, a sperm protein having specific affinity for ZP3. *J. Cell Biol.* **125**, 867–878.
Dietl, J., ed. (1989). "The Mammalian Egg Coat. Structure and Function." Springer-Verlag, Berlin.
Dodd, J., and Jessell, T. (1985). Lactoseries carbohydrates specify subsets of dorsal root ganglion neurons projecting to the superficial dorsal horn of rat spinal cord. *J. Neurosci.* **5**, 3278–3294.
Florman, H., and Wassarman, P. (1985). O-Linked oligosaccharides of mouse egg ZP3 account for its sperm receptor activity. *Cell* **41**, 313–324.
Florman, H., Bechtol, K., and Wassarman, P. (1984). Enzymatic dissection of the functions of the mouse egg's receptor for sperm. *Dev. Biol.* **106**, 243–255.
Greve, J., and Wassarman, P. (1985). Mouse extracellular coat is a matrix of interconnected filaments possessing a structural repeat. *J. Mol. Biol.* **181**, 253–264.
Gwatkin, R. (1977). "Fertilization Mechanisms in Man and Mammals." Plenum Press, New York.

1. Sperm–Egg Recognition Mechanisms in Mammals

Hynes, M., Buck, L., Gitt, M., Barondes, S., Dodd, J., and Jessell, T. (1989). Carbohydrate recognition in neuronal development: Structure and expression of surface oligosaccharides and β-galactoside-binding lectins. *Ciba Found. Symp.* **145**, 189–223.

Kalab, P., Visconti, P., Leclerc, P., and Kopf, G. (1994). p95, the major phosphotyrosine-containing protein in mouse spermatozoa, is a hexokinase with unique properties. *J. Biol. Chem.* **269**, 3810–3817.

Kinloch, R., and Wassarman, P. (1989a). Profile of a mammalian sperm receptor gene. *New Biol.* **1**, 232–238.

Kinloch, R., and Wassarman, P. (1989b). Nucleotide sequence of the gene encoding zona pellucida glycoprotein ZP3—The mouse sperm receptor. *Nucleic Acids Res.* **17**, 2861–2863.

Kinloch, R., Roller, R., Fimiani, C., Wassarman, D., and Wassarman, P. (1988). Primary structure of the mouse sperm receptor's polypeptide chain determined by genomic cloning. *Proc. Natl. Acad. Sci. USA* **85**, 6409–6413.

Kinloch, R., Ruiz-Seiler, B., and Wassarman, P. (1990). Genomic organization and polypeptide primary structure of zona pellucida glycoprotein, hZP3, and hamster sperm receptor. *Dev. Biol.* **142**, 414–421.

Kinloch, R., Mortillo, S., Stewart, C., and Wassarman, P. (1991). Embryonal carcinoma cells transfected with ZP3 genes differentially glycosylate similar polypeptides and secrete active mouse sperm receptor. *J. Cell Biol.* **115**, 655–664.

Kinloch, R., Lira, S., Mortillo, S., Schickler, M., Roller, R., and Wassarman, P. (1993). Regulation of expression of mZP3, the sperm receptor gene, during mouse development. *In* "Molecular Basis of Morphogenesis" (M. Bernfield, ed.), pp. 19–33. Wiley-Liss, New York.

Kinloch, R., Sakai, Y., and Wassarman, P. (1994). *Proc. Natl. Acad. Sci. USA* **91**.

Kopf, G., and Gerton, G. (1991). The mammalian sperm acrosome and the acrosome reaction. *In* "Elements of Mammalian Fertilization" (P. Wassarman, ed.), Vol. 1, pp. 153–204. CRC Press, Boca Raton, Florida.

Lasky, L. (1993). A "roll" in acute inflammation. *Curr. Biol.* **3**, 680–683.

Lee, Y.C. (1992). Biochemistry of carbohydrate–protein interaction. *FASEB J.* **6**, 3193–3200.

Leyton, L., and Saling, P. (1989a). Evidence that aggregation of mouse sperm receptors by ZP3 triggers the acrosome reaction. *J. Cell Biol.* **108**, 2163–2168.

Leyton, L., and Saling, P. (1989b). 95 kd sperm proteins bind ZP3 and serve as tyrosine kinase substrates in response to zona binding. *Cell* **57**, 1123–1130.

Lillie, F. (1919). "Problems of Fertilization." University of Chicago Press, Chicago.

Litscher, E., and Wassarman, P. (1993). Carbohydrate-mediated adhesion of eggs and sperm during mammalian fertilization. *Trends Glycosci. Glycotech.* **5**, 369–388.

Litscher, E., Wassarman, P., Juntunen, K., Seppo, A., Niemela, R., Penttila, L., and Renkonen, O. (1993). Use of oligosaccharide constructs to evaluate carbohydrate determinants involved in gamete interactions in mice. *Mol. Biol. Cell* **4**, 140a (abstract).

Millar, S., Chamow, S., Baur, A., Oliver, C., Robey, F., and Dean, J. (1989). Vaccination with a synthetic zona pellucida peptide produces long-term contraception in female mice. *Science* **246**, 935–938.

Miller, D., Macek, M., and Shur, B. (1992). Complementarity between sperm surface β-1,4-galactosyltransferase and egg-coat ZP3 mediates sperm–egg binding. *Nature* **357**, 589–593.

Moller, C., Bleil, J., Kinloch, R., and Wassarman, P. (1990). Structural and functional relationships between mouse and hamster zona pellucida glycoproteins. *Dev. Biol.* **137**, 276–286.

Mortillo, S., and Wassarman, P. (1991). Differential binding of gold-labeled zona pellucida glycoproteins mZP2 and mZP3 to mouse sperm membrane compartments. *Development* **113**, 141–150.

Noguchi, S., Hatanaka, Y., Tobita, T., and Nakano, M. (1992). Structural analysis of the N-linked carbohydrate chains of the 55-kDa glycoprotein family (PZP3) from porcine zona pellucida. *Eur. J. Biochem.* **204**, 1089–1100.

Palumbi, S., and Metz, E. (1991). Strong reproductive isolation between closely related tropical sea urchins (genus *Echinometra*). *Mol. Biol. Evol.* **8,** 227–239.

Philpott, C., Ringuette, M., and Dean, J. (1987). Oocyte-specific expression and developmental regulation of ZP3, a sperm binding protein of the mouse zona pellucida. *Dev. Biol.* **121,** 568–575.

Roller, R., Kinloch, R., Hiraoka, B., Li, S-L., and Wassarman, P. (1989). Gene expression during mammalian oogenesis and early embryogenesis: Quantification of three messenger-RNAs abundant in fully-grown oocytes. *Development* **106,** 251–261.

Rosiere, T., and Wassarman, P. (1992). Identification of a region of mouse zona pellucida glycoprotein mZP3 that possesses sperm receptor activity. *Dev. Biol.* **154,** 309–317.

Rothschild, L. (1956). "Fertilization." Methuen, London.

Sadler, J. (1984). Biosynthesis of glycoproteins: Formation of O-linked oligosaccharides. In "Biology of Carbohydrates" (V. Ginsburg and P. Robbins, eds.), Vol. 2, pp. 199–288. Wiley-Interscience, New York.

Salzmann, G., Greve, J., Roller, R., and Wassarman, P. (1983). Biosynthesis of the sperm receptor during oogenesis in the mouse. *Eur. Mol. Biol. Org. J.* **2,** 1451–1456.

Schachner, M. (1989). Families of neural adhesion molecules. *Ciba Found. Symp.* **145,** 156–172.

Shalgi, R., Bleil, J., and Wassarman, P. (1990). Carbohydrate-mediated sperm–egg interactions in mammals. In "Advances in Assisted Reproductive Technologies" (Z. Ben-Rafael, ed.), pp. 437–441. Plenum Press, New York.

Shimizu, S., Tsuji, M., and Dean, J. (1983). *In vitro* biosynthesis of three sulfated glycoproteins of murine zonae pellucidae by oocytes grown in follicle culture. *J. Biol. Chem.* **258,** 5858–5863.

Shur, B., and Hall, N. (1982). A role for mouse sperm surface galactosyl-transferase in sperm binding to the egg zona pellucida. *J. Cell Biol.* **95,** 574–579.

Springer, T. (1991). Sticky sugars for selectins. *Nature* **349,** 196–197.

Thillai-Koothan, P., van Duin, M., and Aitken, R. (1993). Cloning, sequencing and oocyte-specific expression of the marmoset sperm receptor protein, ZP3. *Zygote* **1,** 93–101.

Vacquier, V., and Lee, Y-H. (1993). Abalone sperm lysin: Unusual mode of evolution of a gamete recognition protein. *Zygote* **1,** 181–196.

van Duin, M., Polman, J., Verkoelen, E., Bunshoten, H., Meyerink, J., Olijve, W., and Aitken, R. (1992). Cloning and characterization of the human sperm receptor ligand ZP3: Evidence for a second polymorphic allele with a different frequency in the Caucasian and Japanese populations. *Genomics* **14,** 1064–1070.

Vazquez, M., Phillips, D., and Wassarman, P. (1989). Interaction of mouse sperm with purified sperm receptors linked to silica beads. *J. Cell Sci.* **92,** 713–722.

Ward, C., and Kopf, G. (1993). Molecular events mediating sperm activation. *Dev. Biol.* **158,** 9–34.

Wassarman, P. (1988). Zona pellucida glycoproteins. *Annu. Rev. Biochem.* **57,** 415–442.

Wassarman, P. (1990). Profile of a mammalian sperm receptor. *Development* **108,** 1–17.

Wassarman, P., ed. (1992). "Elements of Mammalian Fertilization," Vols. 1 and 2. CRC Press, Boca Raton, Florida.

Wassarman, P. (1993a). Mammalian eggs, sperm, and fertilization: Dissimilar cells with a common goal. *Sem. Dev. Biol.* **4,** 189–197.

Wassarman, P. (1993b). Mammalian fertilization: Sperm receptor genes and glycoproteins. *Adv. Dev. Biochem.* **2,** 159–199.

Wassarman, P., and Mortillo, S. (1991). Structure of the mouse egg extracellular coat, the zona pellucida. *Int. Rev. Cytol.* **130,** 85–109.

Wassarman, P., Bleil, J., Florman, H., Greve, J., Roller, R., Salzmann, G., and Samuels, F. (1985). The mouse egg's receptor for sperm: What is it and how does it work? *Cold Spring Harbor Symp. Quant. Biol.* **50,** 11–19.

Weetall, M., Litscher, E., and Wassarman, P. (1993). Structure–function relationships between mouse and hamster sperm receptors. *Mol. Biol. Cell* **4,** 248a (abstract).

Wilson, I, Gavel, Y., and von Heijne, G. (1991). Amino acid distributions around O-linked glycosylation sites. *Biochem. J.* **275,** 529–534.

Yanagimachi, R. (1977). Specificity of sperm–egg interaction. *In* "Immunobiology of Gametes" (M. Edidin and M. Johnson, eds.), pp. 255–295. Cambridge University Press, Cambridge.

Yanagimachi, R. (1994). Mammalian fertilization. *In* "The Physiology of Reproduction" (E. Knobil and J. Neill, eds.), Vol. 1, pp. 189–317. Raven Press, New York.

Yurewicz, E., Pack, B., and Sacco, A. (1991). Isolation, composition, and biological activity of sugar chains of porcine oocyte zona pellucida 55K glycoproteins. *Mol. Reprod. Dev.* **30,** 126–134.

2
Molecular Basis of Mammalian Egg Activation

Richard M. Schultz
Department of Biology
University of Pennsylvania
Philadelphia, Pennsylvania 19104

Gregory S. Kopf
Department of Obstetrics and Gynecology
University of Pennsylvania
Philadelphia, Pennsylvania 19104

    I. Introduction
   II. Description of Fertilization of Mammalian Eggs
      A. Events of Fertilization prior to Sperm–Egg Plasma Membrane Interaction
      B. Interaction of Sperm and Egg Membranes prior to Fertilization
      C. Events of Fertilization after Sperm–Egg Plasma Membrane Interaction
   III. Intracellular Second Messengers that Regulate Egg Activation
      A. Calcium
      B. IP_3
      C. Diacylglycerol
   IV. Maturation-Associated Changes in CG Exocytosis in Response to Second Messengers
      A. Calcium Oscillations during Oocyte Maturation
      B. Effect of Calcium and IP_3 on CG Exocytosis in Oocytes
   V. Signal Transduction Pathways Leading to Egg Activation
      A. Introduction
      B. Receptor Hypothesis of Egg Activation
      C. Fusion Hypothesis of Egg Activation
   VI. Receptors on the Egg that Mediate Sperm-Induced Egg Activation
      A. General Approaches to Studying Egg Surface Sperm Receptors
      B. Putative Egg Receptors for Sperm
   VI. Sperm Surface Ligands Involved in Egg Activation
   VII. Is There a Need for an Egg-Specific Receptor and a Sperm-Specific Ligand in Egg Activation?
      References

I. Introduction

Gametogenesis results in the formation of two highly and terminally differentiated, nonproliferative haploid cells—the sperm and the egg. In the absence of fertilization, these cells have a relatively short life-span: eggs typically die within

48 hr after ovulation. The interaction of these gametes at fertilization restores the ability of the fertilized egg to proliferate and develop into a new organism.

In animals, the interaction of sperm and egg membranes ultimately gives rise to a series of cellular responses in the egg that are required to initiate embryonic development. In most organisms, these responses occur in a temporal fashion and are classified as "early" and "late" events. Early events include the transient rise in intracellular Ca^{2+} and the consequent cortical granule (CG) exocytosis. The contents of these CGs modify the extracellular coats surrounding eggs and result in a block to polyspermy.[1] Late events include resumption of meiosis (and cell cycle), changes in intracellular pH, recruitment of maternal mRNAs, pronucleus formation, initiation of DNA synthesis, and cleavage.

Many of these egg activation events (e.g., ionic changes, exocytosis, mitogenesis) also occur in somatic cells in response to soluble ligands or extracellular matrix (ECM) ligands. In somatic cells, many of these responses are mediated by cell surface receptors that are coupled to specific intracellular effector systems that amplify the initial cell surface interactions via the production of second messengers. Similar to somatic cells, recent work in mammalian and nonmammalian eggs suggests that egg activation is mediated by signal transduction pathways. It should be noted, however, that unlike the case in somatic cells where the ligand is presented to a significant fraction of the cell surface, signal transduction at the time of fertilization emanates from a localized and restricted interaction of a specific region of the sperm head and egg plasma membrane. For example, assuming that the sperm head is spherical (and of average diameter of 2 μm) and that one hemisphere interacts with an egg (of average diameter of 80 μm), then less than 0.05% of the egg surface interacts with the sperm. This value is likely to be an overestimate, since it does not take into account the additional surface area of the egg generated by the presence of microvilli, nor the fact that only a very restricted region of the sperm head (and not an entire hemisphere) interacts with the egg. Since signal transduction and effector activation in the egg are likely to occur via highly conserved mechanisms, eggs may utilize very efficient signal amplification pathways distal to sperm–egg interaction to achieve the desired cellular response. This efficient amplification could result from the activation of multiple parallel signal transduction effector pathways and/or "cross-talk" between such pathways. For purposes of discussion in this chapter, we will define a **ligand** as a soluble or immobilized molecule that interacts with an extracellular domain of a specific cell surface receptor; a **receptor** as a transmembrane protein whose activation by ligand binding results in receptor activation and signal transduction; a **signal transducer** as a protein that interacts with the activated receptor and, in turn, is activated; and an **effector** as a protein

---

[1]There are several instances of physiological polyspermy established at the cytoplasmic level (Elinson, 1986; Grandin and Charbonneau, 1992). For example, urodeles (salamanders and newts) display polyspermic fertilization, followed by elimination of the supernumerary sperm in the cytoplasm by cytoplasmic blebbing. Fertilization of these eggs, which do not have CGs, is not associated with a rise in cytoplasmic calcium.

2. Molecular Basis of Mammalian Egg Activation

whose interaction with the activated signal transducer results in the generation of an **intracellular messenger**.

The discussion in this chapter will be restricted to what is known about signal transduction mechanisms in the activation of mammalian eggs, with emphasis on the mouse, although relevant information obtained in other species will be drawn upon. The discussion will first focus on the sequence of events that accompanies fertilization in the mouse. This information will provide a framework within which to examine known and proposed signaling mechanisms regulating egg activation. The discussion will then focus on the role of several possible second messengers in egg activation, and will conclude by examining how the sperm–egg interaction may result in the production of these second messengers. The reader should consult the following reviews that explore egg activation in additional species, with different focuses and perspectives (Nuccitelli, 1991; Foltz and Lennarz, 1993; Miyazaki *et al.*, 1993; Myles, 1993; Whitaker and Swann, 1993).

II. Description of Fertilization of Mammalian Eggs

A. Events of Fertilization prior to Sperm–Egg Plasma Membrane Interaction

In the mammal, the first interaction between the sperm and egg is at the level of the zona pellucida (ZP), an extracellular coat that surrounds all mammalian eggs. The ZP is a unique product of the growing oocyte and is secreted and assembled during the period of oocyte growth (Bleil and Wassarman, 1980a; Shimizu *et al.*, 1983; Wassarman, 1988, and references therein). In the mammals studied to date, the ZP is composed of 2–4 glycoproteins, the number of which is species dependent (Dunbar *et al.*, 1991).

In the mouse, the ZP is composed of three glycoproteins called ZP1 (M_r = 200,000), ZP2 (M_r = 120,000), and ZP3 (M_r = 83,000) (Bleil and Wassarman, 1980b; Wassarman, 1988). Although acrosome-intact sperm establish primary binding with ZP3 via an O-linked carbohydrate moiety, the identity of this moiety (terminal α-galactose, terminal N-acetylglucosamine) is controversial (Bleil and Wassarman, 1988; Miller *et al.*, 1992). Following sperm binding, ZP3 induces the acrosome reaction of the bound sperm (Bleil and Wassarman, 1983). The acrosome-reacted sperm, which can no longer interact with ZP3, then binds to ZP2 (Bleil *et al.*, 1988); this interaction is termed secondary binding. Secondary binding is thought to be required for maintaining the association of the sperm with the ZP as the sperm progresses through the ZP. The interactions of acrosome-intact sperm with ZP3 and acrosome-reacted sperm with ZP2 are likely mediated by specific receptors and/or binding proteins on the plasma membrane overlying the intact acrosome and on the inner acrosomal membrane, respectively (Bleil and Wassarman, 1986; Vazquez *et al.*, 1989; Mortillo and Wassarman, 1991). The identity of these binding receptors/binding proteins has not been resolved (Leyton and Saling, 1989; Bleil and Wassarman, 1990; Leyton et al., 1992; Miller et al., 1992; Cheng et al., 1994; Kalab et al., 1994).

B. Interaction of Sperm and Egg Membranes prior to Fertilization

Following passage of the acrosome-reacted sperm through the ZP, the sperm gain access to the space between the egg plasma membrane and the inner aspect of the ZP (i.e., the perivitelline space). The initial contact between the membranes of the respective gametes apparently occurs on a microvillar-containing domain of the egg (Phillips, 1991); metaphase II-arrested eggs possess both microvillar-containing and microvillar-free domains. The microvillar-free domain overlies the metaphase spindle and may arise as a consequence of the limited release of CGs during oocyte maturation (Ducibella et al., 1990). It is interesting to note that in *Xenopus laevis* eggs, although sperm preferentially interact with the animal hemisphere, there are no overt morphological differences between the plasma membranes of the animal and vegetal hemispheres (Elinson, 1986).

In all mammals studied to date, the initial interaction of the acrosome-reacted sperm with the egg's plasma membrane occurs at the postacrosomal region of the sperm head (Yanagimachi, 1994). The geometry of this interaction suggests that there is a regionalization on the sperm surface of molecules that mediate sperm–egg membrane recognition, and it is interesting to note that the sperm surface is regionalized with respect to both lipid and protein domains (Primakoff and Myles, 1983, and references therein). Consistent with this hypothesis of localized functional domains on the sperm surface is that sea urchin sperm interact with the egg plasma membrane via the protein bindin, which is localized to the exposed sperm acrosomal vesicle (Vacquier and Moy, 1977).

C. Events of Fertilization after Sperm–Egg Plasma Membrane Interaction

The earliest detected event of mammalian egg activation is an increase in intracellular Ca^{2+} (Cuthbertson et al., 1981; Cuthbertson and Cobbold, 1985; Miyazaki et al., 1986,1992a; Fissore et al., 1992; Kline and Kline, 1992; Sun et al., 1992; Tombes et al., 1992). In lower species, the intracellular Ca^{2+} increase is causally related to CG exocytosis; a similar situation presumably occurs in mammalian eggs. A consequence of CG exocytosis is the release of CG constituents that modify the ZP and ultimately result in a ZP block to polyspermy. Two of the ZP proteins are modified. First, ZP3 is converted to a form called $ZP3_f$. A consequence of this conversion is that $ZP3_f$ loses its ability to bind acrosome-intact sperm and induce the acrosome reaction (Bleil and Wassarman, 1983). This modification is thought to be due to the release of a CG-derived glycosidase

2. Molecular Basis of Mammalian Egg Activation

(Miller et al., 1993), since O-linked carbohydrates are implicated in ZP3 interaction with sperm and since there is no apparent change in the electrophoretic mobility of $ZP3_f$ (Bleil and Wassarman, 1980c,1983). ZP2 is converted to a form called $ZP2_f$, and this conversion is thought to be mediated by a CG-derived protease (Moller and Wassarman, 1989). This proteolytic cleavage is detected by a shift in electrophoretic mobility (from $M_r = 120,000$ to $90,000$) under reducing conditions (Bleil and Wassarman, 1981). A biological consequence of this conversion is that $ZP2_f$ fails to bind to acrosome-reacted sperm (Bleil and Wassarman, 1986). These ZP modifications constitute the ZP block to polyspermy, since acrosome-intact sperm cannot establish primary binding with $ZP3_f$ and acrosome-reacted sperm cannot maintain secondary binding with $ZP2_f$. Consequently, sperm fail to bind to and penetrate the ZP.

In contrast to lower species, there is no experimental evidence for a rapid electrical block (msec) to polyspermy at the level of the plasma membrane in mammals (Jaffe et al., 1983), but this does not exclude the possibility of a plasma membrane polyspermy block (mouse, Horvath et al., 1993, hamster, Stewart-Savage and Bavister, 1988; rabbit, Wolf, 1981). A teleological argument could be made for the absence of a need for a rapid electrical block to polyspermy, since the sperm-to-egg ratio at the site of fertilization (the ampullary region of the oviduct) is extremely low, on the order of 1:1 to 10:1 (Zamboni, 1972; Cummins and Yanagimachi, 1982; Kopf and Gerton, 1991).

In addition to establishing a block to polyspermy, fertilization also results in the resumption of meiosis and entry into the first cell cycle. In the mouse, these events entail, at the morphological level, the extrusion of the second polar body, decondensation of the sperm head and its transformation into a male pronucleus, formation of the female pronucleus, and pronuclear migration and syngamy. At the biochemical level, fertilization triggers the recruitment of maternal mRNAs (Cascio and Wassarman, 1982) and numerous changes in the pattern of protein synthesis (Latham et al., 1991), many of which result from post-translational modifications (van Blerkom, 1981). These changes in protein synthesis occur throughout the first cell cycle and are composed of specific protein subsets that are temporally expressed, but to date they have not been shown to be causally related to any of the aforementioned morphological changes. Since fertilization results in resumption of the cell cycle, there are changes in the activity of cell-cycle-related regulatory proteins (e.g., a decrease in the amount of a cdc2 kinase activity is observed shortly after insemination; Aoki et al., 1992, Verlhac et al., 1994). In the mammal, the mechanism(s) coupling fertilization to the modulation of the activity of these regulatory proteins is poorly characterized.

Collectively, these early and late morphological and biochemical events result in a characteristic set of "fertilization-associated" changes that serve as morphological and molecular markers of egg activation. Moreover, many of these events can be monitored or followed in individual eggs (e.g., changes in intracellular Ca^{2+}, ZP modifications, and pronuclear formation).

III. Intracellular Second Messengers That Regulate Egg Activation

A common approach used to implicate a potential signal transduction pathway is to examine the effect of specific intracellular messengers that would be generated by signal transduction–effector interactions on a specific cellular response. This approach has been used to study mammalian egg activation, and three second messengers—Ca^{2+}, inositol 1,4,5-trisphosphate (**IP$_3$**) and *sn*-1,2-diacylgylcerol (**DAG**)—have been studied in detail.

A. Calcium

1. Sperm-Induced Calcium Oscillations in Mammalian Eggs

As observed in nonmammalian eggs, fertilization of mammalian eggs results in a transient increase in the concentration of free intracellular Ca^{2+} (Miyazaki *et al.*, 1993, and references therein). Initial experiments conducted in mouse eggs injected with aequorin revealed an increase in intracellular Ca^{2+} following fertilization (Cuthbertson *et al.*, 1981; Cuthbertson and Cobbold, 1985). More extensive experiments using hamster eggs injected with aequorin (Miyazaki *et al.*, 1986) demonstrated that the increase in intracellular Ca^{2+} originates at the site of sperm binding to the egg's plasma membrane, approximately 10–30 sec following sperm attachment, and then spreads in a wavelike fashion throughout the egg. Subsequent to the sperm-induced Ca^{2+} wave is a series of additional Ca^{2+} waves. Although several subsequent increases in intracellular Ca^{2+} occur following these waves, these later rises in Ca^{2+} occur synchronously throughout the egg. With the exception of ascidians (Speksnijder *et al.*, 1989), these multiple increases in intracellular Ca^{2+} are usually not observed in lower species.

These observations have been refined using Ca^{2+} imaging methods (Fissore *et al.*, 1992; Kline and Kline, 1992; Miyazaki *et al.*, 1992a; Sun *et al.*, 1992; Tombes *et al.*, 1992) that provide temporal, spatial, and quantitative information regarding the changes in intracellular Ca^{2+} concentration following sperm attachment. The initial Ca^{2+} transient in the mouse egg following fertilization is rather prolonged (about 3–4 min), and rises from a basal concentration of 50–100 nM to concentrations that can reach 1 μM (Kline and Kline, 1992). In the hamster, this initial transient, which occurs in the absence of extracellular Ca^{2+} (Igusa and Miyazaki, 1983) and reaches concentrations of 500–600 nM (Miyazaki *et al.*, 1992b), is followed by several subsequent transients that are shorter in duration (30–60 sec) and that occur at relatively constant intervals spaced 2–4 min apart. In contrast to the initial Ca^{2+} wave in response to sperm, each of these subsequent Ca^{2+} transients is preceded by a slow increase in intracellular Ca^{2+} concentration that is followed by an abrupt increase once an apparent threshold in

2. Molecular Basis of Mammalian Egg Activation

intracellular Ca^{2+} is reached. In addition, although the amplitude of the increase in Ca^{2+} decreases with time, these oscillations can occur for several hours following fertilization (mouse, Kline and Kline, 1992; cow, Fissore et al., 1992; pig, Sun et al., 1992). Finally, as shown with aequorin-injected eggs, these subsequent Ca^{2+} transients arise simultaneously throughout the egg cytoplasm.

Although the initial intracellular Ca^{2+} rise in response to sperm is independent of extracellular Ca^{2+}, these later Ca^{2+} transients require extracellular Ca^{2+}. A possible explanation for this later dependence on extracellular Ca^{2+} is that the Ca^{2+} that gives rise to the initial increase is derived initially from intracellular Ca^{2+} stores and is efficiently pumped into the extracellular milieu. The influx of extracellular Ca^{2+} is then required to refill these depleted internal stores that are then used for the subsequent Ca^{2+} oscillations (Igusa and Miyazaki, 1983).

Although repetitive Ca^{2+} transients are observed in all mammalian eggs studied to date, and therefore are likely to be a general phenomenon in fertilized mammalian eggs, species differences in these oscillations are seen. For example, in the hamster the initial Ca^{2+} transient is composed of 3-4 discrete increases that are separated by a very short time interval (Miyazaki et al., 1992b), whereas in the mouse the initial Ca^{2+} transient is not composed of such discrete increases. In the cow, both the duration of the Ca^{2+} transients (about 100 sec) and the time interval between the Ca^{2+} transients (about 15 min) are longer (Fissore et al., 1992) than in the rodent. Similarly, a longer interval (18 min) between Ca^{2+} oscillations is observed in the pig (Sun et al., 1992). The basis for these differences and any biological sequelae are not known at this time (see Section III,A,3).

2. Role for Calcium in Egg Activation

As in lower species (e.g., Steinhardt and Epel, 1974), changes in intracellular Ca^{2+} concentrations are pivotal to successful mammalian egg activation. Microinjection of eggs with Ca^{2+} results in egg activation (Fulton and Whittingham, 1978). Likewise, treatment of eggs with an ionophore (e.g., A23187), ethanol, or electrical stimulation results in an increase in intracellular Ca^{2+} and egg activation (Miyazaki et al., 1993). In this regard, activation by these treatments appears to mimic fully the activation induced by sperm. A block to polyspermy is mounted, since CGs are released and the resultant modifications to the ZP occur (Ducibella, 1991, and references therein). The cell cycle resumes and the parthenogenotes can develop to the blastocyst stage.

In a reciprocal fashion, blocking the fertilization or ionophore-induced increase in intracellular Ca^{2+} inhibits egg activation, and this is similar to that observed in lower species (e.g., Kline, 1988). Introduction of the Ca^{2+} chelating agent 1,2-bis-(O-aminophenoxy)-ethane-N,N,N',N'-tetraacetic acid (BAPTA) into eggs by incubating the eggs in medium containing the membrane-permeable acetoxymethyl ester of BAPTA (BAPTA AM) results in a concentration-dependent inhibition of the fertilization or ionophore-induced Ca^{2+} transients, CG exocytosis, and second polar body emission (Kline and Kline, 1992). Interestingly, the inhibition of CG exocytosis appears less sensitive than the inhibition of second polar body emission. Thus, a lower Ca^{2+} concentration may promote CG exocytosis, when compared to the Ca^{2+} concentration required for cell cycle resumption. This possibility could be examined by buffering intracellular Ca^{2+} concentrations to set values by microinjecting Ca^{2+}–BAPTA buffers and determining, in the absence of sperm, the Ca^{2+} threshold for each of these events.

3. Biological Implications of Fertilization-Induced Calcium Transients

Although fertilization-induced Ca^{2+} transients are observed in mammalian (and ascidian) eggs—they have not been reported in eggs of other species—the biological significance of such transients is unresolved. These changes may be related to events that occur subsequent to the initial transient increase, and may require an increase in intracellular Ca^{2+}. For example, in the mouse and hamster, Ca^{2+}-stimulated CG exocytosis is completed approximately 15 min following fertilization (Fukuda and Chang, 1978; Stewart-Savage and Bavister, 1991). Since the duration of the initial Ca^{2+} transient is much shorter than this, the Ca^{2+} oscillations may be coupled to the periodic release of CGs, and this has been demonstrated in hamster eggs (Kline and Stewart-Savage, 1994). In a similar vein, the development of mouse eggs subjected to electrical pulses increases with an increase in the number of such pulses (Vitullo and Ozil, 1992). Thus, the Ca^{2+} transients that occur at later times may somehow be coupled to events that occur during the first cell cycle that promote later developmental events.

The Ca^{2+} oscillations in these cells could, in principle, permit the cell to maintain an elevated and sustained Ca^{2+}-dependent response in the face of oscillating Ca^{2+} concentrations, without the associated cellular toxicity that might occur if Ca^{2+} concentrations were continuously maintained at elevated concentrations. The rationale for this is that proteins whose activities are modulated by Ca^{2+} will bind Ca^{2+} during the transient. Despite the rapid decrease in the intracellular free Ca^{2+} concentration that occurs during these transients, the modulation of these proteins will continue as long as Ca^{2+} is bound, and this is dictated by the dissociation constant of Ca^{2+} for each particular target protein. As long as the affinity of the protein for Ca^{2+} is sufficiently high, the protein will continue to be modulated even though the concentration of intracellular Ca^{2+} has returned to basal levels.

Caution should be exercised in adopting this scenario, however, since treatment with Ca^{2+} ionophore induces only a single Ca^{2+} transient that is of a duration and magnitude similar to that occurring after fertilization (Kline and Kline, 1992) but elicits the full complement of early and late events of egg activation, as well as promotes parthenogenetic development. What is common to both fertilization and ionophore activation of eggs is an increase in the basal intracellular concentration of free Ca^{2+} following the initial Ca^{2+} transient. The

2. Molecular Basis of Mammalian Egg Activation

repetitive Ca^{2+} transients that occur following fertilization are associated with a slow but steady increase in the basal concentration of intracellular Ca^{2+} following the initial Ca^{2+} transient. Likewise, the transient increase in intracellular Ca^{2+} concentration following treatment of eggs with Ca^{2+} ionophore is followed by a decrease that does not reach the initial Ca^{2+} concentration. Thus, although some events subsequent to the initial Ca^{2+} transient may depend on these later Ca^{2+} oscillations (e.g., enhanced embryonic development), the resetting of the intracellular Ca^{2+} concentration to a higher steady-state value may also contribute to the modulation of such later Ca^{2+}-modulated events.

4. Potential Targets for Calcium Action

The proteins that are Ca^{2+} regulated and are involved in mammalian egg activation are not known at this time. Results of experiments in *Xenopus laevis*, however, suggest that the calmodulin-dependent protein kinase II ($CaMK_{II}$) may be responsible for the inactivation of maturation-promoting factor (MPF) that occurs following fertilization (Lorca *et al.*, 1993).

MPF, a heterodimer composed of p34^{cdc2} and cyclin B, possesses an intrinsic protein kinase activity that is essential for entry into M phase (Murray, 1993). The activity of MPF dramatically increases with entry into M phase and then abruptly decreases during the metaphase-to-anaphase transition; this decrease is due to cyclin proteolysis. In metaphase II arrested eggs, the activity of MPF remains elevated, and following the fertilization-induced Ca^{2+} transient, there is a rapid loss of MPF activity (Verlhac *et al.*, 1994).

Addition of a constitutively activated form of $CaMK_{II}$ to metaphase II-arrested *Xenopus laevis* egg extracts results in both the degradation of cyclin and the loss of p34^{cdc2} kinase activity (Lorca *et al.*, 1993). Moreover, microinjection of this form of $CaMK_{II}$ into metaphase II-arrested *Xenopus laevis* eggs also inactivates the p34^{cdc2} kinase. These changes, which are required for egg activation, occur in the absence of a Ca^{2+} transient. Finally, microinjection of the autoinhibitory peptide of $CaMK_{II}$ into metaphase II-arrested eggs inhibits cyclin destruction following parthenogenetic activation, and hence inhibits egg activation. The role of $CaMK_{II}$ in mammalian egg activation is not known.

Another potential target for Ca^{2+} action is the Ca^{2+} and phospholipid-dependent protein kinase, protein kinase C (see Section II,C, for further discussion).

B. IP$_3$

1. Implication of IP$_3$ in Egg Activation

The observation that treatment of eggs in Ca^{2+}-free medium with Ca^{2+} ionophores results in an increase in intracellular Ca^{2+} concentration indicates that the source for this released cation is intracellular. There are at least two intracellular Ca^{2+} stores, namely, mitochondrial and endoplasmic reticular. The release of Ca^{2+} from the endoplasmic reticulum is mediated in somatic cells by either the IP_3 or the ryanodine receptor (Sorrentino and Volpe, 1993). Results of immunoblotting experiments indicate that hamster (Miyazaki et al., 1992b) and mouse (T. Ayabe, R. Schultz, and G. Kopf, unpublished results) eggs possess the IP_3 receptor (see Section III,B,4 for discussion of the ryanodine receptor).

Microinjection of IP_3 elicits a regenerative Ca^{2+} transient that mimics closely the first transient that occurs following fertilization with respect to both its temporal and its spatial components (Miyazaki, 1988; Fujiwara et al., 1993). Repetitive Ca^{2+} transients are observed in mouse eggs when IP_3 is microinjected continuously (Swann et al., 1989) or generated following light-induced release from caged IP_3 (Peres et al., 1991). Repetitive Ca^{2+} transients are also observed when high concentrations of inositol 1,4,5-trisphosphorothioate, which is poorly metabolized, are injected into rabbit eggs (Fissore and Robl, 1993). It should be noted that in *Xenopus laevis* eggs, the Ca^{2+} wave that follows fertilization requires IP_3 production, and that this IP_3 is most likely generated from phospholipase C-stimulated hydrolysis of phosphatidylinositol 4,5-bisphosphate, since microinjected antibodies against phosphatidylinositol 4,5-bisphosphate inhibit the sperm-induced Ca^{2+} wave (Nuccitelli et al., 1993).

These Ca^{2+} transients are due to the IP_3-mediated release of Ca^{2+} from intracellular Ca^{2+} stores, since microinjection of hamster eggs with a monoclonal antibody (18A10) that recognizes the IP_3 receptor inhibits IP_3-mediated Ca^{2+} release in a concentration-dependent manner (Miyazaki et al., 1992b); the antibody binds to a domain that does not inhibit binding of IP_3 to the receptor but does inhibit IP_3-induced Ca^{2+} release. In contrast, another monoclonal antibody that also binds to the IP_3 receptor but does not inhibit IP_3-mediated Ca^{2+} release does not inhibit the Ca^{2+} transients induced in response to microinjected IP_3. The mechanism by which IP_3 induces the regenerative release of Ca^{2+} from intracellular stores appears to be by a Ca^{2+}-sensitized, IP_3-mediated Ca^{2+} release mechanism (see Miyazaki et al., 1993, for detailed discussion).

2. Effects of IP_3 on Early and Late Events of Egg Activation

Microinjection of IP_3 into mouse eggs results in modifications of both ZP2 and ZP3 (Kurasawa et al., 1989). ZP2 is converted to $ZP2_f$; the EC_{50} for microinjected IP_3 is 5–10 nM, which is similar to that required to induce a Ca^{2+} transient in hamster eggs (Miyazaki, 1988) and CG exocytosis in sea urchin (Whitaker and Irvine, 1984), and *Xenopus laevis* (Picard et al., 1985) eggs. The extent of the $ZP2$-to-$ZP2_f$ conversion is similar to that following fertilization, and is observed in the absence of extracellular Ca^{2+} (Kurasawa et al., 1989). Moreover, inositol phosphates that are ineffective in releasing Ca^{2+} from intracellular stores do not elicit this conversion.

Microinjection of IP_3 also results in the loss of both biological activities of ZP3, that is, its ability to interact with acrosome-intact sperm and to induce the acrosome reaction (Kurasawa et al., 1989). When assessing the ability of ZP3 to interact with sperm, it is important to assay "primary binding" that is mediated by ZP3 in the absence of "secondary binding" that is mediated by ZP2. The reason for this is that treatment of eggs with agents known to induce the ZP2-to-$ZP2_f$ conversion may result in a decrease in the number of sperm bound to these eggs in the absence of a change in the biological activity of ZP3, which is assayed by determining the number of sperm bound to eggs. Any decrease in the number of bound sperm could result from the binding of acrosome-intact sperm to unmodified ZP3 and their subsequent ability to undergo the acrosome reaction. These acrosome-reacted sperm will then not be able to establish secondary binding with ZP2 due to its conversion to $ZP2_f$, and hence will not be able to maintain their association with the ZP. The use of sperm that can bind to the ZP but not undergo the acrosome reaction is required to distinguish between primary and secondary binding (Kurasawa et al., 1989).

Although microinjected IP_3 promotes the full complement of ZP modifications, it does not induce either the recruitment of maternal mRNAs that occurs following fertilization/ionophore-induced egg activation or cell cycle resumption, as manifested by emission of the second polar body and pronucleus formation (Kurasawa et al., 1989). One possible explanation for the dissociation of these different egg activation events is that only a single Ca^{2+} transient is required for CG exocytosis, but multiple transients are required for cell cycle resumption and progression (see Section III,A,3, for cautionary note), and microinjection of physiological concentrations of IP_3 results in a single Ca^{2+} transient (Miyazaki, 1988). Nevertheless, it is still puzzling that ionophore treatment elicits the full complement of egg activation events although it induces only a single Ca^{2+} transient that is very similar in its amplitude to that induced by microinjected IP_3. At present, there is no explanation for this apparent paradox.

3. Implication of IP_3 during Fertilization

A fertilization-induced increase of IP_3 is detected in eggs of lower species by either examining polyphosphoinositide turnover (Turner et al., 1984; Ciapa and Whitaker, 1986; Ciapa et al., 1992) or measuring IP_3 mass (Stith et al., 1993, 1994). The paucity of biological material and the inherent asynchrony in fertilization of mammalian eggs have prohibited direct measurement of changes in IP_3 concentration during the course of fertilization. Nevertheless, microinjected IP_3 receptor antibody (18A10) inhibits sperm-induced mouse egg activation (Xu et al., 1994). The ZP2-to-$ZP2_f$ conversion is inhibited in a concentration-dependent manner by this antibody, and, as anticipated, these eggs are highly polyspermic. Moreover, a similar concentration dependence is observed for inhibiting both emission of the second polar body and pronucleus formation. Thus, Ca^{2+} released by an IP_3-mediated mechanism appears necessary, but not sufficient, to elicit both early and late events of egg activation, since mouse eggs microinjected with IP_3 undergo the early but not the late events (see Section III,B,2).

4. Ryanodine Receptor and Cyclic ADP Ribose

Calcium can also be released from intracellular stores via the ryanodine receptor (Sorrentino and Volpe, 1993). There are conflicting reports regarding the presence of functional ryanodine receptors in mouse eggs. The results of one study reported that microinjection of ryanodine (μM final intracellular concentrations) into mouse eggs results in an increase in intracellular Ca^{2+} concentration (Swann, 1992), whereas in another study no such increase was observed (Kline and Kline, 1994). The basis for these differences has not been resolved nor was it documented that ryanodine treatment of mouse eggs results in any event of activation that occurs following an increase in Ca^{2+} concentration, for example, CG exocytosis or emission of the second polar body.

It should be noted that ryanodine receptors display biphasic responses with regard to Ca^{2+} release when exposed to different concentrations of ryanodine (Meissner, 1986); that is, nanomolar ryanodine concentrations release Ca^{2+}, whereas micromolar concentrations inhibit Ca^{2+} release. In fact, mouse eggs microinjected with nanomolar final concentrations of ryanodine undergo the ZP2-to-$ZP2_f$ conversion, which is indicative of a rise in intracellular Ca^{2+} and CG exocytosis; this is not observed with micromolar concentrations of ryanodine (T. Ayabe, R. Schultz, and G. Kopf, unpublished results). No evidence for cell cycle resumption (e.g., no decrease in histone H1 kinase activity), however, is observed at any of the ryanodine concentrations tested. Moreover, the ZP2-to-$ZP2_f$ conversion elicited by nanomolar ryanodine concentrations is not inhibited by the antibody 18A10, which inhibits IP_3-mediated Ca^{2+} release, and micromolar concentrations of ryanodine do not inhibit the IP_3-induced ZP2-to-$ZP2_f$ conversion. These data suggest that IP_3 and ryanodine act separately on their receptors. Although mouse eggs appear to have a functional ryanodine receptor, it does not appear that Ca^{2+} released from this receptor is involved in sperm-induced egg activation. The basis for this is that whereas microinjection of the IP_3 receptor antibody 18A10 inhibits early and late events of sperm-induced egg activation (Xu et al., 1994), microinjection of ryanodine (micromolar intracellular concentrations) does not (T. Ayabe, R. Schultz, and G. Kopf, unpublished results).

Cyclic ADP ribose (cADPR), which is synthesized from NAD, can release Ca^{2+} in sea urchin egg extracts in an IP_3-independent fashion (Galione et al., 1991), and although the ryanodine receptor is a possible candidate for the cADPR receptor, the identity of this receptor has not been established. Inhibiting either the IP_3 receptor by heparin injection or the ryanodine receptor by ruthenium red or 8-amino cADPR does not prevent sperm-induced sea urchin egg activation. Thus, the capacity of sea urchin eggs to raise intracellular Ca^{2+} concentrations in response to either cADPR or IP_3 may provide a redundancy in

2. Molecular Basis of Mammalian Egg Activation 33

Ca^{2+} release mechanisms that operate during fertilization (Galione et al., 1993; Lee et al., 1993; Buck et al., 1994).

Mouse eggs microinjected with cADPR undergo the ZP2-to-ZP2$_f$ conversion, and this conversion is inhibited by high concentrations of ryanodine (T. Ayabe, R. Schultz, and G. Kopf, unpublished results); cell cycle resumption is not observed following microinjection of cADPR. Although cADPR is likely to release Ca^{2+} from ryanodine-sensitive stores, the role of cADPR in sperm-induced mouse egg activation is unlikely, since high concentrations of ryanodine do not inhibit fertilization (see above).

C. Diacylglycerol

Diacylglyerol (sn-1,2 diacylglycerol) and its mimetics, 4β-phorbol diesters, activate protein kinase C (PKC). Activation of PKC by DAG or phorbol diester treatment can result in exocytosis in many systems. Treatment of mouse eggs by either a continuous exposure to a DAG or by biologically active phorbol diesters results in both CG exocytosis and ZP modifications (Endo et al., 1987a,b; Ducibella et al., 1993). The ZP2-to-ZP2$_f$ conversion is apparently normal and occurs to an extent similar to that following fertilization. The effect of PKC activators on the ZP3-to-ZP3$_f$ conversion, however, is more complicated.

ZP3 isolated from phorbol diester-treated eggs possesses the same ability to inhibit sperm binding to eggs in sperm-binding assays as ZP3 isolated from untreated eggs (Endo et al., 1987b), suggesting that the sperm-binding activity of the ZP3 from these eggs is not altered. The ability of ZP3 from such treated eggs to induce the acrosome reaction, however, is altered. In the mouse, discrete stages of the acrosome reaction can be observed following incubation of the sperm with chlortetracycline, which is fluorescent when bound to Ca^{2+} in a hydrophobic environment (Saling and Storey, 1979). Capacitated, acrosome-intact sperm give rise to a B pattern in which a bright fluorescence is seen over the head and midpiece while no fluorescence is observed over the postacrosomal region. With the initiation of the acrosome reaction, this pattern converts to the S pattern, which represents an intermediate stage that appears prior to completion of the acrosome reaction and correlates with the loss of ionic gradients (Lee and Storey, 1985). The S pattern is characterized by bright fluorescence over the midpiece and irregular patterns over the anterior head and postacrosomal regions. The S pattern then gives rise to the AR pattern, which is characterized by a complete loss of fluorescence over the anterior portion of the head while the midpiece remains fluorescent. Sperm treated with ZP3 isolated from phorbol diester-treated eggs undergo the B-to-S transition but then accumulate in the S pattern (Endo et al., 1987a,b); that is, they do not complete the acrosome reaction. This inhibition is reversible, since treatment of these sperm with Ca^{2+} ionophore or solubilized ZP, each of which induces the acrosome reaction, initiates the S-to-AR transition (Kligman et al., 1991). The molecular basis for how phorbol diester treatment of eggs results in dissociating the ability of ZP3 to interact with sperm while inducing a partial acrosome reaction is not understood, especially in light of the fact that the extent of CG exocytosis is similar to that observed following fertilization. This issue is discussed in greater detail elsewhere (Kopf and Gerton, 1991).

A brief treatment of hamster eggs with activators of PKC rapidly induces emission of the second polar body (Gallicano et al., 1993). In contrast, continuous treatment of mouse eggs with these activators results in CG exocytosis, but emission of the second polar body and pronucleus formation are not observed (Endo et al., 1987a). This apparent discrepancy has been recently addressed (G. Moore, G. Kopf, and R. Schultz, unpublished results). A brief treatment of metaphase II-arrested hamster eggs with phorbol diesters or a DAG results in the formation of second polar body-like structures commencing 5 min after treatment, a remarkable increase in filamentous actin in the region of these polar body-like structures, and the disassembly of spindle microtubules. Cell cycle resumption, as assessed by a decrease in H1 kinase activity (i.e., cdc2/cyclin B1), and chromosome separation are not detected. Treatment of mouse eggs with these activators does not result in the formation of these polar body-like structures, does not cause an increase in filamentous actin, and does not result in cell cycle resumption; it does, however, induce disassembly of the spindle microtubules. Thus, the "apparent" activation of hamster eggs by activators of PKC is due to the effect of these agents on the cytoskeleton, which gives rise to structures that appear similar to polar bodies, but without any evidence of cell cycle resumption. The different responses seen in mouse and hamster eggs are likely due to differences in the sensitivity of the cytoskeleton to rearrangements induced by these agents, since there appear to be differences in the cytoskeletal architecture between the eggs of these two species (Gallicano et al., 1994).

Based on the aforementioned results, targets for PKC action in mammalian eggs are likely to include the exocytotic machinery, including the CGs, and the cytoskeleton. It appears unlikely that resumption of the cell cycle, that is, the metaphase-to-anaphase transition, is regulated by PKC.

IV. Maturation-Associated Changes in CG Exocytosis in Response to Second Messengers

A. Calcium Oscillations during Oocyte Maturation

Fully grown oocytes are arrested in the first meiotic prophase. Ovulation is coupled to the resumption of meiosis, which is characterized by chromosome condensation, breakdown of the germinal vesicle, and emission of the first polar body and arrest at metaphase II. Calcium oscillations also occur during mouse and hamster oocyte maturation *in vitro* (Carroll and Swann, 1992; Fujiwara et al., 1993). The temporal aspects of these transients are similar to those that

2. Molecular Basis of Mammalian Egg Activation

follow fertilization in that each transient is preceded by a slow rise in Ca^{2+} concentration that is followed by a rapid increase and decrease. The increases, however, occur uniformly throughout the cytoplasm (Fujiwara *et al.*, 1993) and therefore spatially resemble the Ca^{2+} transients that occur subsequent to the initial sperm-induced transients. Moreover, the amplitude of the increase in both species is less than that which occurs following fertilization. For example, whereas the peak intracellular free Ca^{2+} concentration is 500–600 nM in the hamster egg following fertilization, peak values of about 200 nM are observed during oocyte maturation in this species (Fujiwara *et al.*, 1993). Likewise, the maximum value reached during mouse oocyte maturation is about 700 nM (Carroll and Swann, 1992), which is less than the value of 1 μM that is reached following fertilization.

In these rodents, the maturation-associated Ca^{2+} transients are apparently mediated by an IP_3-receptor-based mechanism (Carroll and Swann, 1992; Fujiwara *et al.*, 1993). The transients are inhibited by heparin, which inhibits IP_3-receptor-mediated Ca^{2+} release, but not by de-N-sulfated heparin, which not have these inhibitory properties. The IP_3 receptor antibody 18A10 also inhibits the transients.

The physiological role of these Ca^{2+} transients during oocyte maturation is unclear. Inhibiting them with either heparin or BAPTA does not inhibit the fraction of oocytes that undergoes germinal vesicle breakdown (GVBD), although the effect on reaching metaphase II has not been closely examined (Carroll and Swann, 1992). Oocyte maturation is accompanied by a precocious release of about 30% of the CGs (Ducibella *et al.*, 1990). This release starts to occur following GVBD and results in the formation of a CG-free domain in the region of the metaphase II spindle. It is possible that these Ca^{2+} transients are involved in the precocious release of CGs, and the partial CG loss is related to the fact that the amplitude of the Ca^{2+} transients is significantly less than that of those that occur following fertilization or egg activation. These Ca^{2+} oscillations could also be involved in the appropriate cytoplasmic maturation required for successful fertilization and subsequent embryonic development.

B. Effect of Calcium and IP_3 on CG Exocytosis in Oocytes

Proper cytoplasmic maturation is required for successful fertilization and subsequent development. For example, the ability of the oocyte cytoplasm to support decondensation of the sperm nucleus develops during oocyte maturation (Perreault *et al.*, 1988). In a similar fashion, the ability of eggs to undergo CG exocytosis also develops during oocyte maturation (Ducibella *et al.*, 1990; Ducibella and Buetow, 1994). Treatment of GV-intact oocytes with A23187 does not result in CG exocytosis, and the ability of CGs to undergo exocytosis in response to A23187 increases subsequent to GVBD and reaches a maximum in the metaphase II-arrested egg. This acquired response to ionophore treatment is also observed with respect to the ability of microinjected IP_3 to induce a Ca^{2+} transient (Fujiwara et al., 1993). Moreover, whereas microinjected IP_3 induces CG exocytosis in mouse eggs, similar concentrations are virtually ineffective when injected into GV-intact oocytes (Ducibella et al., 1993) or meiotically incompetent oocytes (S. Kurasawa, R. Schultz, and G. Kopf, unpublished observations). It should be noted that a similar increase in response to microinjected IP_3 occurs during starfish oocyte maturation with respect to CG exocytosis (Chiba et al., 1990).

In the hamster, the Ca^{2+} response to microinjected IP_3 increases in a biphasic manner during oocyte maturation, and the first increase occurs following GVBD (Fujiwara et al., 1993). A maturation-associated decrease in oocyte cAMP occurs during oocyte maturation and is implicated in the resumption of meiosis (Schultz et al., 1983). In principle, this decrease could result in an increase in IP_3 receptor sensitivity to IP_3, since phosphorylation of this receptor by protein kinase A (PKA) can result in a diminished sensitivity to IP_3 (Supattapone et al., 1988). This does not seem to be the case, however, since incubation of metaphase II-arrested eggs in medium containing a membrane-permeable cAMP analog that should activate PKA, and hence result in the phosphorylation of the endogenous IP_3 receptor, does not suppress the ability of these eggs to elevate intracellular Ca^{2+} concentrations (Fujiwara et al., 1993) or undergo CG exocytosis in response to microinjected IP_3 (C. Williams, G. Kopf, and R. Schultz, unpublished results).

The second phase of the response to microinjected IP_3 occurs between metaphases I and II. During this time, the maturing oocytes develop a regenerative IP_3-induced Ca^{2+} release (Fujiwara et al., 1993). As described in Section III,B,1, this mechanism operates in the fertilized egg and is due to the positive feedback of Ca^{2+} on the IP_3 receptor. In other words, Ca^{2+} that is mobilized due to activation of the IP_3 receptor in turn sensitizes the receptor to further IP_3-induced Ca^{2+} release (Miyazaki et al., 1992b). Although the molecular basis for this sensitization is not known, it is interesting to note that an increase in the internal stores of Ca^{2+} occurs during oocyte maturation, with the largest change occurring between metaphase I and II (Tombes et al., 1992). Perhaps this change in the size of the internal Ca^{2+} pool is related to the maturation-associated acquisition of a regenerative IP_3-induced Ca^{2+} release mechanism, as well as to the ability of ionophore treatment to stimulate CG exocytosis at later maturational stages.

The Ca^{2+} sensitivity of CG exocytosis may account for the inability of microinjected IP_3 or ionophore to induce CG exocytosis in GV-intact oocytes, since the amount of Ca^{2+} released in such treated oocytes is significantly less than that released in eggs microinjected with IP_3. Alternatively, the maturation-associated acquisition of CGs to undergo exocytosis could be due to the maturation of the exocytotic apparatus. It should be noted, however, that treatment of either GV-intact oocytes or metaphase II-arrested eggs with PKC activators results in CG exocytosis and ZP modifications (Ducibella et al., 1993). Thus, the CGs that are present in the oocyte are capable of undergoing exocytosis in response to an appropriate stimulus. It will be interesting to determine whether buffering intracellular Ca^{2+} levels in GV-intact oocytes to concentrations similar to those detected during the peak of the fertilization-induced Ca^{2+} transient will result in CG exocytosis.

V. Signal Transduction of Pathways Leading to Egg Activation

A. Introduction

Integral to any discussion regarding the identity and role of specific intracellular effector pathways that modulate egg activation is a consideration of the mechanism by which the egg perceives signals from the sperm and transduces those signals via intracellular effectors. As stated earlier (Section I), unlike intercellular signaling that occurs via endocrine or paracrine pathways, the interaction of the sperm and egg occurs as a restricted and localized interaction between a specific region of the sperm head and the egg plasma membrane. Following this highly restricted interaction, any signal(s) generated at this site of interaction must then be efficiently amplified to give rise to the proper temporal and spatial changes required to regulate egg activation. Such an efficient signal amplification that arises through the activation of parallel and/or intersecting signal transduction pathways must, therefore, remain an important consideration in any discussion of signal transduction mechanisms mediating egg activation.

As in lower vertebrates and invertebrates (Kline *et al.*, 1990; Foltz and Lennarz, 1993), two hypotheses have been proposed to explain the signaling mechanisms by which sperm–egg interaction in the mammal leads to egg activation. The first is that the fertilizing sperm interacts with a specific egg surface receptor, and that this interaction leads to signal transduction and effector activation (receptor hypothesis). The second hypothesis is that following fusion of the sperm and egg plasma membranes, a soluble sperm-derived factor enters the egg's cytoplasm and activates pathways leading to egg activation (fusion hypothesis). As outlined below, these two hypotheses may not necessarily be mutually exclusive in all cases. It must be emphasized that there is at present no overwhelming experimental evidence to support either of these hypotheses. Nevertheless, each possibility can be tested experimentally.

B. Receptor Hypothesis of Egg Activation

The notion that egg activation occurs in response to receptor–effector activation evolved from the fact that many responses of the egg following fertilization are similar to responses of somatic cells following their interaction with specific ligands. Fertilization in the sea urchin egg is accompanied by an increased turnover of polyphosphoinositides, resulting in the generation of IP_3 and DAG (Turner et al., 1984; Ciapa and Whitaker, 1986; Ciapa et al., 1992), and fertilization of *Xenopus laevis* eggs results in an increase in mass of IP_3 (Stith et al., 1993, 1994). These results suggest that sperm activate an egg receptor that is coupled to phospholipid turnover, since such changes in polyphosphoinositide metabolism can occur in somatic cells in response to activation of either heterotrimeric guanine nucleotide-binding regulatory protein (G protein)-coupled receptors, growth factor receptor tyrosine kinases, or integrins (Nishizuka, 1988; Cybulsky et al., 1993).

1. G Protein Receptor Model

A model for how sperm–egg interaction results in the stimulation of G protein-coupled pathways that lead to egg activation is shown in Fig. 1. The interaction of a sperm-derived ligand(s) with an egg-associated sperm receptor results in the coupling to and activation of G proteins. The subunit dissociation that occurs upon activation of these G proteins results in the formation of two potential signal transducers— the α subunit and $\beta\gamma$ heterodimer. Each of these molecules can, in principle, interact with a variety of effector molecules (e.g., ion channels, adenylyl cyclase, phospholipase C-β; Birnbaumer, 1992; Clapham and Neer, 1993). Activation of such effectors could give rise to the generation of multiple second messengers that regulate early and late events of egg activation. For example, activation of a phospholipase C would result in the generation of IP_3 and DAG, each of which can elicit specific egg activation events (see Section III,B and C).

Several lines of investigation implicate G proteins in mammalian egg activation. Microinjection of mammalian eggs with GTPγS, a hydrolysis-resistant G protein activator, results in the release of intracellular Ca^{2+} to variable extents and the likely exocytosis of CGs (Cran et al., 1988; Miyazaki, 1988; Swann et al., 1989; Miyazaki et al., 1990; Swann, 1992). The ability to mimic these two events of egg activation in the absence of sperm supports the notion that G protein activation is an integral regulatory process mediating this event. This possibility is further supported by the observation that the microinjection of hamster eggs with the G protein antagonist GDPβS blocks the release of intracellular Ca^{2+} that occurs in response to the fertilizing sperm, but does not affect the subsequent Ca^{2+} release in response to IP_3 microinjection. At present, it is not known whether the effects of these agents are at the level of phosphoinositide turnover.

An additional approach to examining the role of G proteins in egg activation is to treat eggs with a variety of agonists that are known to activate G protein-coupled receptors, and to assess the effect of such agonists on egg activation. Although treatment of hamster eggs with 5-hydroxytryptamine (Miyazaki et al.,

2. Molecular Basis of Mammalian Egg Activation

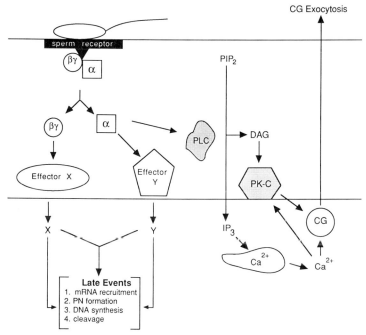

Fig. 1 Schematic depiction of sperm-induced activation utilizing a G protein-coupled receptor. In this model, the sperm interacts with a receptor that couples to effector systems via a G protein. Receptor occupancy by the sperm-associated ligand results in the activation of a G protein(s), as depicted by the dissociation of the α and βγ subunits. In this model, the α subunit activates a phospholipase C (**PLC**), which in turn hydrolyzes PIP_2 to produce a sn-1,2-diacylglycerol (**DAG**) and IP_3. Each of these molecules can bring about CG exocytosis, which is an early event of egg activation. The α subunit, as well as the βγ subunit, could also modulate other effectors (X and Y). Activation of these effectors gives rise to second messengers or ionic changes that, either separately or in cooperation, lead to late events of egg activation. Although not depicted, late events of egg activation may also require the convergence of X, Y, and PIP_2 turnover pathways.

1990) or mouse eggs with carbachol (Swann, 1992) leads to Ca^{2+} oscillations, suggesting a functional coupling to their cognate G protein-coupled receptors, the magnitude of these oscillations does not approach that of those seen following fertilization or microinjection of IP_3. Moreover, no signs of egg activation were noted in these studies, nor was it demonstrated that these effects were mediated by G proteins (i.e., inhibited by G protein antagonists; see subsequent discussion). Finally, the concentration of carbachol (100 μM) used in one of these studies is not physiological.

The inability to observe egg activation in these aforementioned experiments may be related to an insufficient level of expression of the appropriate endogenous receptor. Although activation of these receptors may be sufficient to generate an intracellular signal, the magnitude of such a signal may not be great enough to bring about biological sequelae. To circumvent this experimental limitation, an alternative experimental design is to overexpress a G protein receptor in eggs by microinjection of an exogenous mRNA that encodes such a receptor and then to challenge these eggs with the appropriate ligand and determine whether egg activation occurs. This latter approach has been used successfully in the mouse (Williams *et al.*, 1992; Moore *et al.*, 1993).

Microinjection of mouse eggs with an mRNA encoding the human m1 muscarinic receptor, which is known to couple to G proteins, results in eggs that respond to acetylcholine in a concentration-dependent manner with the conversion of ZP2 to $ZP2_f$ (Williams *et al.*, 1992). The acetylcholine effect appears to be mediated through the expressed receptor since this effect is blocked by atropine, a physiological receptor antagonist. Moreover, receptor activation appears to occur through G proteins, since the acetylcholine-induced $ZP2$-to-$ZP2_f$ conversion is blocked by microinjection of GDPβS, an inhibitor of G protein activation. The G protein that mediates this effect appears to be a member of the pertussis toxin-insensitive family, since the acetylcholine effect on the $ZP2$-to-$ZP2_f$ conversion is not blocked by pertussis toxin treatment; this toxin catalyzes the ADP-ribosylation and functional inactivation of specific classes of G proteins. Similar experiments carried out in *Xenopus laevis* eggs also demonstrate that early events of fertilization (e.g., fertilization potential) are completely insensitive to pertussis toxin (Kline *et al.*, 1991). Although the identity of this pertussis toxin-insensitive G protein is not known, it is unlikely to be G_s (Williams *et al.*, 1992). The pertussis toxin-insensitive G protein G_q is present in mouse eggs and is therefore a candidate for the role of mediating these aforementioned effects.

As described in Sections III,B and C, treatment of eggs with specific second messengers (e.g., IP_3) results only in early events of egg activation. Since the stimulation of multiple second messenger systems regulated by multiple effectors could be required to elicit a totally integrated response essential for both early and late events of egg activation, experimental intervention upstream of effectors may be more likely to initiate a complete response (Fig. 1). The immediate bifurcation of the signaling pathways inherent in such a signaling hierarchy would severely restrict the ability of experimental manipulations of specific second messenger pathways (e.g., IP_3) to bring about a full response (Kurasawa *et al.*, 1989), and this is what is observed. The receptor overexpression approach circumvents this limitation; in fact, complete egg activation occurs in the absence of sperm following treatment with acetylcholine in eggs expressing the human m1 muscarinic receptor (Williams *et al.*, 1992; Moore *et al.*, 1993). When the receptor mRNA-microinjected and acetylcholine-treated mouse eggs are cultured for several hours, ~30% form a pronucleus that supports DNA synthesis. This result is in contrast to 100% of the treated eggs displaying a $ZP2$-to-$ZP2_f$ conversion. In addition, this treatment results in the recruitment and translation of a set of maternal mRNAs very similar to those that are translated following fertiliza-

2. Molecular Basis of Mammalian Egg Activation

tion. Moreover, 95% of the treated eggs that form a pronucleus also cleave to the 2-cell stage and activate their genome. It should also be emphasized that the timing of each of these events is very similar to that observed following fertilization.

Two additional points regarding these experiments require comment. First, every acetylcholine-treated receptor-expressing egg undergoes the ZP2-to-ZP2$_f$ conversion, whereas only 30% of these eggs form a pronucleus. If the requirements needed to initiate CG exocytosis are less stringent than those required for late events of egg activation, then differences in the degree of translation of the receptor mRNA and the subsequent targeting of the protein to the membrane may account for the differential responsiveness of the eggs in this experimental paradigm. Consistent with this proposal is that CG exocytosis appears less sensitive than polar body emission to modulation of intracellular Ca^{2+} concentrations (Kline and Kline, 1992). Second, in contrast to the pertussis toxin insensitivity of the acetylcholine-induced ZP2-to-ZP2$_f$ conversion, pronucleus formation is inhibited ~50% by this toxin (Moore et al., 1993). The molecular basis for this difference remains an enigma.

Although results of these types of experiments support the role of G proteins in mediating egg activation, they do not address the issue of whether sperm-induced egg activation occurs in this manner. It is now essential that experimental approaches be taken to answer this particular question. Although the observations that an inhibitory effect of microinjected GDPβS on the release of intracellular Ca^{2+} in response to the fertilizing sperm support such a notion (Miyazaki, 1988), such experiments must be interpreted with caution since this treatment could block the function of monomeric low molecular weight G proteins, as well as of heterotrimeric G proteins. Moreover, effects of these poorly hydrolyzable GDP analogs on events other than those activated by the fertilizing sperm should also be considered (Crossley et al., 1991).

A number of different investigations are currently under way to answer this particular question. Microinjected GDPβS blocks with a similar concentration dependence both sperm-induced ZP2-to-ZP2$_f$ conversion and the subsequent formation of pronuclei (Moore et al., 1994). The failure to mount a polyspermy block is manifested by an increased number of sperm in the perivitelline space and the presence of multiple sperm heads in the egg cytoplasm. It is unlikely that these GDPβS effects are attributed to inhibitory effects on low molecular weight G proteins of the *ras* family, since microinjection of antibodies against *ras* or microinjection of either constitutively active or dominant negative *ras* has no effect on fertilization (Moore et al., 1994). It should be noted, however, that microinjection of eggs with botulinum ADP-ribosyltransferase C3, which selectively ADP ribosylates and inactivates low molecular weight G proteins of the *rho* family (Aktories et al., 1990), has no effect on sperm-induced cell cycle resumption and CG exocytosis, but does inhibit the actin-based processes of emission of the second polar body and cleavage to the 2-cell stage (Moore et al.,

1994). In the ascidian *Halocynthia roretzi*, rho has been indirectly implicated in events of egg activation (Toratani *et al.*, 1993).

Consistent with a role for heterotrimeric G proteins mediating sperm-induced egg activation is the observation that microinjected phosducin, which is a retinal protein that has high affinity for free βγ subunits, inhibits fertilization (Moore *et al.*, 1994). These data are consistent with the notion that following fertilization, activation of G proteins and subsequent generation of free α and free βγ heterodimer are required for regulating specific aspects of egg activation. The formation of a phosducin–βγ complex would prevent βγ subunits from modulating intracellular effectors in the egg, as well as preventing multiple rounds of G protein association and dissociation that may be required for signaling; these modes of regulation are involved in phototransduction.

Another approach to examining the potential role of G proteins in sperm-induced egg activation is to inhibit selectively specific egg G proteins and then to determine whether specific egg activation events are perturbed after fertilization. For example, overexpression of an mRNA encoding a dominant-negative G protein α subunit construct (Gupta *et al.*, 1990) and microinjection of G protein antibodies that can inhibit function are two possible approaches. As a prerequisite to performing such experiments, however, a systematic determination of the types of G proteins must be undertaken. To date, $G_{s\alpha}$, $G_{i\alpha 1}$, $G_{i\alpha 2}$, $G_{i\alpha 3}$, $G_{o\alpha}$, and $G_{q\alpha}$ have been demonstrated to be present in mouse eggs (Jones and Schultz, 1990; Williams *et al.*, 1992).

2. Receptor Tyrosine Kinase Model

As stated earlier, many of the late events of egg activation can be considered mitogenic events (i.e., stimulation of metabolism, completion of meiosis, initiation of mitosis, activation of the embryonic genome). Based on these end points of cellular activation, as well as on the fact that fertilization in lower species is accompanied by changes in polyphosphoinositide turnover and an alkalinization in internal pH, both of which accompany many mitogenic responses in somatic cells, any receptor-mediated model of sperm-induced egg activation must consider the possibility that receptor tyrosine kinases and/or tyrosine phosphorylation plays a pivotal regulatory role in egg activation. A model for sperm activation of the egg-associated receptor tyrosine kinase is shown in Fig. 2. It should be noted that activation of such a receptor could, in principle, lead to the production of a set of second messengers similar to that produced by activation of a G protein.

In sea urchin eggs, tyrosine kinase activity is stimulated following fertilization and is accompanied by changes in the phosphorylation of a variety of egg proteins (Dasgupta and Garbers, 1983; Ribot *et al.*, 1984; Peaucellier *et al.*, 1988; Ciapa and Epel, 1991; Abassi and Folz, 1994; Moore and Kinsey, 1994). The pH shift that occurs following fertilization in these eggs may be responsible for the activation of a specific tyrosine kinase activity and subsequent protein

2. Molecular Basis of Mammalian Egg Activation 43

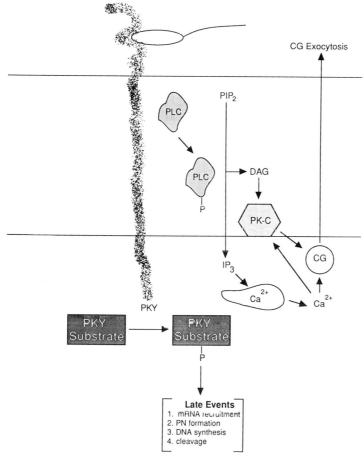

Fig. 2 Schematic depiction of sperm-induced egg activation utilizing a receptor tyrosine kinase. Sperm interaction with this class of receptors results in receptor tyrosine phosphorylation and subsequent interaction with downstream effectors. As with growth factor receptors in somatic cells, these interactions may occur through SH2 and SH3 domains on receptor and effector molecules. In this model, receptor activation results in the activation of a phospholipase C (**PLC**), which in turn hydrolyzes PIP_2 to produce a sn-1,2-diacylglycerol (**DAG**) and IP_3. Each of these molecules can bring about CG exocytosis, which is an early event of egg activation. Tyrosine phosphorylation of other intracellular effectors/substrates (**PKY substrate**) gives rise to second messengers or ionic changes that, either separately or in cooperation, lead to late events of egg activation. Note that activation of this signaling pathway could converge with G protein-mediated signaling pathways at several points.

phosphorylation in the cortex of these cells (Jiang et al., 1989,1991). To date, little is known about the role of tyrosine kinases and tyrosine protein phosphorylation in mammalian egg activation. Although changes in tyrosine phosphorylation of proteins occur following fertilization in mouse eggs (Endo et al.,

1986), the identity and role of such phosphoproteins in egg activation are not known. It should also be stressed that changes in tyrosine phosphorylation of proteins following fertilization do not necessarily implicate a receptor tyrosine kinase in egg activation. For example, tyrosine phosphorylation of the $M_r = 125{,}000$ focal adhesion kinase (pp125 FAK), which normally occurs during integrin-mediated signal transduction (Schaller et al., 1992), can occur in response to the occupancy of G protein-coupled neuropeptide growth factor receptors possessing seven-transmembrane-domain topologies (Zachary et al., 1992).

Microinjection and overexpression of specific receptor tyrosine kinases (e.g., epidermal growth factor receptor, EGFR; platelet-derived growth factor receptor, PDGFR) could be used to explore the potential roles of this class of receptors in egg activation. This approach has been used to study activation of *Xenopus laevis* eggs (Yim et al., 1994). An alternative approach is based on the fact that receptor tyrosine kinases use *src* homology domains (SH2 and SH3) to recruit specific effectors (Koch et al., 1991; Mayer and Baltimore, 1993). In addition, these domains are used by effectors to promote effector–effector interactions. Experimental approaches based on the specificity of these interactions could be exploited to study egg activation. For example, dominant-negative approaches utilizing microinjection of specific SH2 peptides (Songyang et al., 1993) might abrogate sperm-induced egg activation events. This approach has already been used successfully to study the mechanisms by which insulin, insulin-like growth factor 1 (IGF-1; Chuang et al., 1993), and progesterone (Muslin et al., 1993) regulate oocyte maturation in *Xenopus laevis*. These approaches may yield important information regarding the nature of putative receptor candidates on the egg surface for sperm ligands.

C. Fusion Hypothesis of Egg Activation

In the fusion hypothesis, egg activation is initiated by the introduction of a soluble sperm-derived factor (Fig. 3). The essential difference between this hypothesis and the receptor hypothesis is that, in the fusion hypothesis, sperm and egg plasma membrane fusion is an absolute prerequisite to signal transduction, whereas activation through a receptor-mediated process does not necessarily require membrane fusion. However, as will be discussed next, there is room for compromise in models that incorporate both of these mechanisms (see Section VI).

The features of sperm and egg plasma membrane fusion in relation to hallmarks of fertilization have been examined in greatest detail in the sea urchin, using both ultrastructural and electrical methods (Whitaker and Swann, 1993, and references therein). Due to the experimental limitations of these particular methods, it has been impossible to date to obtain agreement on the timing between sperm–egg plasma membrane interaction and membrane fusion. Results

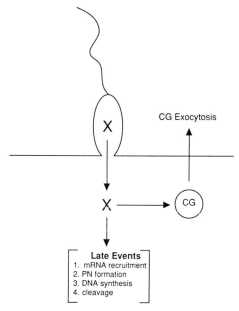

Fig. 3 Schematic depiction of sperm-induced egg activation by introduction of a soluble sperm activator. In this model, membrane fusion and cytoplasmic continuity are absolute prerequisites for egg activation. The introduction of sperm activator X, of unknown identity, in some fashion initiates signaling that leads to the coordinate events of early and late egg activation.

from such experiments suggest, however, that there is a significant time delay of 10–15 sec between the experimental determination of fusion and the initial onset of the Ca^{2+} wave preceding CG exocytosis (as reviewed by Whitaker and Swann, 1993). This delay, which is termed the "latent" period of egg activation (Allen and Griffen, 1958; Whitaker et al., 1989), is the basis for the hypothesis that egg activation might occur as a consequence of the release and diffusion of an egg activator from the sperm into the egg cytoplasm following sperm–egg plasma membrane fusion.

The presence of a latent period should not necessarily be construed as inconsistent with a receptor-mediated process in which intracellular effector activation is assumed to lead to immediate intracellular signaling. For example, it could be argued that there is a delay (latent period) between receptor–effector coupling and experimental detection of a response (e.g., Ca^{2+} transient). This delay may result from the localized nature of an initial ligand–receptor interaction. Such an interaction may have an intrinsic time dependence due to the need for the recruitment of effectors into a signal transduction complex. Moreover, there may be a time-dependent requirement to achieve threshold concentrations of the second messengers necessary to initiate signaling pathways and global responses that can be detected experimentally. To date, the presence of a latent period in mammalian eggs has not been examined, and therefore it is not clear whether this is a characteristic of the fertilization response of all eggs.

The concept of a diffusible egg activating factor was first tested by injecting a soluble sea urchin sperm extract into sea urchin eggs (Dale, 1985). Although this procedure results in the cortical reaction, there was no characterization of this extract, and it is clear that the amounts injected into the egg represent concentrations that would exceed those that would presumably diffuse into the egg cytoplasm following sperm–egg fusion.

Subsequent to this initial observation, two intracellular second messengers— IP_3 and cGMP—were proposed as sperm-derived diffusible activators of sea urchin eggs (Whitaker and Swann, 1993, and references therein). However, the effective concentrations of these second messengers in the sperm are not high enough to cause egg activation (see Whitaker and Swann, 1993). In the mammal, diffusion of these two second messengers from the sperm to the egg is also unlikely to be involved in egg activation, since microinjection of micromolar concentrations of IP_3 does not result in complete egg activation (Kurasawa et al., 1989) and since mammalian sperm do not contain detectable levels of guanylyl cyclase activity (Garbers and Kopf, 1980).

Ca^{2+} has also been proposed as a diffusible sperm factor responsible for egg activation (Jaffe, 1983). In the sea urchin, however, manipulation of the intracellular Ca^{2+} levels of the egg by either microinjection of this cation or manipulation of its endogenous concentration with chelating buffers does not initiate a Ca^{2+} wave similar to that seen following fertilization (Whitaker and Swann, 1993). This result is in contrast to observations made in hamster eggs, in which microinjection of Ca^{2+} mimics the Ca^{2+} transients and Ca^{2+} waves normally seen after fertilization (Miyazaki et al., 1992a,b). Nevertheless, multiple injections of Ca^{2+} do not induce a subsequent release of Ca^{2+} (Igusa and Miyazaki, 1983; Swann, 1992). This result suggests that Ca^{2+}-induced Ca^{2+} release mechanisms become desensitized, a situation not seen following fertilization when transient spikes of Ca^{2+} release continue. Other than these observations, little evidence exists for this cation to be the primary diffusible sperm factor responsible for egg activation.

Mammalian sperm extracts apparently contain a proteinaceous factor(s) that, when microinjected into eggs, can result in egg activation. Crude extracts of rabbit, hamster, or boar sperm (Stice and Robl, 1990; Swann, 1990,1992), when microinjected into rabbit, mouse, and hamster eggs, result in a series of membrane hyperpolarization responses and Ca^{2+} transients (Swann, 1990,1992) as well as in CG exocytosis, pronucleus formation, and cleavage (Stice and Robl, 1990). The activity is present in a high-speed soluble fraction of the sperm and is heat and trypsin labile, suggesting that the factor(s) is a protein (Stice and Robl, 1990). In contrast, extracts prepared in a similar manner from mouse sperm do not activate mouse eggs under conditions in which rabbit sperm extract activates mouse eggs. A concentration of the extract that corresponds to 2–5 whole sperm is required for maximal effects on egg activation (as assessed by pronucleus formation and cleavage). This concentration could be problematic, since it is anticipated that only a small fraction of the total sperm protein would be present in a diffusible compartment in the head. Moreover, the degree of egg activation in the experiments performed with this extract is quite variable. This variability will likely hinder the purification and characterization of the factor(s).

A similar set of experiments has been performed with crude soluble extracts from boar and hamster sperm (Swann, 1990). Microinjection of these extracts into hamster eggs initiates a repetitive series of hyperpolarizing responses and repetitive rises in intracellular Ca^{2+}, the kinetics of which are similar to those seen following fertilization. The boar sperm extract displays similar effects when microinjected into mouse eggs (Swann, 1992). The effect of the sperm extract on the hyperpolarization responses in the egg is sperm specific, since cytosol from brain or liver extracts fails to trigger these responses. The question of specificity is of interest, since Currie et al. (1992) subsequently demonstrated, using a boar sperm extract prepared in the same manner used by Swann (1990), that this extract could trigger Ca^{2+}-dependent currents in rat dorsal root ganglion neurons when applied intracellularly. Fractionation and preliminary characterization of the boar sperm extract suggest that the activity is associated with a proteinaceous fraction of $M_r > 100,000$. Although the responses of these eggs to the sperm extracts are very similar to those seen following fertilization, and are therefore of interest, little consideration is given to the relevance of the final concentration of protein injected into these eggs when compared to the concentration that would normally be delivered to the egg cytoplasm at the time of fertilization. Purification of the aforementioned factor(s) will be essential before an appreciation of its relevance to egg activation can be realized.

VI. Receptors on the Egg That Mediate Sperm-Induced Egg Activation

A. General Approaches to Studying Egg Surface Sperm Receptors

Any working hypothesis of egg activation that assumes transmembrane signal transduction tacitly assumes an egg surface-associated receptor for sperm. The properties of such a receptor must include, at a minimum, the ability to interact with specific components of the sperm and the ability to couple to transmembrane signal transducers and/or intracellular effectors. Moreover, sperm–egg binding and fusion may be mediated by multiple receptors or binding proteins. A number of different experimental approaches have been taken to characterize sperm-binding activity on eggs. To date, there is no consensus on the nature of mammalian egg-associated sperm receptors.

One approach to identify a potential receptor is to examine the effect of enzymatic digestion of the egg plasma membrane on sperm binding and fusion to ZP-free eggs. Treatment of ZP-free mouse eggs with proteases results in a decrease in sperm penetration, whereas treatment with phospholipases or glycosidases does not (Boldt et al., 1988). Although one interpretation of these data is that the loss of sperm binding and penetrating ability is due to the proteolytic modification of a putative sperm receptor(s) on the egg surface, such conclusions must be made with caution since these proteolytic enzymes could grossly perturb the egg membrane and lead to nonspecific inhibitory effects in the sperm-binding assays. In a similar vein, iodination of egg surface proteins reveals a number of radiolabeled proteins that are modified following either fertilization or treatment of the eggs with trypsin or chymotrypsin under conditions that abolish sperm binding and fusion (Boldt et al., 1989). Using this approach, an egg surface protein of $M_r = 94,000$ was identified as a putative sperm receptor (Boldt et al., 1989). However, these conclusions are based solely on correlations between the ability of protease-treated eggs to bind and fuse with sperm and changes in the molecular weights of the radiolabeled egg surface proteins.

Experiments were also designed to correlate the ability of protease-treated eggs to recover the capacity to be penetrated by sperm with the reappearance of the egg surface protein of $M_r = 94,000$ (Kellom et al., 1992). Protease treatment results in the loss of this protein, but during a recovery period a protein of similar M_r reappears; this reappearance correlates with the recovery of sperm penetration. This recovery is cycloheximide insensitive, whereas the synthesis of the egg surface protein of $M_r = 94,000$ is cycloheximide sensitive. One obvious interpretation of these data is that this protein is being replenished to the egg surface from a preexisting store. It must be emphasized in these studies that the recovery of sperm-penetration ability following protease treatment is never complete; that is, the degree to which sperm–egg fusion takes place over time never approaches the level of the untreated controls. An additional consideration is that an interaction between the sperm and the $M_r = 94,000$ protein has not been demonstrated. Moreover, multiple proteins, some of which may not be radioiodinated, might be involved in sperm binding.

B. Putative Egg Receptors for Sperm

1. Complement Receptors

The role of the complement system in human reproduction is of increasing interest, since some forms of infertility are associated with the presence of antisperm antibodies that, in theory, could limit the delivery of viable sperm to the site of fertilization by complement-mediated damage to the sperm. Although

2. Molecular Basis of Mammalian Egg Activation

components of the complement cascade are present in the female reproductive tract (Rooney et al., 1993) and are secreted under hormonal control by the uterine epithelium (Sundstrom et al., 1989; Isaacson et al., 1990), components of this system are present on both the male and the female gametes. Thus, these components not only might function in clearing gametes from the reproductive tract but also may be involved in sperm–egg interaction. The inner acrosomal region of sperm contains the complement-binding proteins membrane cofactor protein (MCP/CD46; Anderson et al., 1989; Fenichel et al., 1990; Cervoni et al., 1992; Okabe et al., 1992) and decay accelerating factor (DAF; Cervoni et al., 1993). MCP functions as a cofactor in the Factor I-mediated inactivation of C3b and C4b, and DAF augments the function of MCP. Thus, these proteins might function to protect complement activation on acrosome-reacted sperm or, alternatively, may have another function. In contrast, the membrane attack complex (MAC) inhibitory protein (CD59) is present on acrosome-intact sperm and may inhibit complement-mediated lysis of acrosome-intact sperm (Rooney et al., 1992).

A monoclonal antibody (MH61) raised against Ca^{2+} ionophore-induced acrosome-reacted human sperm (Okabe et al., 1990) recognizes the sperm MCP/CD46 (Okabe et al., 1992). This monoclonal antibody inhibits the fusion of human sperm with ZP free hamster eggs, suggesting that MCP plays a role in this process (Okabe et al., 1990). More recently, it was demonstrated that dimeric complement C3b binds to the acrosomal and equatorial regions of acrosome-reacting and acrosome-reacted human sperm, respectively (Anderson et al., 1993). This binding is presumed to occur via MCP, since the complement receptors CR1, CR2, and CR3 are not present on sperm. In contrast, CR1 and CR3 are present on the surface of human oocytes; human oocytes do not contain immunoreactive CR2 and MCP (Anderson et al., 1993). Addition of low concentrations of dimeric C3b to culture media during the hamster egg penetration test enhances the sperm penetration rate and the number of sperm that enter the eggs; at higher concentrations (>10 μM), these parameters of egg penetration are inhibited. Antibodies against C3 also inhibit sperm–egg fusion and penetration, although these experiments were performed with IgG fractions and not Fab fragments. It is postulated that circulating complement C3 functions as a "bridging" ligand, following its activation to C3b by acrosin released from acrosome-reacted sperm, by binding to MCP on the inner acrosomal membrane and to CR1/CR3 on the oocyte (Anderson et al., 1993), but this hypothesis has not been rigorously tested. The question of the specificity of interaction between C3/C3b and these complement-binding proteins to mediate this very specialized cellular interaction must be considered carefully in the design of future experiments. This experimental paradigm, however, remains an interesting one since complement is implicated in cell–cell interactions (Springer, 1990) and CR3 is an integrin of the β_2 class (Wright et al., 1987; see subsequent discussion).

2. Integrins

The restricted and localized interaction between the sperm head and the egg plasma membrane that gives rise to egg activation suggests that the sperm may present an immobilized ligand to the egg. In this regard, this interaction is similar to many cell-to-cell recognition and ECM interactions that are mediated by cell surface receptors of the integrin superfamily (Albelda and Buck, 1990; Ruoslahti, 1991; Hynes, 1992). Members of the integrin family are also involved in transmembrane signaling events that give rise to changes in tyrosine phosphorylation (Golden et al., 1990; Kornberg et al., 1991; Lipfert et al., 1992), changes in intracellular Ca^{2+} (Ng-Sikorski et al., 1991; Pelletier et al., 1992; Schwartz, 1993), and activation of phospholipase C (Cybulsky et al., 1993), all of which may be involved in sperm-induced egg activation. These parallels have led to investigations examining the potential role of egg integrins in fertilization.

Peptides containing the RGD (Arg-Gly-Asp) sequence that can be involved in integrin binding (Pierschbacher and Ruoslahti, 1984a,b) inhibit the binding of either human or hamster sperm to ZP-free hamster eggs (Bronson and Fusi, 1990). Moreover, beads coated with either RGD peptide or anti-$\alpha 5$ antibody bind to the plasma membrane of human oocytes (Fusi et al., 1992). These data suggest that ligand–integrin interactions play a role in sperm–egg interaction and egg activation, although the nature of the sperm ligand and the egg receptor are not known.

A number of laboratories have described the complement of integrins present on and within the mammalian egg. Hamster egg extracts contain α_v, α_5, and $\beta 1$ integrin subunits as determined by enzyme-linked immunosorbent assay (ELISA) (Fusi et al., 1993). Moreover, α_2, α_5, and α_6 are detected in human and hamster eggs by following the rosetting of ZP-free eggs incubated with immunobeads coupled to specific monoclonal antibodies (Fusi et al., 1993), suggesting that the epitopes recognized by these antibodies are externally exposed. In the mouse, α_3, α_5, α_6, and $\beta 1$ integrin subunits have been identified following immunoprecipitation from surface-labeled eggs using subunit-specific antisera (Tarone et al., 1993), and α_6 and $\beta 1$ are localized to the egg plasma membrane (Hierck et al., 1993; Tarone et al., 1993). The presence of α_2, α_5, α_v, and $\beta 1$ integrin subunits, as well as the fibronectin (FNR) and vitronectin (VNR) receptors in mouse eggs, is detected by immunoprecipitation (Evans et al., 1995). Localization of these subunits within the egg by indirect immunofluorescence, immunoelectron microscopy (Tarone et al., 1993), and confocal microscopy (Evans et al., 1995) has yielded variable results depending on the antibody used. For example, $\beta 1$ has been localized to either the microvillar (Tarone et al., 1993) or the amicrovillar (Evans et al., 1995) region of the egg, and this difference depends on the antibody used. The distribution to the amicrovillar region occurs as a consequence of oocyte maturation, since a uniform distribution pattern is observed in GV-intact oocytes (Evans et al., 1995). Why such differences are seen is not clear, but it is possible that the difference relates to the epitopes recognized by the respective antibodies and/or the source of the antigen used for the production of the antibodies. Although the presence of β1 in the microvillar region of the egg, which is the region that preferentially binds sperm, is consistent with its potential role in sperm–egg interaction, such conclusions must be interpreted with caution based on the aforementioned discrepancies of localization. It should also be noted that two integrins, α_5 and VNR, appear to reside predominantly in the cytoplasm (Evans et al., 1995). The presence of multiple integrin subunits with varying localizations within the egg therefore demonstrates the potential complexity of identifying an integrin(s) that participates in sperm–egg interaction. At this juncture, a complete characterization of the members of this receptor family in eggs is warranted before a stepwise evaluation of the role of these receptors in egg activation can be fully undertaken.

A variety of experimental approaches can be taken to determine whether integrins mediate egg activation. Incubation of eggs with antibodies against extracellular domains of specific integrins and/or microinjection of antibodies against intracellular domains of specific integrins could be performed to determine whether these treatments inhibit specific events of egg activation. A dominant-negative approach might also be a valid way to test this hypothesis once the proper integrin constructs are available.

VII. Sperm Surface Ligands Involved in Egg Activation

The identification of a sperm ligand that mediates the binding and fusion reactions at the time of fertilization will represent a major step toward an understanding of how this and any subsequent signal transduction processes regulate egg activation. It must be emphasized that in developing experimental approaches to identify such sperm ligands, one must consider the possibility that multiple sperm and egg components function together at the egg plasma membrane to bring about egg activation. This cautionary note arises from the fact that, under a number of different experimental conditions, sperm binding and fusion can be separated (Yanagimachi, 1978,1984,1988b) and can therefore be considered separate events. If one adopts the receptor hypothesis of egg activation, fusion does not necessarily have to lead to signal transduction, although it is essential to successful fertilization. Such a sperm-associated ligand must also be present on the sperm surface following the acrosome reaction and also in the equatorial segment, since all mammalian sperm studied to date interact with the egg plasma membrane at this region (Yanagimachi, 1988b;1994).

Although several potential sperm-associated proteins have been proposed to be involved in the sperm–egg binding and/or fusion process, most of these proteins have not been thoroughly characterized (see Myles, 1993, for further discussion). One protein, PH-30 (now called fertilin), does possess the requisite properties of a potential sperm-specific ligand (Primakoff *et al.*, 1987; Blobel *et al.*, 1990,1992; Myles *et al.*, 1994). PH-30 is an integral membrane protein of guinea pig sperm and exists as a heterodimer. Its proposed role in sperm binding and/or fusion derives from sequence homology of the cloned subunits that reveals a disintegrin-like domain within the β subunit and a putative viral fusion domain within the α subunit (Blobel *et al.*, 1992). Disintegrins represent a family of soluble peptides from the venom of vipers and contain an amino acid sequence that resembles the RGD-containing sequence present in several ECM integrin ligands (Gould *et al.*, 1990). Disintegrins can therefore antagonize specific integrin–ECM interactions. The presence of such a domain in PH-30 has led to the speculation that this region of the molecule interacts with an egg-associated integrin (Blobel *et al.*, 1992). Likewise, the presence of a putative viral fusion domain has led to the speculation that this region is responsible for sperm–egg plasma membrane fusion. Consistent with this hypothesis is the observation that cyclic peptides that contain the disintegrin domain of fertilin inhibit sperm–egg fusion in the guinea pig (Myles *et al.*, 1994).

It will be of interest to ascertain whether a PH-30-like protein is present on the surface of other mammalian sperm. If PH-30 is the sole molecule involved in sperm–egg binding and fusion, it remains to be seen whether this molecule can trigger signal transduction that leads to egg activation and, if so, what receptor–effector signaling pathways are utilized. Alternatively, different sperm-associated ligands that mediate signal transduction might require PH-30 for their proper presentation to the egg's surface.

VIII. Is There a Need for an Egg-Specific Receptor and a Sperm-Specific Ligand in Egg Activation

A tacit underlying assumption that pervades the literature on egg activation is that the sperm presents to the egg a sperm-specific molecule that is recognized by an egg-specific receptor. This assumption has been borne out for the sea urchin. Bindin is a unique sperm-specific ligand (Vacquier and Moy, 1977) and interacts with a unique transmembrane egg receptor (Foltz and Lennarz, 1993). It should be noted, however, that there is no reason that unique sperm- and egg-specific molecules must be involved in sperm-induced egg activation (e.g., the use of a novel egg-specific G protein-coupled receptor, as opposed to the use of a previously characterized G protein receptor that is present on somatic cells).

The rationale for this position is based on the known highly specific interactions of the sperm and the ZP. The ZP is composed of unique glycoproteins that are synthesized solely by the oocyte (Wassarman, 1988). In the mouse, the ZP presents to the sperm immobilized ligands (ZP3 and ZP2) that are responsible for the species-specific interaction of sperm with the egg. This species-specific interaction implies that sperm possess complementary unique receptors for the ZP ligands (see Ward and Kopf, 1993, for further discussion). The specificity of sperm–egg interaction, therefore, resides at the level of the ZP. Consistent with this suggestion is the observation that ZP-free hamster eggs are penetrated by sperm of a number of species (Yanagimachi, 1984). *In vivo*, sperm are the only cells that can penetrate the ZP and thereby gain access to the perivitelline space. Thus, the need for unique sperm ligands and egg receptors in sperm-induced egg activation is not apparent, and sperm-induced egg activation could employ ligands and receptors used in signaling pathways of somatic cells. The identification of the sperm ligand(s) and egg receptor(s) that lead to egg activation will clarify this issue.

References

Abassi, Y. A., and Foltz, K. R. (1994) Tyrosine phosphorylation of the egg receptor for sperm at fertilization. *Dev. Biol.* **164**, 430–443.

Aktories, K., Braun, U., Habermann, B., and Rösener, S. (1990). Botulinum ADP-ribosyltransferase C3. *In* "ADP Ribosylating Toxins and G Proteins" (J. Moss and M. Vaughan, eds.), pp. 97–115. American Society of Microbiology, Washington, DC.

Albelda, S. M., and Buck, C. A. (1990). Integrins and other cell adhesion molecules. *FASEB J.* **4**, 2868–2880.

Allen, R. D., and Griffin, J. L. (1958). The time sequence of early events in the fertilization of sea urchin eggs. 1. The latent period and the cortical reaction. *Exp. Cell Res.* **15**, 163–173.

Anderson, D. J., Michaelson, J. S., and Johnson, P. M. (1989) Trophoblast/leukocyte-common antigen is expressed by human testicular germ cells and appears on the surface of acrosome reacted sperm. *Biol. Reprod.* **41**, 285–293.

Anderson, D. J., Abbott, A. F., and Jack, R. M. (1993). The role of complement component C3b and its receptors in sperm-oocyte interaction. *Proc. Natl. Acad. Sci. USA* **90**, 10051–10055.

Aoki, F., Choi, T., Mori, M., Yamashita, M., Nagahama, T., and Kohmoto, K. (1992). A deficiency in the mechanism for p34^{cdc2} protein kinase activation in mouse embryos arrested at 2-cell stage. *Dev. Biol.* **154**, 66–72.

Birnbaumer, L. (1992). Receptor-to-effector signaling through G proteins: Roles of βγ dimers as well as α subunits. *Cell* **71**, 1069–1072.

Bleil, J. D., and Wassarman, P. M. (1980a). Synthesis of *zona pellucida* proteins by denuded and follicle-enclosed mouse oocytes during culture *in vitro*. *Proc. Natl. Acad. Sci. USA* **77**, 1029–1033.

Bleil, J. D., and Wassarman, P. M. (1980b). Structure and function of the *zona pellucida*: Identification and characterization of the proteins of the mouse oocyte's *zona pellucida*. *Dev. Biol.* **76**, 185–202.

Bleil, J. D., and Wassarman, P. M. (1980c). Mammalian sperm–egg interaction: Identification of a glycoprotein in mouse egg *zonae pellucidae* possessing receptor activity for sperm. *Cell* **20**, 873–882.

Bleil, J. D., and Wassarman, P. M. (1981). Mammalian sperm–egg interaction: Fertilization of mouse eggs triggers modifications of the major *zona pellucida* glycoprotein, ZP2. *Dev. Biol.* **86**, 189–197.

Bleil, J. D., and Wassarman, P. M. (1983). Sperm–egg interactions in the mouse: Sequence of events and induction of the acrosome reaction by a *zona pellucida* glycoprotein. *Dev. Biol.* **95**, 317–324.

Bleil, J. D., and Wassarman, P. M. (1986). Autoradiographic visualization of the mouse egg's sperm receptor bound to sperm. *J. Cell Biol.* **102**, 1363–1371.
Bleil, J. D., and Wassarman, P. M. (1988). Galactose at the nonreducing terminus of O-linked oligosaccharides of mouse *zona pellucida* glycoprotein ZP3 is essential for the glycoprotein's sperm-receptor activity. *Proc. Natl. Acad. Sci. USA* **85**, 6778–6782.
Bleil, J. D., and Wassarman, P. M. (1990). Identification of a ZP3-binding protein on acrosome-intact mouse sperm by photoaffinity crosslinking. *Proc. Natl. Acad. Sci. USA* **87**, 5563–5567.
Bleil, J. D., Greve, J. M., and Wassarman, P. M. (1988). Identification of a secondary sperm receptor in the mouse egg *zona pellucida*: Role in maintenance of binding of acrosome-reacted sperm to eggs. *Dev. Biol.* **128**, 376–385.
Blobel, C. P., Myles, D. G., Primakoff, P., and White, J. M. (1990). Proteolytic processing of a protein involved in sperm–egg fusion correlates with acquisition of fertilization competence. *J. Cell. Biol.* **111**, 69–78.
Blobel, C. P., Wolfsberg, T. G., Turck, C. W., Myles, D. G., Primakoff, P., and White, J. M. (1992). A potential fusion peptide and an integrin ligand domain in a protein active in sperm–egg fusion. *Nature (London)* **356**, 248–252.
Boldt, J., Howe, A. M., and Preble, J. (1988). Enzymatic alteration of the ability of mouse egg plasma membrane to interact with sperm. *Biol. Reprod.* **39**, 19–27.
Boldt, J., Gunter, L. E., and Howe, A. M. (1989). Characterization of cell surface polypeptides of unfertilized, fertilized, and protease-treated *zona*-free mouse eggs. *Gamete Res.* **23**, 91–101.
Bronson, R. A., and Fusi, F. (1990). Evidence that an Arg–Gly–Asp adhesion sequence plays a role in mammalian fertilization. *Biol. Reprod.* **43**, 1019–1025.
Buck, W. R., Hoffman, E. E., Rakow, T. L., and Shen, S. S. (1994). Synergistic calcium release in the sea urchin egg by ryanodine and cyclic ADP ribose. *Dev. Biol.* **163**, 1–10.
Carroll, J., and Swann, K. (1992). Spontaneous cytosolic calcium oscillations driven by inositol trisphosphate occur during *in vitro* maturation of mouse oocytes. *J. Biol. Chem.* **267**, 11196–11201.
Cascio, S. M., and Wassarman, P. M. (1982). Program of early development in the mammal: Post-transcriptional control of a class of proteins synthesized by mouse oocytes and early embryos. *Dev. Biol.* **89**, 397–408.
Cervoni, F., Oglesby, T. J., Adams, E. M., Milesifluet, C., Nickells, M., Fenichel, P., Atkinson, J. P., and Hsi, B. L. (1992). Identification and characterization of membrane cofactor protein of human spermatozoa. *J. Immunol.* **148**, 1431–1437.
Cervoni, F., Oglesby, T. J., Fenichel, P., Dohr, G., Rossi, B., Atkinson, J. P., and Hsi, B. L. (1993). Expression of decay accelerating factor (CD55) of the complement system on human spermatozoa. *J. Immunol.* **151**, 939–948.
Cheng, A., Le, T., Palacios, M., Bookbinder, L. H., Wassarman, P. M., Suzuki, F., and Bleil, J. D. (1994). Sperm–egg recognition in the mouse: Characterization of sp56, a sperm protein having specific affinity for ZP3. *J. Cell Biol.* **125**, 867–878.
Chiba, K., Kado, R. T., and Jaffe, L. A. (1990). Development of calcium release mechanisms during starfish oocyte maturation. *Dev. Biol.* **146**, 300–306.
Chuang, L. M., Myers, M. G., Backer, J. M., Shoelson, S. E., White, M. F., Birnbaum, M. J., and Kahn, C. R. (1993). Insulin-stimulated oocyte maturation requires insulin receptor substrate 1 and interaction with the SH2 domains of phosphatidylinositol 3-kinase. *Mol. Cell. Biol.* **13**, 6653–6660.
Ciapa, B., and Epel, D. (1991). A rapid change in phosphorylation on tyrosine accompanies fertilization of sea urchin eggs. *FEBS Lett.* **295**, 167–170.
Ciapa, B., and Whitaker, M. J. (1986). Two phases of inositol polyphosphate and diacylglycerol production at fertilization. *FEBS Lett.* **195**, 347–351.
Ciapa, B., Borg, B., and Whitaker, M. (1992). Polyphosphoinositide metabolism during the fertilization wave in sea urchin eggs. *Development* **115**, 187–195.

Clapham, D. E., and Neer, E. J. (1993). New roles for G protein βγ-dimers in transmembrane signalling. *Nature (London)* **365**, 403–406.
Cran, D. G., Moor, R. M., and Irvine, R. F. (1988). Initiation of the cortical reaction in hamster and sheep oocytes in response to inositol trisphosphate. *J. Cell Sci.* **91**, 139–144.
Crossley, I., Whalley, T., and Whitaker, M. (1991). Guanosine 5'-thiotriphosphate may stimulate phosphoinositide messenger production in sea urchin eggs by a different rout than the fertilizing sperm. *Cell Reg.* **2**, 121–133.
Cummins, J. M., and Yanagimachi, R. (1982). Sperm-egg ratios and the site of the acrosome reaction during *in vitro* fertilization in the hamster. *Gamete Res.* **5**, 239–256.
Currie, K. P. M., Swann, K., Galione, A., and Scott, R. H. (1992). Activation of Ca^{2+}-dependent currents in cultured rat dorsal root ganglion neurones by a sperm factor and cyclic ADP-ribose. *Mol. Biol. Cell* **3**, 1415–1425.
Cuthbertson, K. S. R., and Cobbold, P. H. (1985). Phorbol ester and sperm activate mouse oocytes by inducing sustained oscillations in cell Ca^{2+}. *Nature (London)* **316**, 541–542.
Cuthbertson, K. S. R., Whittingham, D. G., and Cobbold, P. H. (1981). Free Ca^{2+} increases in exponential phases during mouse oocyte activation. *Nature (London)* **294**, 754–757.
Cybulsky, A. V., Carbonetto, S., Cyr, M.-D., McTavish, A. J., and Huang, Q. (1993). Extracellular matrix-stimulated phospholipase activation is mediated by $β_1$-integrin. *Ann. J. Physiol.* **264**, C323–C332.
Dale, B., DeFelice, L. J., and Ehrenstein, G. (1985). Injection of a soluble sperm fraction into sea urchin eggs triggers the cortical reaction. *Experientia* **41**, 1068–1070.
Dasgupta, J. D., and Garbers, D. L. (1983). Tyrosine protein kinase activity during embryogenesis. *J. Biol. Chem.* **258**, 6174–6178.
Ducibella, T. (1991). Mammalian egg cortical granules and the cortical reaction. *In* "Elements of Mammalian Fertilization" (P. M. Wassarman, ed.), Vol. 1, pp. 205–231. CRC Press, Bocca Raton, Florida.
Ducibella, T., and Buetow, T. (1994). Competence to undergo normal fertilization-induced cortical activation develops after metaphase I of meiosis in mouse oocytes. *Dev. Biol.* **165**, 95–104.
Ducibella, T., Kurasawa, S., Rangarajan, S., Kopf, G. S., and Schultz, R. M. (1990). Changes in distribution of mouse egg cortical granules during meiotic maturation and correlation with an egg-induced modification of the *zona pellucida*. *Dev. Biol.* **137**, 46–55.
Ducibella, T., Kurasawa, S., Duffy, P., Kopf, G. S., and Schultz, R. M. (1993). Regulation of the polyspermy block in the mouse egg: Maturation-dependent differences in cortical granule exocytosis and *zona pellucida* modifications induced by inositol 1,4,5-trisphosphate and an activator of protein kinase C. *Biol. Reprod.* **48**, 1251–1257.
Dunbar, B. S., Prasad, S. V., and Timmons, T. M. (1991). Comparative structure and function of mammalian zonae pellucidae. *In* "A Comparative Overview of Mammalian Fertilization" (B. S. Dunbar, and M. G. O'Rand, eds.), pp. 97–114. Plenum Press, New York.
Elinson, R. P. (1986). Fertilization in amphibians: The ancestry of the block to polyspermy. *Int. Rev. Cytol.* **101**, 59–100.
Endo, Y., Kopf, G. S., and Schultz, R. M. (1986). Stage-specific changes in protein phosphorylation accompanying meiotic maturation of mouse oocytes and fertilization of mouse eggs. *J. Exp. Zool.* **239**, 401–409.
Endo, Y., Schultz, R. M., and Kopf, G. S. (1987a). Effects of phorbol esters and a diacylglycerol on mouse eggs: Inhibition of fertilization and modification of the *zona pellucida*. *Dev. Biol.* **119**, 199–209.
Endo, Y., Mattei, P., Kopf, G. S., and Schultz, R. M. (1987b). Effects of phorbol esters on mouse eggs: Dissociation of sperm receptor activity from acrosome-inducing activity of the mouse *zona pellucida* protein, ZP3. *Dev. Biol.* **123**, 574–577.
Evans, J. P., Schultz, R. M., and Kopf, G. S. (1995). Identification and localization of integrin subunits in oocytes and eggs of the mouse. *Mol. Reprod. Dev.* **20**, 211–220.
Fenichel, P., Dohr, G., Grivaux, C., Cervoni, F., Donzeau, M., and Hsi, B. L. (1990). Localiz-

ation and characterization of the acrosomal antigen recognized by GB24 on human spermatozoa. *Mol. Reprod. Dev.* **27**, 173–178.

Fissore, R. A., and Robl, J. M. (1993). Sperm, inositol trisphosphate, and thimerosal-induced intracellular Ca^{2+} elevations in rabbit eggs. *Dev. Biol.* **159**, 122–130.

Fissore, R. A., Dobrinsky, J. R., Balise, J. J., Duby, R. T., and Robl, J. M. (1992). Patterns of intracellular calcium concentrations in fertilized bovine eggs. *Biol. Reprod.* **47**, 960–969.

Foltz, K. R., and Lennarz, W. J. (1993). The molecular basis of sea urchin gamete interactions at the egg plasma membrane. *Dev. Biol.* **158**, 46–61.

Fujiwara, T., Nakada, K., Shirakawa, H., and Miyazaki, S. (1993). Development of inositol trisphosphate-induced calcium release mechanism during maturation of hamster oocytes. *Dev. Biol.* **156**, 69–79.

Fukuda, Y., and Chang, M. C. (1978). The time of cortical granule breakdown and sperm penetration in mouse and hamster eggs inseminated *in vitro*. *Biol. Reprod.* **19**, 261–266.

Fulton, B. P., and Whittingham, D. G. (1978). Activation of mammalian oocytes by intracellular injection of calcium. *Nature (London)* **273**, 149–151.

Fusi, F. M., Vignali, M., Busacca, M., and Bronson, R. A. (1992). Evidence for the presence of an integrin cell adhesion receptor on the oolemma of unfertilized human oocytes. *Mol. Reprod. Dev.* **36**, 212–219.

Fusi, F. M., Vignali, M., Gailit, J., and Bronson, R. A. (1993). Mammalian oocytes exhibit specific recognition of RGD (Arg-Gly-Asp) tripeptide and express oolemmal integrins. *Mol. Reprod. Dev.* **36**, 212–219.

Galione, A., Lee, H. C., and Busa, W. (1991). Ca^{2+}-induced Ca^{2+} release in sea urchin egg homogenates: Modulation by cyclic ADP-ribose. *Science* **253**, 1143–1146.

Galione, A., McDougall, A., Busa, W. B., Willmott, N., Gillot, I., and Whitaker, M. (1993). Redundant mechanisms of calcium-induced calcium release underlying calcium waves during fertilization in sea urchin eggs. *Science* **261**, 348–352.

Gallicano, G. I., Schwarz, S. M., McGaughey, R. W., and Capco, D. G. (1993). Protein kinase C, a pivotal regulator of hamster egg activation, functions after elevation of intracellular free calcium. *Dev. Biol.* **156**, 94–106.

Gallicano, G. I., Larabell, C. A., McGaughey, R. W., and Capco, D. G. (1994). Novel cytoskeletal elements in mammalian eggs are composed of a unique arrangement of intermediate filaments. *Mech. Dev.* **45**, 211–226.

Garbers, D. L., and Kopf, G. S. (1980). The regulation of spermatozoa by calcium and cyclic nucleotides. *Adv. Cyclic Nuc. Res.* **13**, 251–306.

Golden, A., Brugge, J. S., and Shattil, S. J. (1990). Role of platelet membrane glycoprotein IIb-IIIa in agonist-induced tyrosine phosphorylation of platelet proteins. *J. Cell Biol.* **111**, 3117–3127.

Gould, R. J., Polokoff, M. A., Friedman, P. A., Huant, T.-F., Holt, J. C., Cook, J. J., and Niewiarowsi, S. (1990). Disintegrins: A family of integrin inhibitory proteins from viper venom. *Proc. Soc. Exp. Biol. Med.* **195**, 168–171.

Grandin, N., and Charbonneau, M. (1992). Intracellular free Ca^{2+} changes during physiological polyspermy in amphibian eggs. *Development* **114**, 617–624.

Gupta, S. K., Diez, E., Heasley, L. E., Osawa, S., and Johnson, G. L. (1990). A G protein mutant that inhibits thrombin and purinergic receptor activation of phospholipase A2. *Science* **249**, 662–666.

Hierck, B. P., Thorsteindottir, S., Niessen, C. M., Freund, E., Iperen, L. V., Feyen, A., Hogervorst, F., Poleman, R. E., Mummery, C. L., and Sonnenberg, A. (1993). Variants of the $\alpha_6\beta_1$ receptor in early murine development: Distribution, molecular cloning and chromosomal localization of the mouse integrin α_6 subunit. *Cell Adh. Commun.* **1**, 33–53.

Horvath, P. M., Kellom, T., Caulfield, J., and Boldt, J. (1993). Mechanistic studies of the plasma membrane block to polyspermy in mouse eggs. *Mol. Reprod. Dev.* **34**, 65–72.

2. Molecular Basis of Mammalian Egg Activation

Hynes, R. O. (1992). Integrins: Versatility, modulation and signaling in cell adhesion. *Cell* **69**, 11–25.

Igusa, Y., and Miyazaki, S. (1983). Effects of altered extracellular and intracellular calcium concentration on hyperpolarizing responses of the hamster egg. *J. Physiol.* **340**, 611–632.

Isaacson, K. B., Galman, M., Coutifaris, C., and Lyttle, C. R. (1990). Endometrial synthesis and secretion of complement component-3 by patients with and without endometriosis. *Fert. Steril.* **53**, 836–841.

Jaffe, L. A., Sharp, A. P., and Wolf, D. P. (1983). Absence of an electrical polyspermy block in the mouse. *Dev. Biol.* **96**, 317–327.

Jaffe, L. F. (1983). Sources of calcium in egg activation: A review and hypothesis. *Dev. Biol.* **99**, 256–276.

Jiang, W., Peaucellier, G., and Kinsey, W. H. (1989). Affinity purification of embryo proteins phosphorylated on tyrosine in vitro. *Dev. Growth Diff.* **31**, 573–580.

Jiang, W., Veno, P. A., Wood, R. W., Peaucellier, G., and Kinsey, W. H. (1991). pH regulation of an egg cortex tyrosine kinase. *Dev. Biol.* **146**, 81–88.

Jones, J., and Schultz, R. M. (1990). Pertussis toxin-catalyzed ADP ribosylation of a G protein in mouse oocytes, eggs, and preimplantation embryos: Developmental changes and possible functional roles. *Dev. Biol.* **139**, 250–262.

Kalab, P., Visconti, P., Leclerc, P., and Kopf, G. S. (1994). p95, the major phosphotyrosine-containing protein in mouse spermatozoa, is a hexokinase with unique properties. *J. Biol. Chem.* **269**, 3810–3817.

Kellom, T., Vick, A., and Boldt, J. (1992). Recovery of penetration ability in protease-treated zona-free mouse eggs occurs coincident with recovery of a cell surface 94 kD protein. *Mol. Reprod. Dev.* **33**, 46–52.

Kligman, I., Glassner, M., Storey, B. T., and Kopf, G. S. (1991). *Zona pellucida*-mediated acrosomal exocytosis in mouse spermatozoa: Characterization of an intermediate stage prior to the completion of the acrosome reaction. *Dev. Biol.* **146**, 344–355.

Kline, D. (1988). Calcium-dependent events at fertilization of the frog egg: Injection of a calcium buffer block ion channel opening, exocytosis, and formation of pronuclei. *Dev. Biol.* **126**, 246–261.

Kline, D., and Kline, J. T. (1992). Repetitive calcium transients and the role of calcium exocytosis and cell cycle activation in the mouse egg. *Dev. Biol.* **149**, 80–89.

Kline, D., and Stewart-Savage, J. (1994). The timing of cortical granule fusion, content dispersal, and endocytosis during fertilization of the hamster egg: An electrophysiological and histochemical study. *Dev. Biol.* **162**, 277–287.

Kline, D., Kado, R. T., Kopf, G. S., and Jaffe, L. A. (1990). Receptors, G proteins, and activation of the amphibian egg. *In* "NATO ASI Series: Mechanisms of Fertilization" (B. Dale, ed.), Vol. H45, pp. 529–541. Springer-Verlag, Berlin.

Kline, D., Kopf, G. S., Muncy, L. F., and Jaffe, L. A. (1991). Evidence of the involvement of a pertussis toxin-insensitive G protein in egg activation of the frog, *Xenopus laevis*. *Dev. Biol.* **143**, 218–229.

Kline, J. T., and Kline, D. (1994). Regulation of intracellular calcium in the mouse egg: Evidence for inositol trisphosphate-induced calcium release, but not calcium-induced calcium release. *Biol. Reprod.* **50**, 193–203.

Koch, C. A., Anderson, D., Moran, M. F., Ellis, C., and Pawson, T. (1991). SH2 and SH3 domains: Elements that control interactions of cytoplasmic signaling proteins. *Science* **252**, 668–674.

Kopf, G. S., and Gerton, G. L. (1991). The mammalian sperm acrosome and the acrosome reaction. *In* "Elements of Mammalian Fertilization" (P. M. Wassarman, ed.), Vol. 1, pp. 153–203. CRC Press, Boca Raton, Florida.

Kornberg, L. J., Earp, H. S., Turner, C. E., Prockop, C., and Juliano, R. L. (1991). Signal transduction by integrins: Increased tyrosine phosphorylation caused by clustering of β1 integrins. *Proc. Natl. Acad. Sci. USA* **88,** 8392–8396.

Kurasawa, S., Schultz, R. M. and Kopf, G. S. (1989). Egg-induced modifications of the *zona pellucida* of the mouse egg: Effects of microinjected inositol 1,4,5-trisphosphate. *Dev. Biol.* **133,** 295–304.

Latham, K. E., Garrels, J. I., Chang, C., and Solter, D. (1991). Quantitative analysis of protein synthesis in mouse embryos. I. Extensive reprogramming at the one- and two-cell stages. *Development* **112,** 921–932.

Lee, H. C., Aarhus, R., and Walseth, T. F. (1993). Calcium mobilization by dual receptors during fertilization of sea urchin eggs. *Science* **261,** 352–355.

Lee, M. A., and Storey, B. T. (1985). Evidence for plasma membrane impermeability to small ions in acrosome-intact spermatozoa bound to mouse *zonae pellucidae*, using an aminoacridine fluorescent probe: Time course of the zona-induced acrosome reaction monitored by both chlortetracycline and pH probe fluorescence. *Biol. Reprod.* **33,** 235–246.

Leyton, L., and Saling, P. (1989). 95 kd sperm proteins bind ZP3 and serve as tyrosine kinase substrates in response to zona binding. *Cell* **57,** 1123–1130.

Leyton, L., LeGuen, P., Bunch, D., and Saling, P. M. (1992). Regulation of mouse gamete interaction by a sperm tyrosine kinase. *Proc. Natl. Acad. Sci. USA* **89,** 11692–11695.

Lipfert, L., Haimovich, M., Schaller, M. D., Cobb, B. S., Parsons, J. T., and Brugge, J. S. (1992). Integrin-dependent phosphorylation and activation of the protein tyrosine kinase pp125^{FAK} in platelets. *J. Cell Biol.* **119,** 905–912.

Lorca, T., Cruzalegui, F. H., Fesquet, D., Cavadore, J.-C., Méry, J., Means, A., and Dorée, M. (1993). Calmodulin-dependent protein kinase II mediates inactivation of MPF and SCF upon fertilization of *Xenopus* eggs. *Nature (London)* **366,** 270–273.

Mayer, B. J., and Baltimore, D. (1993). Signaling through SH2 and SH3 domains. *Trends Cell Biol.* **3,** 8–13.

Meissner, G. (1986). Ryanodine activation and inhibition of Ca^{2+} release channel of sarcoplasmic reticulum. *J. Biol. Chem.* **261,** 6300–6306.

Miller, J. D., Macek, M. B., and Shur, B. D. (1992). Complementarity between sperm surface β-1,4-galactosyltransferase and egg-coat ZP3 mediates sperm-egg binding. *Nature (London)* **357,** 589–593.

Miller, J. D., Gong, X., Decker, G., and Shur, B. D. (1993). Egg cortical granule N-acetylglucosaminidase is required for the mouse zona block to polyspermy. *J. Cell Biol.* **123,** 1431–1440.

Miyazaki, S. (1988). Inositol 1,4,5-trisphosphate-induced calcium release and guanine nucleotide-binding protein-mediated periodic calcium rises in golden hamster eggs. *J. Cell Biol.* **106,** 345–353.

Miyazaki, S., Katayama, Y., and Swann, K. (1990). Synergistic activation by serotonin and GTP analogue and inhibition by phorbol ester of cyclic Ca^{2+} rises in hamster eggs. *J. Physiol. (London)* **426,** 209–227.

Miyazaki, S., Hashimoto, N., Yoshimoto, Y., Kishimoto, T., Igusa, Y., and Hiramoto, Y. (1986). Temporal and spatial dynamics of the periodic increase in intracellular free calcium at fertilization of golden hamster eggs. *Dev. Biol.* **118,** 259–267.

Miyazaki, S., Shirakawa, H., Nakada, K., Honda, Y., Yuzaki, Y., Nakade, S., and Mikoshiba, K. (1992a). Antibody to the inositol trisphosphate receptor blocks thimerosal-enhanced Ca^{2+}-induced Ca^{2+} release and Ca^{2+} oscillations in hamster eggs. *FEBS Lett.* **309,** 180–184.

Miyazaki, S., Yuzaki, M., Nakada, K., Shirakawa, H., Nakanishi, S., Nakade, S., and Mikoshiba, K. (1992b). Block of Ca^{2+} wave and Ca^{2+} oscillation by antibody to the inositol 1,4,5-trisphosphate receptor in fertilized hamster eggs. *Science* **257,** 251–255.

Miyazaki, S., Shirakawa, H., Nakada, K., and Honda, Y. (1993). Essential role of inositol 1,4,5-trisphosphate receptor/Ca^{2+} release channel in Ca^{2+} waves and Ca^{2+} oscillations at fertilization of mammalian eggs. *Dev. Biol.* **158,** 62–78.

2. Molecular Basis of Mammalian Egg Activation

Moller, C. C., and Wassarman, P. M. (1989). Characterization of a proteinase that cleaves *zona pellucida* glycoprotein ZP2 following activation of mouse oocytes. *Dev. Biol.* **132**, 103–112.

Moore, G. D., Kopf, G. S., and Schultz, R. M. (1993). Complete mouse egg activation in the absence of sperm by stimulation of an exogenous G protein-coupled receptor. *Dev. Biol.* **159**, 669–678.

Moore, G. D., Ayabe, T., Visconti, P. E., Schultz, R. M., and Kopf, G. S. (1994). Roles of heterotrimeric and monomeric G proteins in sperm-induced activation of mouse eggs. *Development* **120**, 3313–3323.

Moore, K. L., and Kinsey, W. H. (1994). Identification of an Abl-related protein tyrosine kinase in the cortex of the sea urchin egg: Possible role at fertilization. *Dev. Biol.* **164**, 444–455.

Mortillo, S., and Wassarman, P. M. (1991). Differential binding of gold-labeled *zona pellucida* glycoproteins mZP2 and mZP3 to mouse sperm membrane compartments. *Development* **113**, 141–149.

Murray, A. W. (1993). Turning on mitosis. *Curr. Biol.* **3**, 291–293.

Muslin, A. J., Klippel, A., and Williams, L. T. (1993). Phosphatidylinositol 3-kinase activity is important for progesterone-induced *Xenopus* oocyte maturation. *Mol. Cell Biol.* **13**, 6661–6666.

Myles, D. G. (1993). Molecular mechanisms of sperm-egg membrane binding and fusion in mammals. *Dev. Biol.* **158**, 35–45.

Myles, D. G., Kimmel, L. H., Blobel, C. P., White, J. M., and Primakoff, P. (1994). Identification of a binding site in the disintegrin domain of fertilin required for sperm–egg fusion. *Proc. Natl. Acad. Sci. USA* **91**, 4195–4198.

Ng-Sikorski, J., Andersson, R., Patarroya, M., and Andersson, T. (1991). Calcium signalling capacity of the CD11b/CD18 integrin on human neutrophils. *Exp. Cell Res.* **195**, 504–508.

Nishizuka, Y. (1988). The molecular heterogeneity of protein kinase C and its implications for cellular regulation. *Nature (London)* **334**, 661–665.

Nuccitelli, R. (1991). How do sperm activate eggs? *Cur. Top. Dev. Biol.* **25**, 1–16.

Nuccitelli, R., Yim, D. L., and Smart, T. (1993). The sperm-induced Ca^{2+} wave following fertilization of the *Xenopus* egg requires the production of Ins (1,4,5) P_3. *Dev. Biol.* **158**, 200–212.

Okabe, M., Nagira, M., Kawai, Y., Matzno, S., Mimura, T., and Mayumi, T. (1990). A human sperm antigen possibly involved in binding and/or fusion with *zona*-free hamster eggs. *Fert. Steril.* **54**, 1120–1121.

Okabe, M., Ying, X., Nagira, M., Ikawa, M., Yauhiro, K., Mimura, T., and Tanaka, K. (1992). Homology of an acrosome-reacted sperm-specific antigen to CD 46. *J. Pharamcobiol. Dynam.* **15**, 455–459.

Peaucellier, G., Veno, P. A., and Kinsey, W. H. (1988). Protein tyrosine phosphorylation in response to fertilization. *J. Biol. Chem.* **263**, 13806–13811.

Pelletier, A. J., Bodary, S. C., and Levinson, A. D. (1992). Signal transduction by the platelet integrin $\alpha_{IIb}\beta_3$: Induction of calcium oscillations required for protein tyrosine phosphorylation and ligand-induced spreading of stably transfected cells. *Mol. Biol. Cell* **3**, 989–998.

Peres, A., Bertollini, L., and Racca, C. (1991). Characterization of Ca^{2+} transients induced by intracellular photorelease of $InsP_3$ in mouse ovarian oocytes. *Cell Calcium* **12**, 457–465.

Perreault, S. D., Barbee, R. R., and Slott, V. L. (1988). Importance of glutathione in the acquisition and maintenance of sperm nuclear decondensing activity in maturing hamster oocytes. *Dev. Biol.* **125**, 181–186.

Phillips, D. M. (1991). Electron microscopy of mammalian fertilization. *In* "Elements of Mammalian Fertilization" (P. M. Wassarman, ed.), Vol. 1, pp. 249–267. CRC Press, Boca Raton, Florida.

Picard, A., Giraud, F., Le Bouffant, F., Sladeczek, F., Le Peuch, C., and Dorée, M. (1985). Inositol 1,4,5-trisphosphate microinjection triggers activation, but not meiotic maturation in amphibian and starfish oocytes. *FEBS Lett.* **182**, 446–450.

Pierschbacher, M. D., and Ruoslahti, E. (1984a). Cell attachment activity of fibronectin can be duplicated by small synthetic fragments of the molecules. *Nature (London)* **309**, 30–33.

Pierschbacher, M. D., and Ruoslahti, E. (1984b). Variants of the cell recognition site of fibronectin that retain attachment-promoting activity. *Proc. Natl. Acad. Sci. USA* **81**, 5985–5988.

Primakoff, P., and Myles, D. G. (1983). A map of the guinea pig sperm surface constructed with monoclonal antibodies. *Dev. Biol.* **98**, 417–428.

Primakoff, P., Hyatt, H., and Tredick-Kline, J. (1987). Identification and purification of a sperm surface protein with a potential role in sperm-egg fusion. *J. Cell Biol.* **104**, 141–149.

Ribot, H. D., Eisenman, E. A., and Kinsey, W. H. (1984). Fertilization results in increased tyrosine phosphorylation of egg proteins. *J. Biol. Chem.* **259**, 5333–5338.

Rooney, I. A., Davies, A., and Morgan, B. P. (1992). Membrane attack complex (MAC)-mediated damage to spermatozoa: Protection of the cells by the presence on their membranes of MAC inhibitory proteins. *Immunology* **75**, 499–506.

Rooney, I. A., Oglesby, T. J., and Atkinson, J. P. (1993). Complement in human reproduction. *In* "Immunological Research: The Year in Complement" (J. M. Cruse, ed.), Karger, Basel.

Ruoslahti, E. (1991). Integrins. *J. Clin. Invest.* **87**, 1–5.

Saling, P. and Storey, B. T. (1979). Mouse gamete interactions during fertilization *in vitro*: Chlortetracycline as a fluorescent probe for mouse sperm acrosome reaction. *J. Cell Biol.* **83**, 544–555.

Schaller, M. D., Borgman, C. A., Bobb, B. S., Vines, R. R., Reynolds, A. B., and Parsons, J. T. (1992). pp125 FAK, a structurally distinctive protein-tyrosine kinase associated with focal adhesions. *Proc. Natl. Acad. Sci. USA* **89**, 51925196.

Schultz, R. M., Montgommery, R. R., and Belanoff, J. R. (1983). Regulation of mouse oocyte meiotic maturation: Implication of a decrease in oocyte cAMP and protein phosphorylation in commitment to resume meiosis. *Dev. Biol.* **97**, 294–304.

Schwartz, M. A. (1993). Spreading of human endothelial cells on fibronectin or vitronectin triggers elevation of intracellular free calcium. *J. Cell Biol.* **120**, 1003–1010.

Shimizu, S., Tsuji, M., and Dean, J. (1983). *In vitro* biosynthesis of three sulfated glycoproteins of murine *zonae pellucidae* by oocyte grown in follicle culture. *J. Biol. Chem.* **258**, 5858–5863.

Songyang, Z., Shoelson, S. E., Chaudhuri, M., Gish, G., Pawson, T., Haser, W. G., King, F., Roberts, T., Ratnofsky, S., Lechleider, R. J., Neel, B. G., Birge, R. B., Fajardo, J. E., Chou, M. M., Hanafusa, H., Schaffhausen, B., and Cantley, L. C. (1993). SH2 domains recognize specific phosphopeptide sequences. *Cell* **72**, 767–778.

Sorrentino, V., and Volpe, P. (1993). Ryanodine receptors: How many, where, and why? *Trends. Pharamcol. Sci.* **14**, 98–103.

Speksnijder, J. E., Corson, D. W., Sardet, C., and Jaffe, L. F. (1989). Free calcium pulses following fertilization in the ascidian egg. *Dev. Biol.* **135**, 182–190.

Springer, T. A. (1990). Adhesion receptors of the immune system. *Nature (London)* **346**, 425–434.

Steinhardt, R. A., and Epel, D. (1974). Activation of sea urchin eggs by a calcium ionophore. *Proc. Natl. Acad. Sci. USA* **71**, 1915–1919.

Stewart-Savage, J., and Bavister, B. D. (1988). A cell surface block to polyspermy occurs in golden hamster eggs. *Dev. Biol.* **128**, 150–157.

Stewart-Savage, J., and Bavister, B. D. (1991). Time course and pattern of cortical granule breakdown in hamster eggs after sperm fusion. *Mol. Reprod. Dev.* **30**, 390–395.

Stice, S. L., and Robl, J. M. (1990). Activation of mammalian oocytes by a factor obtained from rabbit sperm. *Mol. Reprod. Dev.* **25**, 272–280.

Stith, B. J., Goalstone, M., Silva, S., and Jaynes, C. (1993). Inositol 1,4,5-trisphosphate mass changes from fertilization through first cleavage in *Xenopus laevis*. *Mol. Biol. Cell* **4**, 435–443.

2. Molecular Basis of Mammalian Egg Activation

Stith, B. J., Espinoza, R., Roberts, D., and Smart, T. (1994). Sperm increase inositol 1,4,5-trisphosphate mass in *Xenopus laevis* eggs preinjected with calcium buffers or heparin. *Dev. Biol.* **165**, 206–215.

Sun, F. Z., Hoyland, J., Haung, X., Maxson, W., and Moor, R. M. (1992). A comparison of intracellular changes in porcine eggs after fertilization and electroactivation. *Development* **115**, 947–956.

Sundstrom, S. A., Komm, B. S., Ponce-de-Leon, H., Yi, Z., Teuscher, C., and Lyttle, C. R. (1989). Estrogen regulation of tissue-specific expression of complement C3. *J. Biol. Chem.* **264**, 16941–16947.

Supattapone, S., Danoff, S. K., Theibert, A., Joseph, S. K., Steiner, J., and Synder, S. H. (1988). Cyclic AMP-dependent phosphorylation of a brain inositol trisphosphate receptor decreases its release of calcium. *Proc. Natl. Acad. Sci. USA* **85**, 8747–8750.

Swann, K. (1990). A cytosolic sperm factor stimulates repetitive calcium increases and mimics fertilization in hamster eggs. *Development* **110**, 1295–1302.

Swann, K. (1992). Different triggers for calcium oscillations in mouse eggs involve a ryanodine-sensitive calcium store. *Biochem. J.* **287**, 79–84.

Swann, K., Igusa, Y., and Miyazaki, S. (1989). Evidence for an inhibitory effect of protein kinase C on the G protein-mediated repetitive calcium transients in hamster eggs. *EMBO J.* **8**, 3711–3718.

Tarone, G., Russo, M. A., Hirsch, E., Odorisio, T., Altruda, F., Silengo, L., and Siracusa, G. (1993). Expression of β1 integrin complexes on the surface of unfertilized mouse oocyte. *Development* **117**, 1369–1375.

Tombes, R. M., Simerly, C., Borisy, G. G., and Schatten, G. (1992). Meiosis, egg activation, and nuclear envelope breakdown are differentially reliant on Ca^{2+}, whereas germinal vesicle breakdown is Ca^{2+} independent in the mouse oocyte. *J. Cell Biol.* **117**, 799–811.

Toratani, S., Katada, T., and Yokosawa, H. (1993). Botulinum ADP-ribosyltransferase C3 induces elevation of the vitelline coat of ascidian eggs. *Biochem. Biophys. Res. Commun.* **193**, 1311–1317.

Turner, P. R., Sheetz, M. P., and Jaffe, L. A. (1984). Fertilization increases the polyphosphoinositide content of sea urchin eggs. *Nature (London)* **310**, 414–415.

Vacquier, V. D., and Moy, G. W. (1977). Isolation of bindin: The protein responsible for adhesion of sperm to sea urchin eggs. *Proc. Natl. Acad. Sci. USA* **74**, 2456–2460.

van Blerkom, J. (1981). Structural relationship and post-translational modification of stage-specific proteins synthesized during preimplantation development in the mouse. *Proc. Natl. Acad. Sci. USA* **78**, 7629–7633.

Vazquez, M. H., Phillips, D. M., and Wassarman, P. M. (1989). Interaction of mouse sperm with purified sperm receptors covalently linked to silica beads. *J. Cell Sci.* **92**, 713–722.

Verlhac, M.-H., Kubiak, J. Z., Clarke, H. J., and Maro, B. (1994). Microtubule and chromatin behavior follow MAP kinase activity but not MPF activity during meiosis in mouse oocytes. *Development* **120**, 1017–1025.

Vitullo, A. D., and Ozil, J.-P. (1992). Repetitive stimuli drive meiotic resumption and pronuclear development during oocyte activation. *Dev. Biol.* **151**, 128–136.

Ward, C. R., and Kopf, G. S. (1993). Molecular events mediating sperm activation. *Dev. Biol.* **158**, 9–34.

Wassarman, P. M. (1988). *Zona pellucida* glycoproteins. *Ann. Rev. Biochem.* **57**, 414–442.

Whitaker, M. J., and Irvine, R. F. (1984). Inositol 1,4,5-trisphosphate microinjection activates sea urchin eggs. *Nature (London)* **312**, 636–639.

Whitaker, M. and Swann, K. (1993). Lighting the fuse at fertilization. *Development* **117**, 1–12.

Whitaker, M. J., Swann, K., and Crossley, I. B. (1989). What happens during the latent period at fertilization in sea urchin eggs? *In* "Mechanisms of Egg Activation" (R. Nuccitelli, G. N. Cherr, and W. H. Clark, eds.), pp. 157–171. Plenum Press, New York.

Williams, C. J., Schultz, R. M., and Kopf, G. S. (1992). Role of G proteins in mouse egg activation: Stimulatory effects of acetylcholine on the ZP2-to-ZP2$_f$ conversion and pronuclear formation in eggs expressing a functional ml muscarinic receptor. *Dev. Biol.* **151,** 288–296.

Wolf, D. P. (1981). The mammalian egg's block to polyspermy. *In* "Fertilization and Embryonic Development *In Vitro*" (L. Mastroianni, Jr. and J. D. Biggers, eds.), pp. 183–196. Plenum Press, New York.

Wright, S. D., Reedy, P. A., Jong, M. T. C., and Erickson, B. W. (1987). C3bi receptor (complement receptor type 3) recognizes a region of complement protein C3 containing the sequence Arg-Gly-Asp. *Proc. Natl. Acad. Sci. USA* **84,** 1965–1968.

Xu, Z., Kopf, G. S., and Schultz, R. M. (1994). Involvement of inositol 1,4,5-trisphosphate-mediated Ca^{2+} release in early and late events of mouse egg activation. *Development* **120,** 1851–1859.

Yanagimachi, R. (1978). Calcium requirement for sperm-egg fusion in mammals. *Biol. Reprod.* **19,** 948–958.

Yanagimachi, R. (1984). Zona-free hamster eggs: Their use in assessing fertilizing capacity and examining chromosomes of human spermatozoa. *Gamete Res.* **19,** 187–232.

Yanagimachi, R. (1988). Sperm–egg fusion. *In* "Current Topics in Membranes and Transport" (N. Duzgunes, and F. Bronner, eds.), Vol. 32, pp. 3–43. Academic Press, San Diego.

Yanagimachi, R. (1994). Mammalian fertilization. *In* "The Physiology of Reproduction" (E. Knobil and J. Neill, eds.), pp. 189–317. Raven Press, New York.

Yim, D. L., Opresko, L. K., Wiley, H. S., and Nuccitelli, R. (1994). Highly polarized EGF receptor tyrosine kinase activity initiates egg activation in *Xenopus*. *Dev. Biol.* **162,** 41–55.

Zachary, I., Sinnett-Smith, J., and Rozengurt, E. (1992). Bombesin, vasopressin, and endothelin stimulation of tyrosine phosphorylation in Swiss 3T3 cells. Identification of a novel tyrosine kinase as a major substrate. *J. Biol. Chem.* **267,** 19031–19034.

Zamboni, L. (1972). Fertilization in the mouse. *In* "Biology of Mammalian Fertilization and Implantation" (K. S. Moghissi and E. S. E. Hafez, eds.), pp. 213–262. Charles Thomas, Springfield, Illinois.

3
Mechanisms of Calcium Regulation in Sea Urchin Eggs and Their Activities during Fertilization

Sheldon S. Shen
Department of Zoology and Genetics
Iowa State University
Ames, Iowa 50011

I. Changes in $[Ca^{2+}]_i$ during Fertilization
   A. Resting Level in the Unfertilized Egg
   B. Transient Rise during Fertilization
II. Regulation of Intracellular Ca^{2+} Release
   A. IP_3 Receptor-Dependent Ca^{2+} Release
   B. Ryanodine Receptor-Dependent Ca^{2+} Release
   C. Cyclic ADP-Ribose-Induced Ca^{2+} Release
III. How Do Sperm Activate the Rise in $[Ca^{2+}]_i$?
   A. Receptor-Mediated Egg Activation
   B. Fusion Mediated Egg Activation
IV. Conclusions
   References

More than 56 years have passed since publication of the paper entitled "The release of calcium in *Arbacia* eggs on fertilization" (Mazia, 1937). This seminal work "contended that a primary effect of a stimulating agent on a cell is to cause a release of calcium from organic combinations in the cortex of the cell into the main body of protoplasm." It also established quantitatively a shift of about 15% from bound Ca^{2+} to free Ca^{2+} upon fertilization, and concluded that "this does not lead much nearer to an understanding of how the activating influence itself acts on the cell colloids to cause the Ca release." Despite an abundance of work that established possible organic combinations in the cortex as pathways of the calcium release, the manner in which sperm cause a release of Ca^{2+} remains elusive. This chapter focuses on the current evidence of Ca^{2+} regulation in sea urchin eggs during fertilization; it is not meant to be all inclusive, nor are its findings necessarily comparable with respect to other species. Studies on calcium regulation during fertilization of mammalian eggs (Miyazaki *et al.*, 1993) and comparative changes in deuterostome eggs during fertilization (Nuccitelli, 1991; Yoshimoto and Hiramoto, 1991; Berger, 1992; Whitaker and Swann, 1993) have been reviewed, as have other events associated with egg membranes during fertilization (Jaffe, 1994).

The regulation of intracellular Ca^{2+} activity ($[Ca^{2+}]_i$) in a wide variety of cells is well established as an important event during signal transduction of numerous extrinsic stimuli. Different sources, including influx of external Ca^{2+} and release from intracellular stores, contribute to the rise in $[Ca^{2+}]_i$. Irrespective of cell type, resting $[Ca^{2+}]_i$ is generally in the range of 50–200 nM and accounts for only a minute fraction of the total cellular Ca^{2+} content, so the bulk of cellular Ca^{2+} is bound and may reside within membrane-bound organelles (Carafoli, 1987). Release of Ca^{2+} from organelles is dependent on two types of intracellular receptor release mechanisms, which are sensitive to inositol 1,4,5-trisphosphate (IP_3) and ryanodine (Berridge, 1993a). A propagated wave of intracellular Ca^{2+} release characterizes the fertilization response of eggs from invertebrates to mammals (Jaffe, 1983, 1991). Some fertilization responses are only dependent on an IP_3-regulated mechanism (Miyazaki et al., 1992a,1993; Nuccitelli et al., 1993); however, two different regulatory mechanisms appear to mediate Ca^{2+} release in sea urchin eggs. An IP_3-sensitive pathway (Swann and Whitaker, 1986) and an IP_3-insensitive pathway (Rakow and Shen, 1990; Crossley et al., 1991) have been described. Both types of receptor release mechanism have been putatively identified immunologically in the sea urchin egg (McPherson et al., 1992; Parys et al., 1994). The Ca^{2+} wave is both necessary and sufficient for metabolic activation of the quiescent urchin egg (Whitaker and Steinhardt, 1985), so the two Ca^{2+} release mechanisms may be redundant to ensure the occurrence of an increase in $[Ca^{2+}]_i$ during fertilization. Ca^{2+} release during fertilization appears to be similar when either release mechanism is blocked (Galione et al., 1993a; Lee et al., 1993b; Shen and Buck, 1993).

Despite more than half a century of research, we still do not understand how sperm initiate the all-important rise in $[Ca^{2+}]_i$ during fertilization. The recent identification of the sperm receptor in the egg membrane (Foltz et al., 1993) has not provided recognition of any immediate transduction pathway. The cytoplasmic domain of the receptor is unlike any other known receptor and its function remains uncertain. Two general concepts of egg activation by sperm have been proposed. One hypothesis originally proposed that Ca^{2+} from the sperm contributed to the activation of the egg (L. F. Jaffe, 1980,1983,1990). This idea was later broadened to encompass an unknown activator in the sperm that must diffuse into the egg (Whitaker et al., 1989), according to studies of egg activation by microinjection of a water-soluble sperm extract (Dale et al., 1985). The other general hypothesis was similar to that of signal transduction pathways of receptor-mediated IP_3 production in somatic cells. Sperm activation of the egg occurred either through a GTP-binding protein (Turner et al., 1986; L. A. Jaffe, 1990) or through protein tyrosine kinase activity (Ciapa and Epel, 1991) in the egg membrane coupled to phospholipase C activity. An alternative proposal was modeled after activation of cells in the immune response (Berger, 1992).

Although more recent experiments might argue that phosphatidylinositol hydrolysis is a consequence rather than a cause of sperm-mediated egg activation, no diffusible activator in sperm has been identified to date. Ample experimental evidence supports either hypothesis; however, significant questions remain unanswered. One difficulty with studies attempting to identify the activating pathway during fertilization is that both pathways may be activated independently. In fact, current studies, suggest sperm activation of separate but redundant pathways regulating cytoplasmic Ca^{2+} levels, so inhibitor studies must be carefully evaluated. Much of this chapter focuses on recent studies of intracellular calcium release mechanisms in the egg and their activities during fertilization.

I. Changes in $[Ca^{2+}]_i$ during Fertilization

A. Resting Level in the Unfertilized Egg

After the pioneering detection by ultrafiltration technique of an increase in $[Ca^{2+}]_i$ in *Arbacia* eggs during fertilization (Mazia, 1937), temporal resolution and quantification of the change in $[Ca^{2+}]_i$ languished for 40 years until the development of improved techniques for Ca^{2+} detection. Renewed interest in Ca^{2+} during fertilization was stimulated by the demonstration that the divalent ionophore A23187 can activate sea urchin eggs (Chambers et al., 1974; Steinhardt and Epel, 1974) and eggs of other species that are widely separated phylogenetically (Steinhardt et al., 1974). Furthermore, the rise in $[Ca^{2+}]_i$ during fertilization was found to be necessary for development (Zucker and Steinhardt, 1978; Hamaguchi and Hiramoto, 1981). A major development in biological Ca^{2+} indicators was the purification of the photoprotein aequorin (Shimomura et al., 1962), which emits photons on Ca^{2+} binding (Shimomura and Johnson, 1970). These aequorin studies of $[Ca^{2+}]_i$ during fertilization were initially made using Medaka eggs (Ridgeway et al., 1977) and sea urchin eggs; sea urchin studies required a field of 10–30 aequorin-injected eggs because of their smaller size (Steinhardt et al., 1977). Since the luminescence of aequorin increases exponentially with increases in $[Ca^{2+}]_i$ (Allen et al., 1977), the resting level of $[Ca^{2+}]_i$ in unfertilized eggs was difficult to observe but, after a brief latency, a large increase in luminescence was observed after insemination. The rise in $[Ca^{2+}]_i$ in the field of sea urchin eggs lasted 2–3 min and was estimated to reach 2.5–4.5 μM throughout the cytoplasm (Steinhardt et al., 1977). The development of improved image intensifer and vidicon units allowed the luminescence increase during fertilization in single sea urchin eggs to be monitored with aequorin; it was estimated to rise from <0.1 μM to ~1 μM about 23 sec after the beginning of membrane depolarization (Eisen et al., 1984). These estimates may be low because of the low ionic strengths of calibration buffers and a varying exponential relationship between light and Ca^{2+} (Allen et al., 1977). A mean peak $[Ca^{2+}]_i$ of 1–2.8 μM during fertilization was estimated with aequorin using calibration solutions resembling the egg cytoplasm (Swann and Whitaker, 1986).

Recombinant semisynthetic aequorins with improved sensitivity to Ca^{2+} have been developed (Shimomura et al., 1993) and may be useful for detecting submicromolar levels of $[Ca^{2+}]_i$ and their changes in animal eggs (Kubota et al., 1993).

More recent measurements of $[Ca^{2+}]_i$ have employed fluorescent Ca^{2+} indicators, particularly the ratiometric chelator fura-2 (Grykiewicz et al., 1985), and the long wavelength fluoroprobes fluo-3 (Minta et al., 1989) and calcium green (Eberhard and Erne, 1991; Haughland, 1992). Since fura-2 exhibits a spectral shift in excitation maxima on binding Ca^{2+}, increasing free Ca^{2+} increases the fluorescence from 350-nm excitation and decreases the fluorescence from 380-nm excitation. The ratio of fluorescence is a measure of $[Ca^{2+}]_i$ without uncertainties in fluoroprobe concentration, egg thickness, and optical efficiency of the instrumentation. Although fluo-3 and calcium green lack a spectral shift on binding Ca^{2+}, normalizing fluorescence intensity changes provides useful information about transient Ca^{2+} changes from basal levels (Cornell-Bell et al., 1990). Two potential artifacts of fluorescent Ca^{2+} indicators are binding and compartmentalization (Blatter and Wier, 1990; Bolsover and Silver, 1991; Connor, 1993). In myoplasm, some 30% of fura-2 appeared to be compartmentalized, and two-thirds of the diffusible dye may have been bound (Blatter and Wier, 1990). Compartmentalization in organelles may be due to incomplete hydrolysis of the membrane-permeant ester form of these fluorescent indicators (Almers and Neher, 1985; Scanlon et al., 1987), but because of the size of eggs and the generally poor loading of ester forms, indicators are introduced into eggs in the free acid form by microinjection. Nonetheless, compartmentalization and binding of fluoroprobes occur in eggs by 20 min after microinjection (Mozingo and Chandler, 1990; Shen and Buck, 1990; Gillot and Whitaker, 1993). Additionally, the internal membrane of eggs is significantly reorganized during fertilization (Poenie and Epel, 1987; Terasaki and Jaffe, 1991), possibly resulting in differential dye distribution during fertilization. A potential solution to compartmentalization may be the use of dextran-coupled fluorescence Ca^{2+} indicators (Gillot and Whitaker, 1993; Harris, 1994). Even less is known about possible binding of fluoroprobes to cellular components (Gillot and Whitaker, 1993), which would compromise the reliability of nonratioed indicators including dextran-coupled dyes. Other problems include microviscosity (Poenie, 1990), which may be corrected by choosing excitation wavelengths that are independent of the viscosity artifact (Busa, 1992), and spatial resolution, which may be minimized with confocal laser scanning microscopy systems (Silver et al., 1992).

Despite inherent imaging problems (Mason et al., 1993), an advantage of ratiometric fluorescent Ca^{2+} indicators is their ability to estimate $[Ca^{2+}]_i$ in the unfertilized egg and its changes during fertilization (Tsien and Poenie, 1986). Estimates of $[Ca^{2+}]_i$ in the unfertilized sea urchin egg have ranged from 50 nM (Whalley et al., 1992) to >350 nM (Galione et al., 1991). Most mean resting

3. Calcium Regulation during Sea Urchin Fertilization 67

$[Ca^{2+}]_i$ in unfertilized eggs of different species have been similar, within the range of 100–150 nM (Poenie *et al.*, 1985; Hafner *et al.*, 1988; Mohri and Hamaguchi, 1991; Buck *et al.*, 1992), but $[Ca^{2+}]_i$ in individual eggs, even from the same female, show remarkable variability (Poenie *et al.*, 1985). We have observed variations in $[Ca^{2+}]_i$ from 50 to 200 nM in eggs from the same female, whereas other batches of eggs have had less than 10% variation of the mean between individual measurements. The variation may be due to microinjection damage, but the total Ca^{2+} concentration determined by X-ray microanalysis also displayed significant variability among eggs (Gillot *et al.*, 1989). The Ca^{2+} content in unfertilized sea urchin eggs has been reported to be as low as 1–2 mM (Lindvall and Carsjö, 1951) or as high as 35 mM (Nakamura and Yasamasu, 1974), but eggs from most different species contain 3–10 mM (Mazia, 1937; Rothschild and Barnes, 1953; Azarnia and Chambers, 1976; Gillot *et al.*, 1989, 1991).

An alternative technique to photometric measurements of Ca^{2+} activity is the use of ion-sensitive intracellular microelectrodes. Ion-sensitive microelectrodes have been used to measure a variety of cation activities in the sea urchin egg, including K^+ (Steinhardt *et al.*, 1971; Shen and Sui, 1989), H^+ (Shen and Steinhardt, 1978), Na^+ (Shen and Burgart, 1985), and Mg^{2+} (Sui and Shen, 1986). These ion activities changed by varying amounts during the fertilization response. The resting level of cytosolic Ca^{2+} measured by intracellular Ca^{2+}-sensitive microelectrode was about 100 nM (de Santis *et al.*, 1987), similar to estimates by photometric techniques. Surprisingly, no changes in $[Ca^{2+}]_i$ during fertilization were detected by these microelectrodes, although electrophysiological and morphological changes of fertilization were observed. A rise in $[Ca^{2+}]_i$ was detected only after the addition of a lethal dose of Ca^{2+} ionophore (de Santis *et al.*, 1987); thus, the absence of detectable changes in Ca^{2+} during fertilization was puzzling. The authors suggested that the rate of the Ca^{2+} transient during fertilization may have been faster than the electrode response time; however, Ca^{2+}-sensitive microelectrodes detected the cytosolic Ca^{2+} wave during fertilization in *Xenopus* eggs (Busa and Nuccitelli, 1985). In preliminary efforts, we found that the tip size necessary for reliable Ca^{2+} measurements precluded insertion into the unfertilized sea urchin egg without significant cytological damage, which included vesiculation of the electrode tip in the cytosol (S. Shen, unpublished observations).

B. Transient Rise during Fertilization

The temporal and spatial characteristics of the Ca^{2+} transient during sea urchin fertilization have been studied in detail. The Ca^{2+} transient, with respect to membrane depolarization as $t = 0$, generally begins 10–25 sec later (Whitaker *et al.*, 1989; Crossley *et al.*, 1991; Shen and Buck, 1993), reaches its peak by 12–

20 sec (Poenie et al., 1985; Swann and Whitaker, 1986; Rakow and Shen, 1990; Mohri and Hamaguchi, 1991), and lasts in excess of 5 min (Poenie et al., 1985; Whitaker et al., 1989; Rakow and Shen, 1990). The delay between sperm–egg interaction and onset of the Ca^{2+} rise is now referred to as the latent period. This latent period was initially defined as the time elapsed between sperm–egg interaction and cortical granule exocytosis (Allen and Griffin, 1958) and the sensitivity of sperm incorporation to spermicides (Baker and Presley, 1969). The latent period could be divided into a minimal or absolute period of 7 sec, followed by a variable or relative period (Shen and Steinhardt, 1984). Electrophysiology has better defined the latent period as the time between sperm–egg membrane fusion and the onset of the Ca^{2+} wave (Whitaker et al., 1989). Measurements of membrane conductance and capacitance have indicated that the two are coincident with the establishment of electrical continuity between the gametes (McCulloh and Chambers, 1992). The presence of a latent period has been interpreted to suggest a requirement for the accumulation of a triggering molecule for Ca^{2+} release. Proposed signal pathways for increasing Ca^{2+} levels that are activated by sperm–egg interaction must account for the latent period.

Variability in the peak $[Ca^{2+}]_i$ during the fertilization response has also been reported. Generally, the mean peak $[Ca^{2+}]_i$ during fertilization is between 1.5 and 2.5 μM (Poenie et al., 1985; Swann and Whitaker, 1986; Hafner et al., 1988; Buck et al., 1992), although individual eggs may have peak Ca^{2+} rises from as low as 1 μM (Swann and Whitaker, 1986) to greater than 3 μM (Mohri and Hamaguchi, 1991). We have also observed similar ranges in peak $[Ca^{2+}]_i$ values and have found decreasing values with increasing durations between microinjection of the indicator dye and fertilization, suggesting an effect of dye compartmentalization. Based on the inhibition of the rise of $[Ca^{2+}]_i$ by injection of known amounts of ethylene glycol-bis (β-amino-ethyl ether) N,N,N',N'-tetraacetic acid (EGTA), a Ca^{2+} chelator, the total increase in Ca^{2+} in the cytosol during sea urchin fertilization is estimated between 150 and 170 μM (Mohri and Hamaguchi, 1991).

The rise in $[Ca^{2+}]_i$ initiates at the point of sperm–egg interaction and traverses across the egg to the antipode with a velocity of 5–7 $\mu m/sec$ (Eisen et al., 1984; Swann and Whitaker, 1986; Hafner et al., 1988). The rise in $[Ca^{2+}]_i$ during fertilization is distributed throughout the egg cytoplasm (Eisen et al., 1984; Swann and Whitaker, 1986; Hafner et al., 1988); the propagation velocity appears to be faster in the cortex than in the central region (Mohri and Hamaguchi, 1991). Using the fluorescent Ca^{2+} indicator fluo-3 and confocal laser scanning microscopy, the rise in Ca^{2+} was shown to be heterogeneous with enhancement in the cortex and pronucleus rather than in the central cytoplasm (Stricker et al., 1992; Shen and Buck, 1993); however, imaging with the fluorescent indicator calcium green showed a homogeneous rise in $[Ca^{2+}]_i$ (Galione et al., 1993a). Whether the different images are due to differential compartmentalization or nonspecific binding of the dyes is unclear (Gillot and Whitaker, 1993); alter-

3. Calcium Regulation during Sea Urchin Fertilization

natively, the fluorescent emission respond of calcium green is 3- to 7-fold less than that of fluo-3 (Haughland, 1992). Normalized fluorescence response of fluo-3 during fertilization was near 4-fold (Shen and Buck, 1993), but was only about 1.6-fold for calcium green (Harris, 1994); thus, the heterogeneous distribution of the rise in $[Ca^{2+}]_i$ may be below the response sensitivity of calcium green.

The resting level and regulation of $[Ca^{2+}]_i$ in the sea urchin egg pronucleus may be different from that in the cytoplasm. Using fluo-3 in confocal laser scanning microscopy, the fluorescence in the pronucleus was shown to be less than the fluorescence detected in the cytosol in unfertilized eggs (Stricker *et al.*, 1992; Shen and Buck, 1993). This difference in fluorescence is not evident in calcium green-loaded eggs (Stricker *et al.*, 1992; Galione *et al.*, 1993a). During fertilization, the increase in fluo-3 fluorescence in the pronucleus is significantly greater than the fluorescence increase measured in the cytoplasm (Stricker *et al.*, 1992; Shen and Buck, 1993). Whereas one research group (Stricker *et al.*, 1992) reported a similar pronuclear enhancement during fertilization in calcium green-loaded eggs, another research group (Galione *et al.*, 1993a) did not detect significant pronuclear enhancement. After the transient rise in Ca^{2+}, the fluorescence in the pronucleus and cytoplasm became similar, in contrast to the situation in parthenogenetically activated eggs, which still displayed a pronuclear-to-cytoplasmic fluorescence ratio of less than 1 (Shen and Buck, 1993). Possible explanations for the differences in fluorescence in the pronucleus and cytoplasm include dye exclusion from the pronucleus in unfertilized eggs, compartmentalization of the dye, and nonspecific binding. Since fluorescence of pronucleus and cytoplasm were similar in calcium green-loaded eggs, it is unlikely that the pronuclear envelope in the unfertilized egg is less permeant. Alternatively, dye entry into Ca^{2+} storage compartments in the cytoplasm would give an apparently higher fluorescence intensity for the cytoplasm in the unfertilized egg and a lower increase during fertilization. An increase in nucleus-associated Ca^{2+} activity during cell activation has been observed in a variety of cell systems, and time-dependent and uneven fluoroprobe distributions have been a problem in studying differential nuclear and cytoplasmic Ca^{2+} changes (Hernández-Cruz *et al.*, 1990; AlMohanna *et al.*, 1994). Although these problems cannot be excluded in the sea urchin egg, differences in pronucleus-associated Ca^{2+} responses to sperm and to parthenogenetic agents (Shen and Buck, 1993) suggest that the differences between pronucleus and cytoplasm may be real.

Using confocal laser scanning microscopy, we have also detected an earlier cortex-localized rise in $[Ca^{2+}]_i$, or a cortical flash, shortly after insemination (Shen and Buck, 1993). This rise occurred nearly coincident with sperm attachment and was blocked when Ca^{2+} influx was blocked with low-Ca^{2+} seawater containing nifedipine (Fujiwara *et al.*, 1988), suggesting that the cortical flash was due to the increase in Ca^{2+} channel conductance during the action potential of membrane depolarization (Okamoto *et al.*, 1977; Chambers and deArmendi, 1979). The cortical flash appeared uniform throughout the egg cortex and was not localized to the sperm–egg binding site; thus, the cortical flash cannot be attributed to Ca^{2+} influx from the sperm. The cortical flash preceded the onset of the main rise in $[Ca^{2+}]_i$ by about 13 sec, corresponding to the latent period between the onset of the sperm-induced inward current (McCulloh and Chambers, 1992) and the main increase in $[Ca^{2+}]_i$ (Swann et al., 1992) that corresponded to the large fertilization current caused by Ca^{2+} activation of egg membrane channels (David et al., 1988). The duration of the cortical flash was very short (3–5 sec); it was present in one image and subsided in the next image or two (Shen and Buck, 1993). The cortical flash appeared to correspond to the small step-like rise detected in fura-2 fluorescence at the start of egg membrane depolarization (Whitaker et al., 1989).

The rise in $[Ca^{2+}]_i$ during fertilization is caused by Ca^{2+} influx and release from intracellular stores. Radioisotope studies show a rapid influx of Ca^{2+} during the first 30–60 sec after insemination (Paul and Johnston, 1978), but the main rise in $[Ca^{2+}]_i$ during fertilization is caused by intracellular Ca^{2+} release (Shen and Buck, 1993). No doubt eggs, like most somatic cells, contain an intracellular Ca^{2+} store; however, some controversy exists over the source of the intracellular Ca^{2+} store. Using aequorin luminescence of organelle-stratified centrifuged *Arbacia* eggs, researchers showed the endoplasmic reticulum (ER) to be the cellular source of Ca^{2+} during fertilization (Eisen and Reynolds, 1985). On the other hand, using chlortetracycline fluorescence of centrifuged *Hemicentrotus* eggs, other researchers localized the source of Ca^{2+} to the mitochondria (Ohara and Sato, 1986). This discrepancy may have been the result of compartmentalization of the dye into the mitochondria (Gillot and Whitaker, 1993). Ultrastructural localization of fluoride ion precipitation of Ca^{2+} in unfertilized *Strongylocentrotus* eggs was to cortical granules, tubular ER, and acidic vesicles, but not to mitochondria (Poenie and Epel, 1987). Furthermore, mitochondrial uncoupling agents, which would release mitochondrial Ca^{2+} (Fiskum and Lehninger, 1980), did not cause a change in $[Ca^{2+}]_i$ in unfertilized eggs (Eisen and Reynolds, 1985; Buck et al., 1991).

Fertilization apparently activates mitochondria to sequester Ca^{2+}, since aequorin luminescence in fertilized stratified eggs shifts to the mitochondrial zone and mitochondrial uncouplers stimulate a transient rise in $[Ca^{2+}]_i$ in fertilized eggs (Eisen and Reynolds, 1985). X-Ray microanalysis showed a transient Ca^{2+} accumulation in the mitochondria during fertilization (Girard et al., 1991). Ca^{2+} must also be resequestered into the ER (Eisen and Reynolds, 1985; Ohara and Sato, 1986; Girard et al., 1991), with a significant region of Ca^{2+} uptake in the egg cortex (Oberdorf et al., 1986; Payan et al., 1986; Terasaki and Sardet, 1991). In a wide variety of cell types, thapsigargin is a specific irreversible inhibitor of Ca^{2+} ATPase activity in the ER (Thastrup et al., 1990) without blocking plasma membrane Ca^{2+} ATPase activity (Lytton et al., 1991). Thapsigargin completely blocked ATP-dependent Ca^{2+} uptake in egg homogenates (Buck et al., 1992; Lee, 1993). Treatment of unfertilized eggs with thapsigargin generally caused only a slight rise in $[Ca^{2+}]_i$, so cellular stores of Ca^{2+} appear not to be leaky (Buck et al., 1992). The duration of the rise in $[Ca^{2+}]_i$ during fertilization was not significantly lengthened by thapsigargin, suggesting that Ca^{2+} efflux is important for returning $[Ca^{2+}]_i$ to near the unfertilized level.

As much as 10–30% of the total Ca^{2+} content in sea urchin eggs may be extruded after fertilization (Azarnia and Chambers, 1976). A consequence of the transient rise in $[Ca^{2+}]_i$ is a wave of cortical granule exocytosis (Vacquier, 1975; Baker and Whitaker, 1978); both precipitation with fluoride (Poenie and Epel, 1987) and X-ray microanalysis (Gillot et al., 1991) showed a significant concentration of Ca^{2+} in the cortical granules, suggesting that the Ca^{2+} loss during fertilization may be due to exocytosis (Gillot et al., 1990). Ca^{2+} release from the egg has been studied with an ion-selective vibrating probe system (Kühtreiber et al., 1993). The Ca^{2+} efflux signal had a duration of 150–200 sec, which is significantly longer than the duration of exocytosis; the authors estimated that cortical granule release accounted for less than 14% of the Ca^{2+} lost. These measurements imply that fertilization activates a plasma membrane Ca^{2+} pump system that has not yet been identified.

II. Regulation of Intracellular Ca^{2+} Release

Two separate intracellular Ca^{2+} release mechanisms regulated by IP_3 and ryanodine have been identified in many somatic cell types (Meldolesi et al., 1990; Berridge, 1993a). The sea urchin egg contains specific proteins recognized by an antibody against the skeletal muscle type 1 ryanodine receptor (McPherson et al., 1992) and an antibody against the type I IP_3 receptor (Parys et al., 1994). Both receptors appear to differ from their counterparts in somatic cells (Mikoshiba, 1993; Sorrentino and Volpe, 1993). The putative ryanodine receptor isoform in eggs has a size of ~380 kDa (McPherson et al., 1992), which is significantly smaller than the somatic receptor isoforms with sizes of ~565 kDa (McPherson and Campbell, 1993). In contrast, the putative IP_3 receptor isoform has a size of ~370 kDa (Parys et al., 1994), which is significantly larger than the IP_3 receptor isoforms of rabbit brain (~270 kDa) and Xenopus oocytes and eggs (~260 kDa). The spatial distribution of antibody bindings to the two receptors in the unfertilized urchin egg appeared to be different. Ryanodine receptor antibody binding was localized to the egg cortex (McPherson et al., 1992), whereas IP_3 receptor antibody binding appeared throughout the cytoplasm and periphery of the female pronucleus (Parys et al., 1994). However, both receptor release mechanisms apparently utilize a common Ca^{2+} store in the unfertilized egg (Whalley et al., 1992; Buck et al., 1994). Ca^{2+} release by agonists of either receptor mechanism was reduced significantly after introduction of a poorly metabolized IP_3 analog. Further evidence of a common Ca^{2+} store was the observation that the ER of unfertilized eggs was a cell-wide interconnected compartment that was discerned using the diffusion of dicarbocyanin dye (Terasaki and Jaffe, 1991). A striking phenomenon was the fractionation and reorganization of the ER within the first minute of fertilization (Jaffe and Terasaki, 1993).

A variety of agents have been shown to act as Ca^{2+} release agonists in sea urchin eggs. The targets of these agonists have been mostly separated by the pharmacological sensitivity of their induced Ca^{2+} release (Palade et al., 1989), especially to the low molecular weight heparin, which is a competitive and potent inhibitor of IP_3-activated Ca^{2+} release (Ghosh et al., 1988) and IP_3 binding (Worley et al., 1987; Ferris et al., 1989). To a lesser extent, ryanodine receptor-mediated Ca^{2+} release has been identified by sensitivity to ruthenium red, which is a specific inhibitor of skeletal and cardiac ryanodine receptor isoforms in sarcoplasmic reticulum (Fleischer and Inui, 1989). For a variety of reasons to be discussed later, ruthenium red is not as useful as a specific inhibitor of sea urchin ryanodine receptor isoform-mediated Ca^{2+} release. Ryanodine receptor-mediated Ca^{2+} release in somatic cells has also been characterized as Ca^{2+}-induced Ca^{2+} release (CICR) because a sudden change in Ca^{2+} can trigger Ca^{2+} release from the sarcoplasmic reticulum (Endo et al., 1970; Ford and Podolsky, 1970). The term "CICR" is misleading since it is now clear that Ca^{2+} may also affect the IP_3-dependent receptor release mechanism (Bezprozvanny et al., 1991; Finch et al., 1991). Furthermore, the ryanodine receptor release mechanism in sea urchin eggs does not appear to act similarly to that described for somatic CICR. Elevation of Ca^{2+} in intact sea urchin eggs (Swann and Whitaker, 1986) or digitonin-permeabilized eggs (Girard et al., 1991) did not generate additional Ca^{2+} release, and the propagation of Ca^{2+} from the injection site was by diffusion (Mohri and Hamaguchi, 1991). Further evidence against occurrence of CICR in egg is the nonpropagated cortical reaction induced by localized application of the divalent ionophore A23187 (Chambers and Hinkley, 1979). CICR was observed in egg homogenates only after excess Ca^{2+} loading of the microsomes (Galione et al., 1991, 1993a).

The multiple mechanisms of intracellular Ca^{2+} release in sea urchin eggs contrast with its release in *Xenopus* oocytes and eggs and in hamster eggs. Although CICR was proposed in *Xenopus* oocytes (Busa et al., 1985) and was supported by observations that the Ca^{2+} wave propagation velocity was similar to the rate of Ca^{2+} diffusion (Lechleiter et al., 1991), *Xenopus* oocytes appear to lack a ryanodine-dependent release mechanism. Rather, IP_3 was absolutely required for Ca^{2+} release and heparin blocked Ca^{2+} wave propagation (DeLisle and Welsh, 1992). Ca^{2+} activity apparently stimulated sensitization of the IP_3 receptor (Lechleiter and Clapham, 1992). The IP_3 receptor, but not the ryanodine receptor, has been immunologically detected in (Parys et al., 1992) and cloned from (Kume et al., 1993) *Xenopus* oocytes. The initiation and propagation of the rise in $[Ca^{2+}]_i$ in *Xenopus* eggs during fertilization were similarly dependent on IP_3 (Nuccitelli et al., 1993). A CICR mechanism was also suggested in hamster eggs (Igusa and Miyazaki, 1983) and mouse oocytes (Peres, 1990). However, a monoclonal antibody against the IP_3 receptor blocked both IP_3-mediated Ca^{2+} release and CICR in hamster eggs (Miyazaki et al., 1992a,b). In hamster eggs, IP_3-mediated Ca^{2+} release also appeared to be essential for generating the rise in $[Ca^{2+}]_i$ during the fertilization response (Miyazaki et al., 1993). Currently, a ryanodine-dependent CICR mechanism has been reported to be both present (Swann, 1992) and absent (Kline and Kline, 1994) in mouse eggs.

A. IP_3 Receptor-Dependent Ca^{2+} Release

IP_3-induced Ca^{2+} release is well established as a cellular mechanism underlying a variety of cellular responses (Berridge and Irvine, 1989). Interest was focused on this pathway during sea urchin fertilization by the observations of an increased turnover of polyphosphoinositides (Turner et al., 1984) and a transient increase in $[Ca^{2+}]_i$ by microinjection of exogenous IP_3 (Whitaker and Irvine, 1984). Within the first 10–30 sec after insemination, a transient rise in IP_3 occurred (Ciapa and Whitaker, 1986). The levels of phosphorylated phosphatidylinositols increased, whereas the precursor phosphatidylinositol decreased (Turner et al., 1984; Kamel et al., 1985), due to a 600-fold increase in the turnover of polyphosphoinositides during this time (Ciapa et al., 1992). The egg cortex contained both phospholipase C (PLC; Whitaker and Aitchison, 1985) and phosphoinositide kinase (Oberdorf et al., 1989) activities, which are stimulated by the rising Ca^{2+} activity (Ciapa et al., 1992). These experiments seem to confirm the earlier suggestion (Swann and Whitaker, 1986) that propagation of the Ca^{2+} rise through the egg from the sperm–egg fusion site to the antipode involves IP_3-induced Ca^{2+} release and Ca^{2+}-stimulated IP_3 production. Unfortunately, further analysis suggests a more complex regulation of IP_3 production during fertilization. Divalent ionophore activation caused an initial decrease in labeling of phosphatidylinositol 4,5-bisphosphate (PIP_2) rather than the increase seen during fertilization, suggesting a sperm-dependent activation of the kinase activity (Ciapa et al., 1992). Furthermore, as noted earlier, Ca^{2+} injection failed to induce additional Ca^{2+} release (Swann and Whitaker, 1986; Mohri and Hamaguchi, 1991), and increase in $[Ca^{2+}]_i$ by exogenous IP_3 (Whalley and Whitaker, 1988) or microinjections of other agonists (Buck et al., 1992; Whalley et al., 1992; Shen and Buck, 1993) also failed to stimulate any apparent IP_3 production. This conclusion has been based on a lack of an increase in cytosol pH (Whalley and Whitaker, 1988) and a nonoccurrence of IP_3-dependent Ca^{2+} release (Shen and Buck, 1993). The regulation of cytosol pH is dependent on Na^+–H^+ antiporter activity that is stimulated by 1,2-diacylglycerol, which is the other product of PIP_2 hydrolysis (Berridge and Irvine, 1984).

Currently it is not possible to test whether a rise in $[Ca^{2+}]_i$ is even necessary for IP_3 production in sea urchin eggs; however, this is not the case in *Xenopus* eggs. Since the mass of IP_3 increased 3- to 5-fold within the first minute after insemination, a rate significantly greater than that detected during artificial activation of the *Xenopus* egg (Stith *et al.*, 1993), a sperm-dependent stimulation of polyphosphoinositide hydrolysis that is not Ca^{2+} dependent was suggested. This stimulation was directly demonstrated when a similar rise in IP_3 mass was measured during fertilization of *Xenopus* eggs preloaded with the Ca^{2+} chelator 1,2-bis (2-aminophenoxy) ethane N,N,N',N'-tetraacetic acid (BAPTA), which would prevent cytosolic elevation in $[Ca^{2+}]_i$ (Stith *et al.*, 1994). Interestingly, the level of IP_3 production was independent of polyspermy. Thus, a sperm dependent event appears to be able to initiate and maintain IP_3 production throughout the egg, independent of a rise in $[Ca^{2+}]_i$.

Using cortical granule exocytosis as an index of egg activation, the half-maximal amount of microinjected IP_3 required is estimated at 3–7.5 nM (Whitaker and Irvine, 1984; Crossley *et al.*, 1988). Complete egg activation was detected with microinjection of IP_3 at an estimated 28 nM (Turner *et al.*, 1986). At higher concentrations of IP_3, the transit time of cortical granule exocytosis coincided with the estimated diffusion of IP_3 (Whitaker and Irvine, 1984; Mohri and Hamaguchi, 1991). An estimated level of 6–10 μM IP_3 is produced during fertilization (Turner *et al.*, 1984; Swann *et al.*, 1987). In sea urchin egg homogenates, the half-maximal concentration for Ca^{2+} release by IP_3 was 50–130 nM; 1 μM was required for full Ca^{2+} release (Clapper and Lee, 1985; Dargie *et al.*, 1990; S. Shen, unpublished observations). The difference in IP_3 receptor sensitivity in intact eggs and homogenates may be due to differences in basal Ca^{2+} activity. Other inositol polyphosphates at 10 to 100-fold greater concentrations were also able to stimulate cortical granule exocytosis, and the effects were independent of external Ca^{2+} (Crossley *et al.*, 1988).

The kinetic characteristics of IP_3-induced Ca^{2+} release in sea urchin eggs have been studied. The peak $[Ca^{2+}]_i$ of ~2 μM induced by exogenous IP_3 is similar to that occurring during fertilization (Swann and Whitaker, 1986; Swann *et al.*, 1987). However, the rise and fall times of the transient change in $[Ca^{2+}]_i$ were significantly faster after IP_3 treatment than during fertilization (Shen and Buck, 1993); thus, the duration of the Ca^{2+} transient was greatly reduced (Swann *et al.*, 1987). The reduction in rise time of the Ca^{2+} transient may be due to the rapid diffusion of IP_3 (Mohri and Hamaguchi, 1991). An enhanced pronucleus-associated rise in fluorescence intensity was observed following IP_3 administration (Shen and Buck, 1993); putative receptors for IP_3 have been localized to the pronuclear membrane (Parys *et al.*, 1994). Unlike in fertilization, the pronucleus/cytoplasm fluorescence ratio did not stabilize near 1 after IP_3 treatment, but returned to the inactivated level of less than 1. Whether the reorganization of the egg ER during fertilization (Terasaki and Jaffe, 1991; Jaffe and Terasaki, 1993) occurs after parthenogenetic activation by IP_3 or whether this might account for the difference in pronucleus/cytoplasm fluorescence ratio remains unknown.

The IP_3-dependent release mechanism is refractory to consecutive applications

3. Calcium Regulation during Sea Urchin Fertilization

of IP_3 in homogenates, permeabilized eggs (Clapper and Lee, 1985), and microinjected intact eggs (Dargie *et al.*, 1986), although successive Ca^{2+} transients elicited by IP_3 injections have also been reported (Swann *et al.*, 1987). IP_3-induced Ca^{2+} release in sea urchin eggs is sensitive to low molecular weight heparin, but not to de-N-sulfated heparin (Rakow and Shen, 1990; Crossley *et al.*, 1991). Specific IP_3 binding activity in egg microsomes was 95% inhibited by heparin at 50 μg/ml (Parys *et al.*, 1994). Heparin at ~0.3 mg/ml was reported to block the Ca^{2+} transient by microinjection of 0.1% egg volume of 50 μM IP_3 (Rakow and Shen, 1990), far exceeding the IP_3 content of sea urchin sperm (Iwasa *et al.*, 1990). Nonetheless, this level of heparin failed to block the increase in Ca^{2+} during fertilization (Rakow and Shen, 1990; Crossley *et al.*, 1991), suggesting that Ca^{2+} release during fertilization may be IP_3 independent. Perhaps because of the structural difference of putative sea urchin IP_3 receptors (Parys *et al.*, 1994), we were unable to block IP_3-induced Ca^{2+} release with a polyclonal antibody (S. Shen and W. Buck, unpublished observations) against the C terminus of the type I IP_3 receptor (Rbt5? was provided by Drs. J. B. Parys and K. P. Campbell, University of Iowa). More recently, ~1 mg/ml heparin or 0.9 mg/ml pentosan polysulfate has been reported to reduce substantially the $[Ca^{2+}]_i$ increase caused by sperm (Mohri *et al.*, 1993; T. Mohri and E. Chambers, personal communication). This finding is in contrast to a report (Lee *et al.*, 1993b) of a sperm-induced rise in $[Ca^{2+}]_i$ in eggs preloaded with 4.7 mg/ml heparin. The source of these apparently contradictory results may be the different heparin compounds used by the two laboratories. Sigma (St. Louis, Missouri) product number H5765, noncleaved heparin, was used by Lee *et al.* (1993b), and H2149, a low molecular weight cleaved heparin preparation, was used by Mohri and Chambers (personal communication). The effectiveness of noncleaved heparin may be less than 20% of that of cleaved heparin (Ghosh *et al.*, 1988); thus, the effective level of heparin inhibition would be much less than 4.7 mg/ml. Nonetheless, the effects of high concentrations of heparin or pentosan polysulfate suggest that the sperm-induced rise in Ca^{2+} is dependent on IP_3 binding or that, at these higher concentrations, the antagonists may also inhibit IP_3-independent Ca^{2+} release mechanisms.

B. Ryanodine Receptor-Dependent Ca^{2+} Release

Ample evidence has accumulated for an IP_3-independent Ca^{2+} release mechanism in the sea urchin egg. In sea urchin egg homogenates, the rate and magnitude of Ca^{2+} release induced by 1–10 mM caffeine (Galione *et al.*, 1991) and 20–500 μM ryanodine (Buck *et al.*, 1994) are dose dependent. Ryanodine-mediated release can be synergized by prior treatment of the homogenate with subthreshold levels of caffeine (Galione *et al.*, 1991), and is sensitive to 1 mM procaine or 30–50 μM ruthenium red (Galione *et al.*, 1991,1993a). The Ca^{2+}

release by caffeine and ryanodine appears to be IP_3 independent because 1 μM IP_3-induced release was not blocked by prior treatment of homogenate with caffeine and ryanodine, nor by the inhibitors procaine and ruthenium red (Galione et al., 1991). These results contrast with other reports in which IP_3-mediated Ca^{2+} release was also procaine and ruthenium red sensitive, and ruthenium red only partially inhibited ryanodine-mediated Ca^{2+} release. $^{45}Ca^{2+}$ release from $^{45}Ca^{2+}$-loaded particulate fractions of sea urchin eggs by either 50 μM ryanodine- and 10 μM IP_3-induced release was inhibited by 1 mM ruthenium red or 0.5 mM procaine (Fujiwara et al., 1990). On the other hand, the discharge of $^{45}Ca^{2+}$ from isolated egg cortices by 100 μM ryanodine was blocked only partially by 100 μM ruthenium red, and ryanodine inhibition of ATP-dependent $^{45}Ca^{2+}$ uptake by the cortices was prevented by 100 μg/ml heparin (Sardet et al., 1992). Thus, in different sea urchin egg fractions and using fluorescence or radioisotope techniques, caffeine and ryanodine appear to mediate a Ca^{2+} release that has variable sensitivities to inhibitors. However, whether the ruthenium red inhibition of IP_3 action or the heparin inhibition of ryanodine was an anomaly of the particular experiment remains unclear.

We have not observed 200–500 μg/ml heparin to show any inhibition of ryanodine action in egg homogenates or in intact eggs (Buck et al., 1992). Heparin at less than 10 μg/ml has been reported to induce Ca^{2+} release from the terminal cysterns of the sarcoplasmic reticulum (SR), probably due to its polyanionic property (Ritov et al., 1985). In most cell systems, IP_3-induced Ca^{2+} release is insensitive to ruthenium red (Palade et al., 1989); in the intact egg, ruthenium red does not prevent IP_3-mediated Ca^{2+} release (Galione et al., 1993a; S. Shen, unpublished observations). A significant problem we have encountered is the quenching of fluorescence signal by ruthenium red in egg homogenates (Penner et al., 1989). In our experience, the fluorescence signals of fura-2 and fluo-3 are quenched by more than 90%, significantly compromising the response of the fluoroprobe. Furthermore, ruthenium red is an undefined mixture of precursor components and variable dye content. The wide range of inhibitory potency, from nanomoles on the ryanodine receptor in skeletal and cardiac muscles to tens of micromoles in egg homogenates, is troubling and may suggest inhibition by different constituents of the dye. We have not examined procaine inhibition in the intact egg extensively because of its weak base action (Shen and Steinhardt, 1978) and the marked receptor-binding sensitivity of Ca^{2+} release agonists to pH between 7 and 8 (Worley et al., 1987; Lee, 1991). Furthermore, eggs are parthenogenetically activated by treatment with procaine (Moy et al., 1977).

Both caffeine and ryanodine mediate a rise in $[Ca^{2+}]_i$ in intact eggs that appears to be independent of the IP_3 receptor release mechanism. A minimal addition to the bath of 10-20 mM caffeine was required to induce fertilization membrane formation (Fujiwara et al., 1990) and a transient rise in $[Ca^{2+}]_i$ (Buck et al., 1992; Harris, 1994). Using fura-2 fluorescence, we have detected an

3. Calcium Regulation during Sea Urchin Fertilization

increase of $\sim$250 nM above the basal level of 120 nM (a level insufficient for fertilization envelope formation), after an initial delay of 15–45 sec, that lasted 2–4 min after 20 mM caffeine treatment of unfertilized eggs under conditions blocking Ca^{2+} influx (Buck et al., 1992). This caffeine-induced effect was insensitive to heparin at 0.3–0.6 mg/ml and to ruthenium red at 1 mM in the bath or at 10–50 µM inside the egg. Using calcium green–dextran (10,000 MW), Harris (1994) showed that 20 mM caffeine treatment produced a sharp rise in $[Ca^{2+}]_i$ approximately two-thirds the height of the normal fertilization response, although no fertilization envelope formation was detected. The difference in rate and magnitude of Ca^{2+} release may be due to the different fluoroprobes used or to differences in time between injection of fluoroprobes and exposure to caffeine. Furthermore, fertilization of caffeine-treated eggs resulted in a reduced rise in Ca^{2+} in fura-2-loaded eggs (Buck et al., 1992) and a nearly normal release in calcium green–dextran-loaded eggs (Harris, 1994). Caffeine did not significantly inhibit Ca^{2+} release by IP_3 (Buck et al., 1994), as has been reported in *Xenopus* oocytes (Berridge, 1991; Parker and Ivorra, 1991) and somatic cells (Brown et al., 1992; Hirose et al., 1993; Bezprozvanny et al., 1994). Interestingly, the washout of caffeine also caused a rise in $[Ca^{2+}]_i$ (Harris, 1994). This transient rise in Ca^{2+} is heparin insensitive but blocked by thapsigargin (P. Harris, personal communication), suggesting intracellular release of Ca^{2+} by an IP_3-independent receptor release mechanism.

Ryanodine at 150 µM in the intact egg was reported to elicit a fertilization-like Ca^{2+} transient to nearly 1 µM (Galione et al., 1991). Addition of ryanodine above 100 µM to the bath stimulated $^{45}Ca^{2+}$ influx and fertilization envelope elevation in a dose-dependent manner, with 50% responses near 1 mM (Fujiwara et al., 1990). Similarly, microinjection of 400 µM ryanodine stimulated full or partial elevation of the fertilization envelope in 62% of the eggs (Sardet et al., 1992). We observed a dose response in Ca^{2+} release to ryanodine injection, 40–60 µM ryanodine induces a rise to about 0.5 µM and 150–200 µM stimulates a release to 0.75 µM (Buck et al., 1992). The observed transient rise in $[Ca^{2+}]_i$ in response to 200 µM ryanodine was significantly slower in rise time and smaller in magnitude than the rise during fertilization (Shen and Buck, 1993). Maximal Ca^{2+} release in egg homogenates required 0.5 mM ryanodine and was significantly slower than release by other Ca^{2+} agonists (Buck et al., 1994). Ryanodine-induced Ca^{2+} rise was not dependent on Ca^{2+} influx, but was enhanced by heparin (0.3–0.6 mg/ml) and only partially blocked by ruthenium red (1 mM in the bath or 10–50 µM inside the egg; Buck et al., 1992). The ryanodine-mediated transient rise in $[Ca^{2+}]_i$ in the unfertilized egg revealed significant kinetic differences from that of Ca^{2+} release in skeletal sarcoplasmic reticulum (McPherson and Campbell, 1993). Agonists of sarcoplasmic reticulum all-or-none Ca^{2+} release (CICR) include Ca^{2+}, caffeine, ryanodine at nanomolar concentrations, and sulfhydryl reagents; inhibitors include ruthenium red, procaine, and ryanodine at micromolar concentrations (Fleischer and Inui, 1989). In contrast, Ca^{2+} release in intact sea urchin eggs is insensitive to Ca^{2+} and is graded in response to caffeine and ryanodine so that their actions are synergistic (Buck et al., 1992). However, the synergistic action of caffeine and ryanodine suggests actions similar to those described in sarcoplasmic reticulum, including the opening of Ca^{2+} channels by caffeine and the enhancement of the duration of channel opening by ryanodine (Rousseau et al., 1987) due to its preferential binding to open channels (Lai and Meissner, 1989). However, hundred of micromoles of ryanodine, which would have closed the channel in sarcoplasmic reticulum (Meissner, 1986), were required to induce Ca^{2+} release in egg. Resolution of these differences will require analysis of the effects of caffeine, ryanodine, and ruthenium red on [^{3}H]ryanodine binding in the sea urchin egg (Pessah and Zimanyi, 1991).

C. Cyclic ADP-Ribose-Induced Ca^{2+} Release

A physiological agonist for the ryanodine receptor has not been identified to date (Berridge, 1993b), but cyclic adenosine diphosphate-ribose (cADPR) has been proposed to act as the endogenous messenger (Galione, 1992,1993). Although cADPR was originally shown to act as a Ca^{2+} release agent in sea urchin microsome preparations (Dargie et al., 1990), cADPR has been reported to mobilize Ca^{2+} in other cell types including pituitary cells (Koshiyama et al., 1991), dorsal root ganglion cells (Currie et al., 1992), pancreatic β cells (Takasawa et al., 1993), cardiac sarcoplasmic reticulum microsomes (Mészáros et al., 1993), and rat brain microsomes (White et al., 1993). Furthermore, cADPR and its synthesizing enzyme have been detected in a wide variety of mammalian and invertebrate cell types (Rusinko and Lee, 1989; Walseth et al., 1991; Lee and Aarhus, 1993). Synthesis of cADPR from β-NAD$^+$ may occur either by a cytosolic 29-kDa enzyme called ADP-ribosyl cyclase (Glick et al., 1991; Lee and Aarhus, 1991) or by a membrane-bound 39-kDa enzyme called NAD glycohydrolyase (Kim et al., 1993a). In the sea urchin egg, the enzyme is associated with the membrane fraction but is still referred to as cADP-ribosyl cyclase (Lee, 1991). The synthesis of cADPR from β-NAD$^+$ involves cleavage of the nicotinamide–ribose bond and cyclization of the ribose to position 1 of the adenine ring (Kim et al., 1993b; Lee et al., 1994) rather than to N^6, as was previously proposed (Lee et al., 1989). Degradation of cADPR to ADPR by cADPR-ribose hydrolase is a part of NAD glycohydrolase activity (Kim et al., 1993a) or ADP-ribosyl cyclase activity (Zocchi et al., 1993). Lymphocyte antigen CD38, which has a 24% sequence identity with ADP-ribosyl cyclase (States et al., 1992), catalyzes formation and hydrolysis of cADPR when CD38 is expressed as a recombinant soluble protein (Howard et al., 1993). Additionally, NAD glycohydrolase is phosphatidylinositol-anchored to the outer surface of erythrocytes (Pekala and Anderson, 1978; Kim et al., 1988) and can convert β-NAD$^+$ to cADPR (Lee et al., 1993a). Thus, cADPR may have cellular functions other than Ca^{2+} mobilization.

Much of the experimental evidence supporting the hypothesis that cADPR is the natural ligand for ryanodine receptors has come from studies of sea urchin eggs and homogenates. Egg homogenates bind cADPR independently of IP_3 and heparin (Lee, 1991) and may have a common Ca^{2+} release mechanism for cADPR, ryanodine, and caffeine (Galione et al., 1991). The half-maximal concentration of cADPR for Ca^{2+} release in egg homogenates was seven times lower than that for IP_3 (Dargie et al., 1990). Subthreshold concentrations of cADPR potentiated Ca^{2+} release by caffeine or Ca^{2+} in egg homogenates (Lee, 1993), or by ryanodine in homogenates and intact eggs (Buck et al., 1994). Further evidence for a common Ca^{2+} release mechanism of cADPR, caffeine, and ryanodine includes mutual desensitization by maximal dosages of these agonists, but not IP_3, in egg homogenates (Dargie et al., 1990; Galione et al., 1991) and inhibition by procaine and ruthenium red (Galione et al., 1991). Ruthenium red inhibition of cADPR action appears to be more potent in homogenates than in intact eggs (Galione et al., 1993a), although we found that quenching was a significant problem in accurate quantification. Ruthenium red at 50 μM reduced Ca^{2+} release by threshold levels of cADPR (Galione et al., 1993a). We have not observed ruthenium red inhibition of Ca^{2+} release by higher concentrations (>100 nM) of cADPR (S. Shen and W. Buck, unpublished observations), a finding that is similar to the effect reported in cultured neurons. Ruthenium red at 100 μM had no effect on the action of cADPR or caffeine in dorsal root ganglion cells (Currie et al., 1992). A more specific inhibitor of cADPR may be the cADPR analog 8-amino cADPR (Walseth and Lee, 1993). Although like the actions of consecutive injections of cADPR, Ca^{2+} release in the recently fertilized egg is refractory to cADPR (Shen and Buck, 1993), the strongest evidence to date for cADPR-mediated Ca^{2+} release during fertilization is the block of Ca^{2+} release during fertilization in eggs preloaded with heparin and 8-amino cADPR (Lee et al., 1993b). However, before cADPR can be conclusively accepted as a natural messenger, a signal pathway and evidence that cADPR level changes rapidly during fertilization must be demonstrated.

cADPR did not appear to bind directly to the ryanodine receptor, but instead bound to two proteins of 100 and 140 kDa (Walseth et al., 1993). The binding sites of the cADPR–protein complex, caffeine, and ryanodine may be separate since FLA365, an inhibitor of the high-affinity binding of ryanodine to both skeletal and cardiac sarcoplasmic reticulum (Mack et al., 1992), inhibited the action of ryanodine but had no effect on the action of cADPR in homogenates or intact eggs (Buck et al., 1994). The synergistic actions of ryanodine and cADPR contrasted with their antagonistic effects in cardiac and brain microsome preparations, where 50 μM ryanodine inhibited 1 μM cADPR-mediated Ca^{2+} release (Mészáros et al., 1993). In the intact egg, the transient rise in $[Ca^{2+}]_i$ caused by exogenous cADPR is similar in kinetics to that induced by IP_3. A rapid increase in $[Ca^{2+}]_i$ from 200 nM to 1 μM occurred in response to a 150 nM cADPR injection, with a half-maximal level of ~60 nM (Dargie et al., 1990). Using confocal laser scanning microscopy, researchers showed that the rise and fall times of cADPR-induced Ca^{2+} transients were significantly faster, with a similar peak magnitude, than the kinetics and peak of Ca^{2+} increase during fertilization (Shen and Buck, 1993). As for the action of IP_3, no delay was observed between the onset of the rise in $[Ca^{2+}]_i$ and the microinjection of cADPR. Heparin at 0.1–0.3 mg/ml had little effect on the kinetics but did reduce the rise in $[Ca^{2+}]_i$ induced by 0.6 μM cADPR. The peak response of heparin-loaded eggs to cADPR was similar to that induced by ryanodine (Shen and Buck, 1993) but, unlike ryanodine, an enhanced pronuclear Ca^{2+} response was seen with cADPR, even in eggs preloaded with heparin.

Lee (1993) offered an alternative possibility that cADPR sensitized the ryanodine receptor to Ca^{2+} in such a way that cADPR may be present at a low constant level and ready to respond to a stimulus that slightly increases $[Ca^{2+}]_i$. The low level of cADPR and the slight rise in $[Ca^{2+}]_i$ would potentiate Ca^{2+} release that would propagate across the egg. Several experiments suggest messenger activation of the ryanodine receptor release mechanism; thus, cADPR would most likely act as a messenger, rather than a modulator of Ca^{2+} release. In eggs preloaded with heparin to block IP_3-mediated Ca^{2+} release, Ca^{2+} release was observed during fertilization (Rakow and Shen, 1990; Crossley et al., 1991; Shen and Buck, 1993). As further evidence that the cellular level of cADPR must rise if it is mediating the ryanodine receptor release mechanism, researchers found that microinjection of Ca^{2+} only caused localized egg activation (Swann and Whitaker, 1986), that the change in Ca^{2+} activity after microinjection of Ca^{2+}–EGTA buffer solution was only a function of diffusion (Mohri and Hamaguchi, 1991), that stimulation of Ca^{2+} influx did not trigger Ca^{2+} release (Zucker et al., 1978), and that a graded rather than an all-or-none Ca^{2+} response occurred to caffeine and ryanodine (Buck et al., 1992). Thus, Ca^{2+} does not appear to initiate Ca^{2+} release during fertilization. Furthermore, Swann and Whitaker (1986) estimated that the Ca^{2+} buffer capacity in the egg cytosol would prevent significant diffusion of released Ca^{2+}, which would restrict the range of Ca^{2+} action (Allbritton et al., 1992). Because of the requirement of at least 480 nM 8-amino cADPR to block Ca^{2+} release during fertilization of eggs also loaded with heparin (Lee et al., 1993b), and because 8-amino cADPR is a reversible, specific antagonist of cADPR (Walseth and Lee, 1993), the level of cADPR must rise to nearly 0.5 μM during fertilization, although the rise may be localized.

Currently, no rapid technique for measuring the level of cADPR is available. The most sensitive and rapid assay for cADPR is the sea urchin egg homogenate system, which has been used to detect cADPR production in other cell systems (Howard et al., 1993). An attempt to label the cADPR precursor, β-NAD^+, in the unfertilized egg would be difficult because of metabolic quiescence (Whitaker and

3. Calcium Regulation during Sea Urchin Fertilization 81

Steinhardt, 1985), and it might otherwise most likely label a variety of other small metabolites. Alternatively, eggs after insemination may be rapidly homogenized and separated by HPLC, and the relevant fraction containing cADPR may be tested for Ca^{2+} release in the egg homogenate system. A preliminary attempt suggested an ~20% increase in cADPR level in eggs at 1 min postinsemination (H. Lee, personal communication). Although the evidence for cADPR-dependent Ca^{2+} release during fertilization is indirect, the sperm-induced rise in $[Ca^{2+}]_i$ appears to be more complex. Like the other exogenous Ca^{2+} release agents (Buck et al., 1992; Whalley et al., 1992), microinjection of cADPR did not trigger IP_3 production (Shen and Buck, 1993). We have not observed an evidence of refractory Ca^{2+} release in response to microinjection of IP_3 after cADPR microinjection or vice versa in intact eggs (S. Shen and W. Buck, unpublished observations).

The level of cADPR in sea urchin eggs may be regulated by a cGMP-dependent pathway (Galione et al., 1993b). Microinjection of 10–20 μM cGMP triggered inositol phospholipid-independent Ca^{2+} release in sea urchin eggs (Whalley et al., 1992). The cGMP-stimulated rise in $[Ca^{2+}]_i$ (Swann et al., 1987) appeared to be indirect, with a latent period between injection and Ca^{2+} release, and with an absence of cGMP response in digitonin-permeabilized eggs (Whalley et al., 1992). The action of cGMP was heparin insensitive and appeared to be independent of PIP_2 hydrolysis. These observations suggested that an important small intermediate necessary for cGMP-mediated Ca^{2+} release had diffused from the egg cytosol. The bioactive molecule is assumed to be β-NAD^+, since its addition to egg homogenates potentiated Ca^{2+} release by cGMP (Galione et al., 1993b). These authors have proposed that cGMP, through cGMP-dependent protein kinase (PKG), regulates the activity of ADP-ribosyl cyclase. Evidence supporting this hypothesis includes mutual desensitization of Ca^{2+} release in homogenates by ryanodine, cADPR, and cGMP but not by IP_3; inhibition of cGMP-mediated Ca^{2+} release by ruthenium red but not by heparin; inhibitors of PKG blocking Ca^{2+} release induced by cGMP but not cADPR; and the stimulation of β-NAD^+ metabolism by cGMP. However, there are several difficulties with the proposal. The level of cGMP necessary for Ca^{2+} release in intact eggs and homogenates was at least 100-fold greater than that necessary for enzyme activation in mammalian preparations (Francis et al., 1988). The actions of 200 μM Rp-cAMPs are not specific to inhibition of PKG. The rate of NAD metabolism by cGMP was quite slow, so very few metabolites had been generated during the first 5 min when the bulk of cGMP-induced Ca^{2+} release occurs. Nonetheless, the idea of the ryanodine receptor—a widely distributed family of intracellular Ca^{2+} release channels (Berridge, 1993a; McPherson and Campbell, 1993), regulated by cADPR, a small β-NAD^+ metabolite of a widely distributed family of NAD glycohydrolases (Lee and Aarhus, 1991; Kim et al., 1993a) that is itself regulated by a cGMP-dependent signal pathway, is an attractive hypothesis (Berridge, 1993b). Further evidence is that 8-amino cADPR blocks cGMP-dependent increase in Ca^{2+} (H. Lee, personal communications).

III. How Do Sperm Activate the Rise in $[Ca^{2+}]_i$?

Despite the wealth of information concerning regulation of $[Ca^{2+}]_i$ in the sea urchin egg, the manner in which "the activating influence itself acts on the cell colloids to cause the Ca release" (Mazia, 1937) is still unknown. For that matter, the nature of the activating influence remains a mystery. Researchers thought that identification of the sperm receptor in the egg would clarify the action of the activating influence and perhaps even reveal its identity (Aldhous, 1993). The sea urchin sperm receptor has been identified and sequenced (Foltz et al., 1993), as well as isolated and characterized (Ohlendieck et al., 1993); however, its cytoplasmic domain is unlike that of any other known receptor and its function remains uncertain. Although the receptor binds sperm species-specifically and bindin, which is the presumed sperm ligand (Vacquier and Moy, 1977; Minor et al., 1993), and the cytoplasmic domain are conserved among echinoderms (Foltz et al., 1993), the receptor does not have any of the obvious characteristics of signal transducing receptors, such as the seven-transmembrane segment or a tyrosine kinase domain. Nonetheless, the receptor appears to play a role in mediating sperm activation of the egg. A glycoprotein extracellular fragment of the receptor bound sperm and inhibited fertilization in a species-specific manner (Foltz and Lennarz, 1990). Antibodies against the receptor fragment also inhibited sperm binding and even activated a subset of eggs (Foltz and Lennarz, 1992). These observations, however, do not help distinguish the proposal of receptor-mediated signaling from that of sperm-delivered activator for egg activation.

A. Receptor-Mediated Egg Activation

Proposals of receptor-mediated egg activation are patterned after messenger systems in somatic cells of either G protein-dependent (Hepler and Gilman, 1992) or tyrosine kinase-dependent (Pazin and Williams, 1992) pathways. Both G protein-dependent and tyrosine kinase-dependent pathways are capable of stimulating IP_3 production through activation of phospholipase C isozymes PLCβ and PLCτ, respectively (Rhee, 1991). Stimulation of G proteins with the nonhydrolyzed analog GTPτS activated sea urchin eggs (Turner et al., 1986) and caused a fertilization-like rise in Ca^{2+} that is heparin sensitive (Rakow and Shen, 1990; Crossley et al., 1991). Putative G proteins have been identified with cholera toxin and pertussis toxin, which catalyzed ADP-ribosylation of a 47-kDa polypeptide and a 40-kDa polypeptide, respectively (Turner et al., 1987). Microinjection of whole cholera toxin and subunit A stimulated cortical granule exocytosis in whole eggs (Turner et al., 1987). Expression of mRNAs encoding hormone receptors has provided further confirmation of a functional G protein-activated PLC pathway in other animal oocytes and eggs, including *Xenopus*

3. Calcium Regulation during Sea Urchin Fertilization   83

(Kline et al., 1988; Lechleiter et al., 1991), starfish (Shilling et al., 1990), and mouse (Williams et al., 1992; Moore et al., 1993). In addition, Ca^{2+} release in hamster eggs by serotonin and GTPτS was synergistic (Miyazaki et al., 1990). Such direct evidence has not been obtained in sea urchin eggs but even so, these experiments do not prove that a G protein-dependent mechanism is activated during fertilization. Inhibition of G proteins by injection of GDPβS in sea urchin eggs failed to block the rise in Ca^{2+} during fertilization (Crossley et al., 1991). The possibility remains that the Ca^{2+} rise was due to a redundant G protein-independent pathway, rather than to an absence of a G protein-dependent pathway. Surprisingly, the injection of GDPβS itself caused a rise to $[Ca^{2+}]_i$ (Crossley et al., 1991) that has not been further pursued. GDPβS apparently inhibited the cortical reaction (Turner et al., 1987), which may be G protein mediated (Mohri and Hamaguchi, 1989). Thus, ample evidence suggests that the egg contains a G protein-mediated activation pathway for Ca^{2+} release, but whether such signal transduction is operating during fertilization has not been tested directly.

In somatic cells, some tyrosine kinase-coupled receptors may activate PLCτ; expression of tyrosine kinase receptors in *Xenopus* (Yim et al., 1994) and starfish (Shilling et al., 1994) eggs resulted in egg activation with the addition of ligands. Tyrosine kinase activation occurs in sea urchin eggs during fertilization (Satoh and Garbers, 1985; Peaucellier et al., 1988). Phosphorylation on these proteins is first detected 30–60 sec after insemination, placing the enzyme activity subsequent to the rise in Ca^{2+}. Tyrosyl phosphorylation of a 350-kDa membrane protein was dependent on cytosol alkalinization (Jiang et al., 1990,1991) that occurred subsequent to the rise in Ca^{2+} (Shen, 1989). Using anti-phosphotyrosine antibody researchers found that residues of 91-kDa and 138-kDa proteins were phosphorylated within 20 sec after insemination (Ciapa and Epel, 1991). Artificial stimulation of intracellular Ca^{2+} release also increased tyrosyl phosphorylation of these two proteins; thus, it is unclear whether phosphorylation precedes or is a consequence of the rise in $[Ca^{2+}]_i$. However, the sperm receptor itself has been found to be phosphorylated on tyrosine residues within 5 sec after insemination (Abassi and Foltz, 1994). Phosphorylation was observed in both intact eggs and an egg plasma membrane preparation. Although the receptor does not have any sequences representing either a tyrosine kinase or *src* homology domains, it could, by analogy with the T-cell antigen receptor (Samelson et al., 1993; Weiss, 1993), activate a tyrosine kinase tightly associated with the egg plasma membrane. One possibility is a 57-kDa protein tyrosine kinase that has been identified in the egg cortex (Peaucellier et al., 1993). Since large numbers of sperm receptors were phosphorylated (Abassi and Foltz, 1994), the receptors were presumably phosphorylated independent of sperm binding, and the receptor is a substrate of the kinase activity.

The proposal that sperm activation of the egg takes place via a T-cell antigen receptor-like mechanism (Berger, 1992) poses several difficulties. Receptor phosphorylation in both intact eggs and the plasma membrane preparation was similarly activated by bindin (Abassi and Foltz, 1994). Bindin is the sperm ligand for the egg (Vacquier and Moy, 1977; Minor et al., 1993), and is recognized species-specifically by the sperm receptor in the egg (Foltz and Lennarz, 1992; Foltz et al., 1993; Ohlendieck et al., 1993). Although there is no documentation that bindin exposure does not cause ionic events in the unfertilized egg, researchers generally agree that bindin treatment does not activate the egg (Minor et al., 1993). Thus, receptor phosphorylation appears not to be sufficient for egg activation. Inhibition of tyrosine phosphorylation prevents T-cell antigen receptor-mediated signaling (June et al., 1990); however, whether tyrosine kinase activity is necessary for egg activation remains unreported. In preliminary experiments with genistein, a tyrosine kinase inhibitor (Akimaya et al., 1987), we observed normal changes in early ionic events during fertilization, although later development, including cleavage, was blocked (S. Shen and W. Buck, unpublished observations). Whether genistein blocked receptor phosphorylation in the egg is unknown. Another possible problem may be that the timing of receptor phosphorylation and the ensuing activation of Ca^{2+} release are longer than the latent period. Although receptor phosphorylation was observed at 5 sec and was maximal by 30 sec after insemination (Abassi and Foltz, 1994), placing the event within the latent period, the ensuing signaling pathway may not fit within the latent period constraint of 10–20 sec. The latent period between receptor activation and rise in $[Ca^{2+}]_i$ in the immune response was 20 sec to 2 min (Randriamampita et al., 1991; Hall et al., 1993), revealing a greater variability than was observed during fertilization (Shen and Steinhardt, 1984). A significant difference with respect to the immune response is the occurrence of bystander effects during fertilization or phosphorylation of multiple sperm receptors (Abassi and Foltz, 1994), including effects that presumably were not directly associated with the site of Ca^{2+} release. In the immune response, receptor-mediated phosphorylation occurred only on receptors that were aggregated as part of cell stimulation or an absence of the bystander effect (Pribluda and Metzger, 1992). Observation of multiple sperm receptor phosphorylation induced by sperm may explain differences between parthenogenetic activation and fertilization. Studies with IP_3-independent Ca^{2+} release agonists have suggested that, despite the rise in $[Ca^{2+}]_i$, IP_3 was not produced (Buck et al., 1992; Whalley et al., 1992). Furthermore, the pattern of phosphatidylinositol hydrolysis was different in fertilization- and ionophore-activated eggs (Ciapa et al., 1992). Hydrolysis of PIP_2 may require phosphorylated sperm receptor-mediated activity as part of its signaling pathway.

B. Fusion-Mediated Egg Activation

The idea that egg activation occurs as a result of sperm–egg membrane fusion preceded the hypothesis of the receptor-mediated activation pathway, but fell into disfavor with the accumulation of experimental parallels to receptor-mediated activation in somatic cells. However, this idea has now renewed its status since attempts to prove receptor-mediated activation appear to have failed. A basic tenet of egg activation by diffusion of a sperm activator is the occurrence of sperm–egg membrane fusion. The molecular basis of sea urchin gamete interactions was recently reviewed (Foltz and Lennarz, 1993). A number of sperm and egg proteins that may play important roles in gamete interactions have been identified, but the two key proteins appear to be bindin, the 30.5-kDa polypeptide of sperm (Lopez *et al.*, 1993), and the egg receptor for sperm, the 350-kDa egg membrane glycopeptide that binds bindin (Foltz *et al.*, 1993). Bindin recognition to the egg receptor would place bindin at the site of sperm–egg attachment (Moy and Vacquier, 1979), where bindin could exhibit fusogenic properties. Bindin has been shown to associate with phospholipids in gel phase (Glabe, 1985a) and to induce fusion of mixed-phase vesicles containing phosphatidylcholine and phosphatidylserine (Glabe, 1985b). Thus, bindin appears to participate in both sperm–egg adhesion and plasma membrane fusion. Another fusogenic sperm protein is lysin (Hong and Vacquier, 1986), although its principal function may be the disruption of the vitelline envelope (Vacquier *et al.*, 1990). Fusion of the sperm and egg has been studied by histochemistry or electrophysiological techniques (Whitaker and Swann, 1993). The electrophysiological studies of correlating membrane capacitance and onset of membrane currents have established time relationships between sperm–egg binding and egg activation (Chambers, 1989). The electrical studies showed that the capacitance increase, which corresponds to sperm–egg membrane fusion, occurred at the same time as the initial current step (McCulloh and Chambers, 1992). This initial current step is insensitive to EGTA injection into the egg and appears to be of sperm origin (Swann *et al.*, 1992). Egg activation was obligatorily linked not only with sperm–egg membrane fusion but with maintenance through the latent period (McCulloh and Chambers, 1992), supporting the hypothesis of activation by diffusion of a sperm activator.

The current idea of an activator that is delivered by the sperm started with the proposal that sperm activate eggs by delivery of Ca^{2+}, causing a localized elevation of $[Ca^{2+}]_i$ in the egg that initiates a subsequent wave of CICR (Jaffe, 1980,1983), thus giving the sperm the anarchist image of "a Ca^{2+} bomb" (Jaffe, 1980; Whitaker and Swann, 1993). Rather than Ca^{2+} from the sperm detonating Ca^{2+} release within the egg, the hypothesis has been modified to propose that the sperm acrosome acts as a conduit for Ca^{2+} entry to fill the cortical cisternum in the egg until a threshold for Ca^{2+} release is exceeded and CICR ensues (Jaffe, 1990). The time for overloading the local egg store would correspond to the latent period. This idea fell into disfavor because of findings that sea urchin eggs can be fertilized in the absence of external Ca^{2+} (Chambers and Angeloni, 1981; Schmidt *et al.*, 1982) and because of developing interest in phosphatidylinositol hydrolysis mechanisms. Some plausible explanations exist for sperm-mediated Ca^{2+} activation of the egg that is independent of Ca^{2+} in the bath. Although the usual pathway of Ca^{2+} influx into the egg would be Ca^{2+} influx into the sperm acrosomal region, an alternative source of Ca^{2+} from the sperm would be its mitochondria (L. F. Jaffe, personal communications). Supportive evidence of a less efficient Ca^{2+} transfer from sperm to egg associated with a Ca^{2+}-free environment includes the low success rates of fertilizations that are polyspermic and delayed (Schmidt et al., 1982). As further evidence for Ca^{2+} loading through the sperm (R. Créton and L. F. Jaffe, personal communications), researchers found that application of 2 mM La^{3+} blocked cell division and membrane elevation completely up to 5 sec after insemination and with decreasing efficiency thereafter. The La^{3+} application became half efficient at about 25 sec. The effect of La^{3+} was independent of sperm concentration, and similar effects were observed using 10 mM BAPTA with Ca^{2+} up to about 1 mM. These experiments alone do not prove that a Ca^{2+} flux from the sperm is necessary for egg activation. La^{3+} may act as a spermicide, disrupting sperm–egg fusion (Whitaker et al., 1989). This possibility may be tested by assaying for dye transfer (Hinkley et al., 1986). Nonetheless, other experiments argue against Ca^{2+} as the sperm activator of eggs. As noted earlier, Ca^{2+} itself does not initiate further Ca^{2+} release in the egg; furthermore, successful fertilization can be increased by reducing Ca^{2+} influx or the rise in $[Ca^{2+}]_i$ at the site of sperm–egg fusion (McCulloh et al., 1990).

Putative activating messengers should satisfy three criteria (Whitaker and Swann, 1993): (1) the activator should be in sufficient quantity in the sperm; (2) it should mimic the fertilization response when microinjected; and (3) its specific antagonist should block fertilization. The notion of a sperm activator triggering egg activation counters earlier dogma based on observation that injection of live spermatozoa into unfertilized eggs neither activated them nor caused the loss of their ability to be fertilized (Hiramoto, 1962). The activator is membrane impermeant, so the injection of cytosolic sperm extract activates the cortical reaction in sea urchin eggs (Dale et al., 1985), a finding that has been extended to ascidians (Dale, 1988), frogs (Whitaker and Swann, 1993), and mammals (Stice and Robl, 1990; Swann, 1990,1992). Further supporting evidence for the requirement of sperm–egg membrane fusion for diffusion of an activator is the cross-species activation of denuded sea urchin eggs by polyethylene glycol-induced fusion of activated starfish sperm (Kyozuka and Osanai, 1989). Recent in vitro fertilization studies have now reported successful human embryogenesis with injection of spermatozoa into the ooplasm (Edwards, 1993). A number of possible activators have been suggested, but none has proven definitive according to the preceding criteria. Two putative activators, IP_3 and cADPR, can be readily dismissed. Although the sperm contains IP_3 (Iwasa et al., 1990) and IP_3 production is stimulated with sperm activation (Domino and Garbers, 1988), microinjection of IP_3 triggers (without a significant latent period) a rapid release of Ca^{2+} that differs markedly in kinetics from that seen during fertilization (Shen and Buck, 1993). cADPR levels have not been reported in sperm but, like IP_3, exogenous cADPR has induced a rapid transient increase in $[Ca^{2+}]_i$ without a latent period.

A third putative activator, cGMP, is less easily discounted. The level of cGMP

3. Calcium Regulation during Sea Urchin Fertilization

in eggs during fertilization has yet to be reported, although cAMP levels during fertilization and early development have been studied extensively (Ishida and Yasumasu, 1982; Browne et al., 1990). In contrast, the level of cGMP has been measured carefully during sperm activation (Kopf and Garbers, 1979; Hansbrough and Garbers, 1981) by egg peptides, which caused a rapid and dramatic activation of guanylate cyclase (Chinkers and Garbers, 1991). Since microinjection of 5–10 μM cGMP is necessary for egg activation (Whalley et al., 1992), application of diffusion kinetics would require a sperm cytosol concentration near 1 M (Whitaker and Crossley, 1990). Alternatively, the activated guanylate cyclase of sperm (Bentley et al., 1986) may generate a sufficient level at the sperm–egg fusion site during the latent period. This most likely is not the case. The activation of sperm could occur in the absence of detectable changes in cGMP in sperm (Shimomura and Garbers, 1986); in this study, sperm guanylate cyclase appeared to be rapidly (within a few seconds) inactivated (Ward et al., 1985). The existence of guanylate cyclase activity in the egg has not been studied to date, but this issue is distinct from the question concerning possible sperm factors for egg activation. Thus, despite the fertilization-like transient increase in $[Ca^{2+}]_i$ following microinjection of cGMP (Whalley et al., 1992), cGMP does not appear to be the putative activator.

Currently, the most promising candidate for an egg activator in sperm is a high molecular weight protein factor that remains uncharacterized. The factor is >100 kDa and triggers Ca^{2+} oscillations in hamster eggs that are similar to those seen during fertilization (Swann, 1990). This sperm cytosol fraction will also activate mouse eggs (Swann, 1992) and Ca^{2+}-dependent currents in cultured rat dorsal root ganglion neurons (Currie et al., 1992). This factor is believed to act by increasing the sensitivity of CICR (Whitaker and Swann, 1993), an effect also observed by treating hamster eggs with the sulfhydryl reagent thimerosal (Swann, 1991). Thimerosal was presumed to act through sulfhydryl groups since its effects were blocked by dithiothreitol, although other sulfhydryl reagents such as N-ethylmaleimide did not affect CICR (Swann, 1991). Thimerosal apparently enhanced CICR of an IP_3-dependent mechanism, because a monoclonal antibody against the IP_3 receptor blocked Ca^{2+} oscillations induced by thimerosal (Miyazaki et al., 1992b). Thimerosal also induced Ca^{2+} release in mouse (Swann, 1992) and sea urchin (McDougall et al., 1993) eggs. Regardless of the effect of sperm factor on sensitizing CICR, the question of how this activating agent initiates the production of endogenous Ca^{2+} release agonists remains unanswered, as do a number of questions, until the putative activator has been identified.

IV. Conclusions

In the nearly six decades that have passed since Mazia reported an increase in unbound calcium with sea urchin egg fertilization, much has been learned about the different mechanisms cells utilize to regulate cytosol calcium levels. One might even be tempted to complain that too many alternatives for $[Ca^{2+}]_i$ regulation are now known. Although the pathway from sperm–egg interactions to the transient increase in ooplasm $[Ca^{2+}]_i$ might be more direct in *Urechis* (Gould and Stephano, 1987) or more restricted in *Xenopus* (Nuccitelli *et al.*, 1993) and hamster (Miyazaki *et al.*, 1993), the multiple pathways in the sea urchin egg present unique challenges. Although IP_3 is well established as a second messenger for regulating $[Ca^{2+}]_i$, it remains unclear how sperm induce the production of IP_3. It now appears that not only is increased $[Ca^{2+}]_i$ insufficient for stimulating PIP_2 hydrolysis, but sperm may stimulate IP_3 production independent of a Ca^{2+} signal. Although the sea urchin egg contains a ryanodine receptor release mechanism, a natural ligand has not been established. cADPR, whose production may be regulated by a cGMP-dependent pathway, is a strong candidate; however, transduction of the fertilizing sperm signal to regulate ryanodine receptors is unknown. In addition, the kinetics of Ca^{2+} release induced by IP_3 and cADPR are significantly faster than the kinetics of increased $[Ca^{2+}]_i$ during fertilization; thus, propagation of the signal for Ca^{2+} release must account for the extended duration of increased $[Ca^{2+}]_i$.

Perhaps the most intriguing issue of Ca^{2+} regulation in sea urchin eggs is why both IP_3 and ryanodine receptor release mechanisms are present. A wide variety of somatic cells also contain both IP_3-sensitive and ryanodine-sensitive Ca^{2+} release mechanisms. Immunolocalization in Purkinje neurons (Walton *et al.*, 1991) and rat hepatocytes (Feng *et al.*, 1992) suggested the possibility of both multiple Ca^{2+} stores controlled by different release mechanisms and a common store with multiple release mechanisms. Such regionalization of Ca^{2+} receptor release mechanisms may play an important role in cellular activities and may permit the generation of a variety of spatiotemporal Ca^{2+} release patterns. The spatial distribution of the antibody binding against the two receptors in the unfertilized sea urchin egg appears to be different. Ryanodine receptors were localized to the egg cortex (McPherson *et al.*, 1992), whereas IP_3 receptors appeared distributed throughout the cytoplasm and periphery of the female pronucleus (Parys *et al.*, 1994). However, both receptor release mechanisms apparently release from a common Ca^{2+} store in the unfertilized egg (Whalley *et al.*, 1992; Buck *et al.*, 1994). Although different spatiotemporal patterns of Ca^{2+} release by parthenogenetic activation were observed (Shen and Buck, 1993), the increase in $[Ca^{2+}]_i$ during fertilization of eggs preloaded with either heparin or 8-amino cADPR appeared similar (Lee *et al.*, 1993). A common store and similar fertilization-induced change in Ca^{2+} suggest that the two mechanisms may be redundant to ensure occurrence of the event that is necessary and sufficient for initiating development. The reorganization of the endoplasmic reticulum is a striking phenomenon during fertilization (Terasaki and Jaffe, 1991). This reorganization may result in separate Ca^{2+} stores with differential regulation in the zygote. In addition to the rise in Ca^{2+} during fertilization, multiple free Ca^{2+} changes occur during the first cell cycle (Poenie *et al.*, 1985). In somatic cells, separate sources for the transient elevation of Ca^{2+} may have different cellular effects, such as control of exocytosis in bovine chromaffin cells (Kim and Westhead, 1989) and differential regulation of gene expression in hippocampal neurons (Bading et al., 1993). The roles of IP_3-sensitive and ryanodine-sensitive receptor release mechanisms during early development remain unexplored.

Acknowledgments

I thank Drs. Ted Chambers and Lionel Jaffe for numerous conversations concerning calcium regulation during fertilization. In addition, I thank Drs. P. Harris, L. A. Jaffe, W. H. Kinsey, H. C. Lee, M. McCloskey, R. Nuccitelli, and B. J. Stith for sharing unpublished observations and/or their perspectives on signal transduction during fertilization. Work in my laboratory has been supported by National Science Foundation Grants DCB 89-03837 and DIR 91-13595.

References

Abassi, Y. A., and Foltz, K. R. (1994). Tyrosine phosphorylation of the egg receptor for sperm at fertilization. *Dev. Biol.* **164**, 430–443.

Akimaya, T., Ishida, J., Nakagawa, S., Ogawara, H., Watanabe, S., Itoh, N., Shibuya, M., and Fukami, Y. (1987). Genistein, a specific inhibitor of tyrosine specific protein kinases. *J. Biol. Chem.* **262**, 5592–5595.

Aldhous, P. (1993). Long search for sea urchin sperm receptor pays off. *Science* **259**, 1403–1404.

Allbritton, N. L., Meyer, T., and Stryer, L. (1992). Range of messenger action of calcium ion and inositol 1,4,5-trisphosphate. *Science* **258**, 1812–1815.

Allen, D. G., Blinks, J. R., and Prendergast, F. G. (1977). Aequorin luminescence: Relation of light emission to calcium concentration-a calcium-independent component. *Science* **195**, 996–998.

Allen, R. D., and Griffin, J. L. (1958). The time sequence of early events in the fertilization of sea urchin eggs. *Exp. Cell Res.* **15**, 163–173.

Almers, W., and Neher, E. (1985). The Ca signal from fura-2 loaded mast cells depends strongly on the method of dye-loading. *FEBS Lett.* **192**, 13–18.

Al-Mohanna, F. A., Caddy, K. W. T., and Bolsover, S. R. (1994). The nucleus is insulated from large cytosolic calcium ion changes. *Nature* **367**, 745–750.

Azarnia, R., and Chambers, E. L. (1976). The role of divalent cations in activation of the sea urchin egg. I. The effect of fertilization on divalent cation content. *J. Exp. Zool.* **198**, 65–78.

Bading, H., Ginty, D. D., and Greenberg, M. E. (1993). Regulation of gene expression in hippocampal neurons by distinct calcium signaling pathways. *Science* **260**, 181–186.

Baker, P. F., and Presley, R. (1969). Kinetic evidence for an intermediate stage in the fertilization of the sea urchin egg. *Nature* **221**, 488–490.

Baker, P. F., and Whitaker, M. J. (1978). Influences of ATP and calcium on the cortical reaction in sea urchin eggs. *Nature* **276**, 513–515.

Bentley, J. K., Tubb, D. J., and Garbers, D. L. (1986). Receptor-mediated activation of spermatozoan guanylate cyclase. *J. Biol. Chem.* **261**, 14859–14862.

Berger, F. (1992). Mechanisms of initiation and propagation of the calcium wave during fertilization in deuterostomes. *Int. J. Dev. Biol.* **36**, 245–262.

Berridge, M. J. (1991). Caffeine inhibits inositol trisphosphate-induced membrane potential oscillations in *Xenopus laevis*. *Proc. R. Soc. London B* **244**, 57–62.

Berridge, M. J. (1993a). Inositol trisphosphate and calcium signalling. *Nature* **361**, 315–325.
Berridge, M. J. (1993b). A tale of two messengers. *Nature* **365**, 388–389.
Berridge, M. J., and Irvine, R. F. (1984). Inositol trisphosphate, a novel second messenger in cellular signal transduction. *Nature* **312**, 315–321.
Berridge, M. J., and Irvine, R. F. (1989). Inositol phosphates and cell signalling. *Nature* **341**, 197–205.
Bezprovzanny, I., Watras, J., and Ehrlich, B. E. (1991). Bell-shaped calcium-response curves of Ins(1,4,5)P_3- and calcium-gated channels from endoplasmic reticulum of cerebellum. *Nature* **351**, 751–754.
Bezprozvanny, I., Bezprozvannaya, S., and Ehrlich, B. E. (1994). Caffeine-induced inhibition of inositol (1,4,5)-trisphosphate-gated calcium channels from cerebellum. *Mol. Biol. Cell* **5**, 97–103.
Blatter, L. A., and Wier, W. G. (1990). Intracellular diffusion, binding, and compartmentalization of the fluorescent calcium indicators indo-1 and fura-2. *Biophys. J.* **58**, 1491–1499.
Bolsover, S., and Silver, R. A. (1991). Artifacts in calcium measurement: Recognition and Remedies. *Trends Cell Biol.* **1**, 71–74.
Brown, G. R., Sayer, L. G., Kirk, C. J., Mitchell, R. H., and Michelangeli, F. (1992). The opening of the inositol 1,4,5-trisphosphate-sensitive Ca^{2+} channel in rat cerebellum is inhibited by caffeine. *Biochem. J.* **282**, 309–312.
Browne, C. L., Bower, W. A., Palazzo, R. E., and Rebhun, L. L. (1990). Inhibition of mitosis in fertilized sea urchin eggs by inhibition of the cyclic AMP-dependent protein kinase. *Exp. Cell Res.* **188**, 122–128.
Buck, W. R., Rakow, T. L., and Shen, S. S. (1991). Calcium release in unfertilized and fertilized sea urchin eggs. *J. Cell Biol.* **115**, 321a.
Buck, W. R., Rakow, T. L., and Shen, S. S. (1992). Synergistic release of calcium in sea urchin eggs by caffeine and ryanodine. *Exp. Cell Res.* **202**, 59–66.
Buck, W. R., Hoffmann, E. E., Rakow, T. L., and Shen, S. S. (1994). Synergistic calcium release in the sea urchin egg by ryanodine and cyclic ADP ribose. *Dev. Biol.* **163**, 1–10.
Busa, W. B. (1992). Spectral characterization of the effect of viscosity on fura-2 fluorescence: Excitation wavelengths optimization abolishes the viscosity artifact. *Cell Calcium* **13**, 313–319.
Busa, W. B., and Nuccitelli, R. (1985). An elevated free cytosolic Ca^{2+} wave follows fertilization in eggs of the frog, *Xenopus laevis*. *J. Cell Biol.* **100**, 1325–1329.
Busa, W. B., Ferguson, J. E., Suresh, K. J., Williamson, J. R., and Nuccitelli, R. (1985). Activation of frog (*Xenopus laevis*) eggs by inositol trisphosphate. I. Characterization of Ca^{2+} release from intracellular stores. *J. Cell Biol.* **101**, 677–682.
Carafoli, E. (1987). Intracellular calcium homeostasis. *Annu. Rev. Biochem.* **56**, 395–433.
Chambers, E. L. (1989). Fertilization in voltage-clamped sea urchin eggs. In "Mechanisms of Egg Activation" (R. Nuccitelli, G. N. Cherr, and W. N. Clark, Jr., eds.), pp. 1–18. Plenum Press, New York.
Chambers, E. L., and Angeloni, S. V. (1981). Is external Ca^{2+} required for fertilization of sea urchin eggs by acrosome-reacted sperm? *J. Cell Biol.* **91**, 181a.
Chambers, E. L., and deArmendi, J. (1979). Membrane potential, action potential and activation potential of eggs of the sea urchin *Lytechinus variegatus*. *Exp. Cell. Res.* **122**, 203–218.
Chambers, E. L., and Hinkley, R. E. (1979). Nonpropagated cortical reactions induced by the divalent ionophore A23187 in eggs of the sea urchin *Lytechinus variegatus*. *Exp. Cell Res.* **124**, 441–446.
Chambers, E. L., Pressman, B. C., and Rose, B. (1974). The activation of sea urchin eggs by the divalent ionophores A23187 and X-537A. *Biochem. Biophys. Res. Commun.* **60**, 126–132.
Chinkers, M., and Garbers, D. L. (1991). Signal transduction by guanylyl cyclases. *Annu. Rev. Biochem.* **60**, 553–575.

3. Calcium Regulation during Sea Urchin Fertilization

Ciapa, B., and Epel, D. (1991). A rapid change in phosphorylation on tyrosine accompanies fertilization of sea urchin eggs. *FEBS Lett.* **295**, 167–170.

Ciapa, B., and Whitaker, M. (1986). Two phases of inositol polyphosphate and diacylglycerol production at fertilization. *FEBS Lett.* **195**, 347–351.

Ciapa, B., Borg, B., and Whitaker, M. (1992). Polyphoinositide metabolism during the fertilization wave in sea urchin eggs. *Development* **115**, 187–195.

Clapper, D., and Lee, H. C. (1985). Inositol trisphosphate induces calcium release from nonmitochondrial stores in sea urchin egg homogenates. *J. Biol. Chem.* **260**, 13947–13954.

Connor, J. A. (1993). Intracellular calcium mobilization by inositol 1,4,5-trisphosphate: Intracellular movements and compartmentalization. *Cell Calcium* **14**, 185–200.

Cornell-Bell, A. H., Finkbeiner, S. M., Cooper, M. S., and Smith, S. J. (1990). Glutamate induces calcium waves in cultured astrocytes: Long-range glial signaling. *Science* **247**, 470–473.

Crossley, I., Swann, K., Chambers, E., and Whitaker, M. (1988). Activation of sea urchin eggs by inositol phosphates is independent of external calcium. *Biochem. J.* **252**, 257–262.

Crossley, I., Whalley, T., and Whitaker, M. (1991). Guanosine 5′-thiotriphosphate may stimulate phosphoinositide messenger production in sea urchin eggs by a different route than the fertilizing sperm. *Cell Regul.* **2**, 121–133.

Currie, K. P. M., Swann, K., Galione, A., and Scott, R. H. (1992). Activation of Ca^{2+}-dependent currents in cultured rat dorsal root ganglion neurones by a sperm factor and cyclic ADP-ribose. *Mol. Biol. Cell* **3**, 1415–1425.

Dale, B. (1988). Primary and secondary messengers in the activation of ascidian eggs. *Exp. Cell Res.* **177**, 205–211.

Dale, B., DeFelice, L. J., and Ehrenstein, G. (1985). Injection of a soluble sperm fraction into sea urchin eggs triggers the cortical reaction. *Experientia* **41**, 1068–1070.

Dargie, P. J., Agre, M. C., and Lee, H. C. (1986). Parthenogenetic activation of sea urchin egg by microinjection of IP3 and GTP-γ-S. *J. Cell Biol.* **103**, 84a.

Dargie, P. J., Agre, M. C., and Lee, H. C. (1990). Comparison of Ca^{2+} mobilizing activities of cyclic ADP-ribose and inositol trisphosphate. *Cell Regul.* **1**, 279–290.

David, C., Halliwell, J., and Whitaker, M. (1988). Some properties of the membrane currents underlying the fertilization potential in sea urchin eggs. *J. Physiol. (London)* **402**, 139–154.

DeLisle, S., and Welsh, M. J. (1992). Inositol trisphosphate is required for the propagation of calcium waves in *Xenopus* oocytes. *J. Biol. Chem.* **267**, 7963–7966.

de Santis, A., Ciccarelli, C., and Dale, B. (1987). Free intracellular cations in echinoderm oocytes and eggs. *Eur. Biophys. J.* **14**, 471–476.

Domino, S. E., and Garbers, D. L. (1988). The fucose-sulfate glycoconjugate that induces an acrosome reaction in spermatozoa stimulates inositol 1,4,5-trisphosphate accumulation. *J. Biol. Chem.* **263**, 690–695.

Eberhard, M., and Erne, P. (1991). Calcium binding to fluorescent calcium indicators: Calcium green, calcium orange and calcium crimson. *Biochem. Biophys. Res. Commun.* **180**, 209–215.

Edwards, R. G., and Van Steirtegham, A. C. (1993). Intracellular sperm injections (ICSI) and human fertilization: Does calcium hold the key to success? *Human Reprod.* **8**, 988–989.

Eisen, A., and Reynolds, G. T. (1985). Source and sinks for the calcium released during fertilization of single sea urchin eggs. *J. Cell Biol.* **100**, 1522–1527.

Eisen, A., Kiehart, D. P., Wiedland, S. J., and Reynolds, G. T. (1984). Temporal sequence and spatial distribution of early events of fertilization in single sea urchin eggs. *J. Cell Biol.* **99**, 1647–1654.

Endo, M., Tanaka, M., and Ogawa, Y. (1970). Calcium induced release of calcium from the sarcoplasmic reticulum of skinned skeletal muscle fibres. *Nature* **228**, 34–36.

Feng, L., Pereira, B., and Kraus-Friedmann, N. (1992). Different localization of inositol 1,4,5-trisphosphate and ryanodine binding sites in rat liver. *Cell Calcium* **13**, 79–87.

Ferris, C. D., Huganir, R. L., Supattapone, S., and Snyder, S. H. (1989). Purified inositol 1,4,5-

trisphosphate receptor mediates calcium flux in reconstituted lipid vesicles. *Nature* **342**, 87–89.

Finch, E. A., Turner, T. J., and Goldin, S. G. (1991). Calcium as a coagonist of inositol 1,4,5-trisphosphate-induced calcium release. *Science* **252**, 443–445.

Fiskum, G., and Lehninger, A. L. (1980). The mechanisms and regulation of mitochondrial calcium transport. *Fed. Proc.* **39**, 2432–2436.

Fleischer, S., and Inui, M. (1989). Biochemistry and biophysics of excitation-contraction coupling. *Annu. Rev. Biophys. Chem.* **18**, 333–364.

Foltz, K. R., and Lennarz, W. J. (1990). Purification and characterization of an extracellular fragment of the sea urchin egg receptor for sperm. *J. Cell Biol.* **111**, 2951–2959.

Foltz, K. R., and Lennarz, W. J. (1992). Identification of the sea urchin egg receptor for sperm using an antiserum raised against a fragment of its extracellular domain. *J. Cell Biol.* **116**, 647–658.

Foltz, K. R., and Lennarz, W. J. (1993). The molecular basis of sea urchin gamete interactions at the egg plasma membrane. *Dev. Biol.* **158**, 46–61.

Foltz, K. R., Partin, J. S., and Lennarz, W. J. (1993). Sea urchin egg receptor for sperm: Sequence similarity of binding domain and hsp70. *Science* **259**, 1421–1425.

Ford, L. E., and Podolsky, R. I. (1970). Regenerative calcium release within muscle cells. *Science* **167**, 58–59.

Francis, S. H., Noblett, B. D., Todd, B. W., Wells, I. N., and Corbin, J. D. (1988). Relaxation of vascular and tracheal smooth muscle by cyclic nucleotide analogs that preferentially activate purified cGMP-dependent protein kinase. *Mol. Pharmacol.* **34**, 506–517.

Fujiwara, A., Sudoh, K., and Yasumasu, I. (1988). Activation of sea urchin by halothane and its inhibition by dantrolene. *Dev. Growth Diff.* **30**, 1–8.

Fujiwara, A., Taguchi, K., and Yasumasu, I. (1990). Fertilization membrane formation in sea urchin eggs induced by drugs known to cause Ca^{2+} release from isolated sarcoplasmic reticulum. *Dev. Growth Diff.* **32**, 303–314.

Galione, A. (1992). Ca^{2+}-induced Ca^{2+} release and its modulation by cyclic ADP-ribose. *Trends Pharmacol. Sci.* **13**, 304–306.

Galione, A. (1993). Cyclic ADP-ribose: A new way to control calcium. *Science* **259**, 325–326.

Galione, A., Lee, N. C., and Busa, W. B. (1991). Ca^{2+}-induced Ca^{2+} release in sea urchin egg homogenates: Modulation by cyclic ADP-ribose. *Science* **253**, 1143–1146.

Galione, A., McDougall, A., Busa, W. B., Willmott, N., Gillot, I., and Whitaker, M. (1993a). Redundant mechanisms of calcium-induced calcium release underlying calcium waves during fertilization of sea urchin eggs. *Science* **261**, 348–352.

Galione, A., White, A., Willmott, N., Turner, M., Potter, B. V. L., and Watson, S. P. (1993b). cGMP mobilizes intracellular Ca^{2+} in sea urchin eggs by stimulating cyclic ADP-ribose synthesis. *Nature* **365**, 456–459.

Ghosh, T. K., Eis, P. S., Mullaney, J. M., Ebert, C. L., and Gill, D. L. (1988). Competitive, reversible, and potent antagonism of inositol 1,4,5-trisphosphate-activated calcium release by heparin. *J. Biol. Chem.* **263**, 11075–11079.

Gillot, I., and Whitaker, M. (1993). Imaging calcium waves in eggs and embryos. *J. Exp. Biol.* **184**, 213–219.

Gillot, I., Ciapa, B., Payan, P., DeRenzis, G., Nicaise, G., and Sardet, C. (1989). Quantitative X-ray microanalysis of calcium in sea urchin eggs after quick-freezing and freeze-substitution. *Histochem.* **92**, 523–529.

Gillot, I., Payan, P., Girard, J. P., and Sardet, C. (1990). Calcium in sea urchin egg during fertilization. *Int. J. Dev. Biol.* **34**, 117–125.

Gillot, I., Ciapa, B., Payan, P., and Sardet, C. (1991). The calcium content of cortical granules and the loss of calcium from sea urchin eggs at fertilization. *Dev. Biol.* **146**, 396–405.

Girard, J. P., Gillot, I., De Renzis, G., and Payan, P. (1991). Calcium pools in sea urchin eggs: Roles of endoplasmic reticulum and mitochondria in relation to fertilization. *Cell Calcium* **12**, 289–299.

Glabe, C. G. (1985a). Interaction of the sperm adhesive protein, bindin, with phospholipid vesicles. I. Specific association of bindin with gel-phase phospholipid vesicles. *J. Cell Biol.* **100**, 794–799.

Glabe, C. G. (1985b). Interaction of the sperm adhesive protein, bindin, with phospholipid vesicles. II. Bindin induces the fusion of mixed-phase vesicles that contain phosphatidylcholine and phosphatidylserine *in vitro*. *J. Cell Biol.* **100**, 800–806.

Glick, D. L., Hellmich, M. R., Beushausen, S., Tempst, P., Bayley, H., and Strumwasser, F. (1991). Primary structure of a molluscan egg-specific NADase, a second messenger enzyme. *Cell Regul.* **2**, 211–218.

Gould, M., and Stephano, J. L. (1987). Electrical responses of eggs to acrosomal protein similar to those induced by sperm. *Science* **235**, 1654–1656.

Grynkiewicz, G., Poenie, M., and Tsien, R. Y. (1985). A new generation of Ca^{++} indicators with greatly improved fluorescence properties. *J. Biol. Chem.* **260**, 3440–3450.

Hafner, M., Petzelt, C., Nobiling, R., Pawley, J. B., Kramp, D., and Schatten, G. (1988). Wave of free calcium at fertilization in the sea urchin egg visualized with fura-2. *Cell Motil. Cytoskel.* **9**, 271–277.

Hall, C. G., Sancho, J., and Terhorst, C. (1993). Reconstitution of T cell receptor *f*-mediated calcium mobilization in nonlymphoid cells. *Science* **261**, 915–918.

Hamaguchi, Y., and Hiramoto, Y. (1981). Activation of sea urchin eggs by microinjection of calcium buffers. *Exp. Cell Res.* **134**, 171–179.

Hansbrough, J. R., and Garbers, D. L. (1981). Sodium-dependent activation of sea urchin spermatozoa by speract and monensin. *J. Biol. Chem.* **256**, 2235–2241.

Harris, P. J. (1994). Caffeine-induced calcium release in sea urchin eggs and the effect of continuous versus pulsed application on the mitotic apparatus. *Dev. Biol.* **161**, 370–378.

Haugland, R. P. (1992). "Handbook of Fluorescent Probes and Research Chemicals." Molecular Probes, Inc., Eugene, Oregon.

Hepler, J. R., and Gilman, A. G. (1992). G proteins. *Trends Biochem. Sci.* **17**, 383–387.

Hernández-Cruz, A., Sala, F., and Adams, P. R. (1990). Subcellular calcium transients visualized by confocal microscopy in a voltage-clamped vertebrate neuron. *Science* **247**, 858–862.

Hinkley, R. E., Wright, B. D., and Lynn, J. W. (1986). Rapid visual detection of sperm–egg fusion using the DNA-specific fluorochrome Hoechst 33342. *Dev. Biol.* **118**, 148–154.

Hiramoto, Y. (1962). An analysis of the mechanism of fertilization by means of enucleation of sea urchin eggs. *Exp. Cell Res.* **28**, 323–334.

Hirose, K., Iino, M., and Endo, M. (1993). Caffeine inhibits Ca^{2+}-mediated potentiation of inositol 1,4,5-trisphosphate-induced Ca^{2+} release in permeabilized vascular smooth muscle cells. *Biochem. Biophys. Res. Commun.* **194**, 726–732.

Hong, K., and Vacquier, V. D. (1986). Fusion of liposomes induced by a cationic protein from the acrosome granule of abalone spermatozoa. *Biochemistry* **25**, 543–549.

Howard, M., Grimaldi, J. C., Bazan, J. F., Lund, F. E., Santos-Argumedo, L., Parkhouse, R. M. E., Walseth, T. F., and Lee, H. C. (1993). Formation and hydrolysis of cyclic ADP-ribose catalyzed by lymphocyte antigen CD38. *Science* **262**, 1056–1059.

Igusa, Y., and Miyazaki, S. (1983). Effects of altered extracellular and intracellular calcium concentration on hyperpolarizing responses of hamster eggs. *J. Physiol. (London)* **340**, 611–632.

Ishida, K., and Yasumasu, I. (1982). The periodic change in adenosine $3',5'$-monophosphate concentration in sea urchin eggs. *Biochim. Biophys. Acta* **720**, 266–273.

Iwasa, K. H., Ehrenstein, G., DeFelice, L. J., and Russell, J. T. (1990). High concentration of inositol 1,4,5-trisphosphate in sea urchin sperm. *Biochem. Biophys. Res. Commun.* **172**, 932–938.

Jaffe, L. A. (1990). First messengers at fertilization. *J. Reprod. Fert. Suppl.* **42,** 107–116.
Jaffe, L. A. (1994). Egg membranes during fertilization. In "Molecular Biology of Membrane Transport Disorders" (S. G. Schultz, T. Andreoli, A. Brown, D. Fambrough, J. Hoffman, and M. Welsh, eds.). Plenum Publishing, New York.
Jaffe, L. A., and Terasaki, M. (1993). Structural changes of the endoplasmic reticulum of sea urchin eggs during fertilization. *Dev. Biol.* **156,** 566–573.
Jaffe, L. F. (1980). Calcium explosions as triggers of development. *Ann. N.Y. Acad. Sci.* **339,** 86–101.
Jaffe, L. F. (1983). Sources of calcium in egg activation: A review and hypothesis. *Dev. Biol.* **99,** 265–276.
Jaffe, L. F. (1990). The roles of intermembrane calcium in polarizing and activating eggs. *NATO ASI Series* **H45,** 389–417.
Jaffe, L. F. (1991). The path of calcium in cytosolic calcium oscillations: A unifying hypothesis. *Proc. Natl. Acad. Sci. USA* **88,** 9883–9887.
Jiang, W., Gottlieb, R. A., Lennarz, W. J., and Kinsey, W. H. (1990). Phorbol ester treatment stimulates tyrosine phosphorylation of a sea urchin egg cortex protein. *J. Cell Biol.* **110,** 1049–1053.
Jiang, W., Veno, P. A., Wood, R. W., Peaucellier, G., and Kinsey, W. H. (1991). pH regulation of an egg cortex tyrosine kinase. *Dev. Biol.* **146,** 81–88.
June, C. H., Fletcher, M. C., Ledbetter, J. A., Schieven, G. L., Siegel, J. N., Phillips, A. F., and Samelson, L. E. (1990). Inhibition of tyrosine phosphorylation prevents T-cell receptor-mediated signal transduction. *Proc. Natl. Acad. Sci. USA* **87,** 7722–7726.
Kamel, L. C., Bailey, J., Schoenbaum, L., and Kinsey, W. (1985). Phosphatidylinositol metabolism during fertilization in the sea urchin eggs. *Lipids* **20,** 350–356.
Kim, H., Jacobson, E. L., and Jacobson, M. K. (1993a). Synthesis and degradation of cyclic ADP-ribose by NAD glycohydrolases. *Science* **261,** 1330–1333.
Kim, H., Jacobson, E. L., and Jacobson, M. K. (1993b). Position of cyclization in cyclic ADP-ribose. *Biochem. Biophys. Res. Commun.* **194,** 1143–1147.
Kim, K. T., and Westhead, E. W. (1989). Cellular responses to Ca^{2+} from extracellular and intracellular sources are different as shown by simultaneous measurements of cytosolic Ca^{2+} and secretion from bovine chromaffin cells. *Proc. Natl. Acad. Sci. USA* **86,** 9881–9885.
Kim, U. H., Rockwood, S. F., Kim, H. R., and Daynes, R. A. (1988). Membrane-associated NAD^+ glycohydrolase from rabbit erythrocytes is solubilized by phosyphatidylinositol-specific phospholipase C. *Biochim. Biophys. Acta* **965,** 76–81.
Kline, D., Simoncini, L., Mandel, G., Maue, R. A., Kado, R. T., and Jaffe, L. A. (1988). Fertilization events induced by neurotransmitters after injection of mRNA in *Xenopus* egg. *Science* **241,** 464–467.
Kline, J. T., and Kline, D. (1994). Regulation of intracellular calcium in the mouse egg: Evidence for inositol 1,4,5-trisphosphate-induced calcium release, but not calcium-induced calcium release. *Biol. Reprod.* **50,** 193–203.
Kopf, G. S., and Garbers, D. L. (1979). A low molecular weight factor from sea urchin eggs elevates sperm cyclic nucleotide concentrations and respiration rates. *J. Reprod. Fert.* **57,** 353–361.
Koshiyama, H., Lee, H. C., and Tashjian, A. H. (1991). Novel mechanism of intracellular calcium release in pituitary cells. *J. Biol. Chem.* **266,** 16985–16988.
Kubota, H. Y., Yoshimoto, Y., and Hiramoto, Y. (1993). Oscillation of intracellular free calcium in cleaving and cleavage-arrested embryos of *Xenopus laevis*. *Dev. Biol.* **160,** 512–518.
Kühtreiber, W. M., Gillot, I., Sardet, C., and Jaffe, L. F. (1993). Net calcium and acid release at fertilization in eggs of sea urchins and ascidians. *Cell Calcium* **14,** 73–86.
Kume, S., Muto, A., Aruga, I., Nakagawa, T., Michikawa, T., Furuichi, T., Nakade, S., Okano,

3. Calcium Regulation during Sea Urchin Fertilization 95

H., and Mikoshiba, K. (1993). The *Xenopus* IP_3 receptor: Structure, function and localization in oocytes and eggs. *Cell* **73**, 555–570.

Kyozuka, K., and Osanai, K. (1989). Induction of cross fertilization between sea urchin eggs and starfish sperm by polyethylene glycol treatment. *Gamete Res.* **22**, 123–129.

Lai, F. A., and Meissner, G. (1989). The muscle ryanodine receptor and its intrinsic Ca^{2+} channel activity. *J. Bioenerg. Biomemb.* **21**, 227–246.

Lechleiter, J. D., and Clapham, D. E. (1992). Molecular mechanisms of intracellular calcium excitability in *X. laevis* oocytes. *Cell* **69**, 283–294.

Lechleiter, J. D., Girard, S., Peralta, E., and Clapham, D. (1991). Spiral calcium wave propagation and annihilation in *Xenopus laevis* oocytes. *Science* **252**, 123–126.

Lee, H. C. (1991). Specific binding of cyclic ADP-ribose to calcium-storing microsomes from sea urchin eggs. *J. Biol. Chem.* **266**, 2276–2281.

Lee, H. C. (1993). Potentiation of calcium- and caffeine-induced calcium release by cyclic ADP-ribose. *J. Biol. Chem.* **266**, 2276–2281.

Lee, H. C., and Aarhus, R. (1991). ADP-ribosyl cyclase: An enzyme that cyclizes NAD^+ into a calcium-mobilizing metabolite. *Cell Regul.* **2**, 203–209.

Lee, H. C., and Aarhus, R. (1993). Wide distribution of an enzyme that catalyzes the hydrolysis of cyclic ADP-ribose. *Biochim. Biophys. Acta* **1164**, 68–74.

Lee, H. C., Walseth, T. F., Bratt, G. T., Haynes, R. N., and Clapper, D. L. (1989). Structural determination of a cyclic metabolite of NAD^+ with intracellular Ca^{2+}-mobilizing activity. *J. Biol. Chem.* **264**, 1608–1615.

Lee, H. C., Zocchi, E., Guida, L., Franco, L., Benatti, I. U., and De Flora, A. (1993a). Production and hydrolysis of cyclic ADP-ribose at the outer surface of human erythrocytes. *Biochem. Biophys. Res. Commun.* **191**, 639–645.

Lee, H. C., Aarhus, R., and Walseth, T. F. (1993b). Calcium mobilization by dual receptors during fertilization of sea urchin eggs. *Science* **261**, 352–355.

Lee, H. C., Aarhus, R., and Levitt, D. (1994). The crystal structure of cyclic ADP-ribose. *Nature Struct. Biol.* **1**, 143–144.

Lindvall, S., and Carsjö, A. (1951). On protein fractions and inorganic ions in sea urchin eggs, unfertilized and fertilized. *Exp. Cell Res.* **2**, 491–498.

Lopez, A., Miraglia, S. J., and Glabe, C. G. (1993). Structure/function analysis of the sea urchin sperm adhesive protein bindin. *Dev. Biol.* **156**, 24–33.

Lytton, J., Westlin, M., and Hanley, M. R. (1991). Thapsigargin inhibits the sarcoplasmic or endoplasmic reticulum Ca-ATPase family of calcium pumps. *J. Biol. Chem.* **266**, 17067–17071.

Mack, M. M., Zimanyi, I., and Pessah, I. N. (1992). Discrimination of multiple binding sites for antagonists of the calcium release channel complex of skeletal and cardiac sarcoplasmic reticulum. *J. Pharmacol. Exp. Therapeut.* **262**, 1028–1037.

Mason, W. T., Hoyland, J., Davison, I., Carew, M., Somasundaram, B., Tregear, R., Zorec, R., Lledo, P. M., Shankar, G., and Horton, M. (1993). Quantitative real-time imaging of optical probes in living cells. *In* "Fluorescent and Luminescent Probes for Biological Activity" (W. T. Mason, ed.), pp. 161–195. Academic Press, San Diego.

Mazia, D. (1937). The release of calcium in Arbacia eggs on fertilization. *J. Cell Comp. Physiol.* **10**, 291–304.

McCulloh, D. H., and Chambers, E. L. (1992). Fusion of membranes during fertilization. *J. Gen. Physiol.* **99**, 137–175.

McCulloh, D. H., Ivonnet, P. I., and Chambers, E. L. (1990). Microinjection of a Ca^{2+} chelator, EGTA and BAPTA promotes sperm entry in sea urchin eggs clamped at negative membrane potentials. *J. Cell Biol.* **111**, 113a.

McDougall, A., Gillot, I., and Whitaker, M. J. (1993). Thimerosal reveals calcium-induced calcium release in unfertilized sea urchin eggs. *Zygote* **1**, 35–42.

McPherson, P. S., and Campbell, K. P. (1993). The ryanodine receptor/Ca^{2+} release channel. *J. Biol. Chem.* **268**, 13765–13768.
McPherson, S. M., McPherson, P. S., Mathews, L., Campbell, K. P., and Longo, F. J. (1992). Cortical localization of a calcium release channel in sea urchin eggs. *J. Cell Biol.* **116**, 1111–1121.
Meissner, G. (1986). Ryanodine activation and inhibition of the Ca^{2+} release channel of sarcoplasmic reticulum. *J. Biol. Chem.* **261**, 6300–6306.
Meldolesi, J., Madeddu, L., and Pozzan, T. (1990). Intracellular Ca^{2+} storage organelles in nonmuscle cells: Heterogeneity and functional assignment. *Biochim. Biophys. Acta* **1055**, 130–140.
Mészáros, L., Bak, J., and Chu, A. (1993). Cyclic ADP-ribose as an endogenous regulator of the nonskeletal type ryanodine receptor Ca^{2+} channel. *Nature* **364**, 76–79.
Mikoshiba, K. (1993). Inositol 1,4,5-trisphosphate receptor. *Trends Pharmacol. Sci.* **14**, 86–89.
Minta, A., Kao, J. P. Y., and Tsien, R. Y. (1989). Fluorescent indicators for cytosolic calcium based on rhodamine and fluorescein chromophores. *J. Biol. Chem.* **264**, 8171–8178.
Minor, J. E., Britten, R. J., and Davidson, E. H. (1993). Species-specific inhibition of fertilization by a peptide derived from the sperm protein bindin. *Mol. Biol. Cell* **4**, 375–387.
Miyazaki, S., Katayama, Y., and Swann, K. (1990). Synergistic activation by serotonin and GTP analogue and inhibition by phorbol ester of cyclic Ca^{2+} rises in hamster eggs. *J. Physiol. (London)* **426**, 209–227.
Miyazaki, S., Yuzaki, M., Nakada, K., Shirakawa, H., Nakanishi, S., Nakade, S., and Mikoshiba, K. (1992a). Block of Ca^{2+} wave and Ca^{2+} oscillation by antibody to the inositol 1,4,5-trisphosphate receptor in fertilized hamster eggs. *Science* **257**, 251–255.
Miyazaki, S., Shirakawa, H., Nakada, K., Honda, Y., Yuzaki, M., Nakade, S., and Mikoshiba, K. (1992b). Antibody to the inositol trisphosphate receptor blocks thimerosal-enhanced Ca^{2+}-induced Ca^{2+} release and Ca^{2+} oscillations in hamster eggs. *FEBS Lett.* **309**, 180–184.
Miyazaki, S., Shirakawa, H., Nakada, K., and Honda, Y. (1993). Essential role of the inositol 1,4,5-trisphosphate receptor/Ca^{2+} release channel in Ca^{2+} waves and Ca^{2+} oscillations at fertilization of mammalian eggs. *Dev. Biol.* **158**, 62–78.
Mohri, T., and Hamaguchi, Y. (1989). Analysis of the breakdown of cortical granules in echinoderm eggs by microinjection of second messengers. *Cell Struct. Func.* **14**, 429–438.
Mohri, T., and Hamaguchi, Y. (1991). Propagation of transient Ca^{2+} increase in sea urchin eggs upon fertilization and its regulation by microinjecting EGTA solution. *Cell Struct. Func.* **16**, 157–165.
Mohri, T., Ivonnet, P., and Chambers, E. L. (1993). Effect on activation current in sea urchin eggs of agents which modify the intracellular concentration and the mobilization of Ca^{2+}. *Mol. Biol. Cell* **4**, 232a.
Moore, G. D., Kopf, G. S., and Schultz, R. M. (1993). Complete mouse egg activation in the absence of sperm by stimulation of an exogenous G protein-coupled receptor. *Dev. Biol.* **159**, 669–678.
Moy, G. W., and Vacquier, V. D. (1979). Immunoperoxidase localization of bindin during the adhesion of sperm to sea urchin eggs. *Curr. Top. Dev. Biol.* **13**, 31–44.
Moy, G. W., Brandriff, B., and Vacquier, V. D. (1977). Cytasters from sea urchin eggs parthenogenetically activated by procaine. *J. Cell Biol.* **73**, 788–793.
Mozingo, N. M., and Chandler, D. E. (1990). The fluorescent probe BDECF has a heterogeneous distribution in sea urchin eggs. *Cell Biol. Int. Rep.* **14**, 689–699.
Nakamura, M., and Yasumasu, I. (1974). Mechanism for increase in intracellular concentration of free calcium in fertilized sea urchin egg. *J. Gen. Physiol.* **63**, 374–388.
Nuccitelli, R. (1991). How do sperm activate eggs? *Curr. Top. Dev. Biol.* **25**, 1–16.
Nuccitelli, R., Yim, D. L., and Smart, T. (1993). The sperm-induced Ca^{2+} wave following fer-

tilization of the *Xenopus* egg requires the production of Ins(1,4,5)P$_3$. *Dev. Biol.* **158**, 200–212.

Oberdorf, J. A., Head, J. F., and Kaminer, B. (1986). Calcium uptake and release by isolated cortices and microsomes from the unfertilized egg of the sea urchin *Strongylocentrotus droebachiensis*. *J. Cell Biol.* **102**, 2205–2210.

Oberdorf, J., Vilar-Rojas, C., and Epel, D. (1989). The localization of PI and PIP kinase activities in the sea urchin egg and their modulation following fertilization. *Dev. Biol.* **131**, 236–242.

Ohara, T., and Sato, H. (1986). Distributional change in membrane-associated calcium in sea urchin eggs during fertilization. *Dev. Growth Diff.* **28**, 369–373.

Ohlendieck, K., Dhume, S. T., Partin, J. S., and Lennarz, W. J. (1993). The sea urchin egg receptor for sperm: Isolation and characterization of the intact, biologically active receptor. *J. Cell Biol.* **122**, 887–895.

Okamoto, H., Takahasi, K., and Yamashita, N. (1977). Ionic currents through the membrane of the mammalian oocyte and their comparisons with those in the tunicate and sea urchin. *J. Physiol. (London)* **267**, 465–495.

Palade, P., Dettbarn, C., Alderson, B., and Volpe, P. (1989). Pharmacologic differentiation between inositol 1,4,5-trisphosphate-induced Ca^{2+} release and Ca^{2+}- or caffeine-induced Ca^{2+} release from intracellular membrane systems. *Mol. Pharmacol.* **36**, 673–680.

Parker, I., and Ivorra, I. (1991). Caffeine inhibits inositol trisphosphate-mediated liberation of intracellular calcium in *Xenopus* oocytes. *J. Physiol. (London)* **433**, 229–240.

Parys, J. B., Sernett, S. W., DeLisle, S., Snyder, P. M., Welsh, M. J., and Campbell, K. P. (1992). Isolation, characterization, and localization of the inositol 1,4,5-trisphosphate receptor protein in *Xenopus laevis* oocytes. *J. Biol. Chem.* **267**, 18776–18782.

Parys, J. B., McPherson, S. M., Mathews, L., Campbell, K. P., and Longo, F. J. (1994). Presence of inositol 1,4,5-trisphosphate receptor, calreticulin, and calsequestrin in eggs of sea urchins and *Xenopus laevis*. *Dev. Biol.* **161**, 466–476.

Paul, M., and Johnston, R. N. (1978). Uptake of Ca^{2+} is one of the earliest responses to fertilization of sea urchin eggs. *J. Exp. Zool.* **203**, 143–149.

Payan, P., Girard, J. P., Sardet, C., Whitaker, M., and Zimmerberg, J. (1986). Uptake and release of calcium by isolated egg corticies of the sea urchin *Paracentrotus lividus*. *Biol. Cell* **58**, 87–90.

Pazin, M. J., and Williams, L. T. (1992). Triggering signaling cascades by receptor tyrosine kinases. *Trends Biochem. Sci.* **17**, 374–378.

Peaucellier, G., Veno, P. A., and Kinsey, W. H. (1988). Protein tyrosine phosphorylation in response to fertilization. *J. Biol. Chem.* **263**, 13806–13811.

Peaucellier, G., Shartzer, K., Jiang, W., Maggio, K., and Kinsey, W. H. (1993). Anti-peptide antibody identifies a 57 kDa protein tyrosine kinase in the sea urchin egg cortex. *Dev. Growth Diff.* **35**, 199–208.

Pekala, P. H., and Anderson, B. M. (1978). Studies of bovine erythrocyte NAD glycohydrolase. *J. Biol. Chem.* **253**, 7453–7459.

Penner, R., Neher, E., Takeshima, H., Nishimura, S., and Numa, S. (1989). Functional expression of the calcium release channel from skeletal muscle ryanodine receptor cDNA. *FEBS Lett.* **259**, 217–221.

Peres, A. (1990). InsP$_3$- and Ca^{2+}-induced Ca^{2+} release in single mouse oocytes. *FEBS Lett.* **275**, 213–216.

Pessah, I. N., and Zimányi, I. (1991). Characterization of multiple [^{3}H]ryanidine binding sites on the Ca^{2+} release channel of sarcoplasmic reticulum from skeletal and cardiac muscle: Evidence for a sequential mechanism in ryanodine action. *Mol. Pharmacol.* **39**, 679–689.

Poenie, M. (1990). Alteration of intracellular fura-2 fluorescence by viscosity: A simple correction. *Cell Calcium* **11**, 85–91.

Poenie, M., and Epel, D. (1987). Ultrastructural localization of intracellular calcium stores by a new cytochemical method. *J. Histochem. Cytochem.* **35**, 939–956.

Poenie, M., Alderton, J., Tsien, R. Y., and Steinhardt, R. A. (1985). Changes of free calcium levels with stages of the cell division cycle. *Nature* **315**, 147–149.

Pribluda, V. S., and Metzger, H. (1992). Transmembrane signaling by the high-affinity IgE receptor on membrane preparations. *Proc. Natl. Acad. Sci. USA* **89**, 11446–11450.

Rakow, T. L., and Shen, S. S. (1990). Multiple stores of calcium are released in the sea urchin egg during fertilization. *Proc. Natl. Acad. Sci. USA* **87**, 9285–9289.

Randriamampita, C., Bismuth, G., Debré, P., and Trautmann, A. (1991). Nitrendipine-induced inhibition of calcium influx in a human T-cell clone: Role of cell depolarization. *Cell Calcium* **12**, 313–323.

Rhee, S. G. (1991). Inositol phospholipid-specific phospholipase C: Interaction of the τ_1 isoform with tyrosine kinase. *Trends Biochem. Sci.* **16**, 297–301.

Ridgeway, E. B., Gilkey, J. C., and Jaffe, L. F. (1977). Free calcium increases exponentially in activating Medaka eggs. *Proc. Natl. Acad. Sci. USA* **74**, 623–627.

Ritov, V. B., Men'shikova, E. V., and Kozlov, Y. P. (1985). Heparin induces Ca^{2+} release from the terminal cysterns of skeletal muscle sarcoplasmic reticulum. *FEBS Lett.* **188**, 77–80.

Rothschild, Lord, and Barnes, H. (1953). The inorganic constituents of the sea urchin egg. *J. Exp. Biol.* **30**, 534–544.

Rousseau, E., Smith, J. S., and Meissner, G. (1987). Ryanodine modifies conductance and gating behavior of single Ca^{2+} release channel. *Am. J. Physiol.* **253**, C364–C368.

Rusinko, N., and Lee, H. C. (1989). Widespread occurrence in animal tissue of an enzyme catalyzing of NAD^+ into a cyclic metabolite with intracellular Ca^{2+}-mobilizing activity. *J. Biol. Chem.* **264**, 11725–11731.

Samelson, L. E., Siegel, J. N., Phillips, A. F., Garcia-Morales, P., Minami, Y., Klausner, R.D., Fletcher, M. C., and June, C. H. (1993). The T-cell antigen receptor: Biochemical aspects of signal transduction. *In* "Molecular Mechanisms of Immunological Self-Recognition" (F. W. Alt and H. J. Vogel, eds.), pp. 55–68. Academic Press, San Diego.

Sardet, C., Gillot, J., Ruscher, A., Payan, P., Girard, J. P., and De Renzis, G. (1992). Ryanodine activates sea urchin eggs. *Dev. Growth Diff.* **34**, 37–42.

Satoh, N., and Garbers, D. L. (1985). Protein tyrosine kinase activity of eggs of the sea urchin *S. purpuratus:* The regulation of its increase after fertilization. *Dev. Biol.* **11**, 515–519.

Scanlon, M., Williams, D. A., and Fay, F. S. (1987). A Ca^{2+}-insensitive form of fura-2 associated with polymorphonuclear leukocytes. *J. Biol. Chem.* **262**, 6308–6312.

Schmidt, T., Patton, C., and Epel, D. (1982). Is there a role for the Ca^{2+} influx during fertilization of the sea urchin egg? *Dev. Biol.* **90**, 284–290.

Sekhar, K. R., Hatchett, R. I., Shabb, J. B., Wolfe, L., Francis, S. H., Wells, J. N., Jastorff,B., Butt, E., Chakinala, M. M., and Corbin, J. D. (1992). Relaxation of pig coronary arteries by new and potent cGMP analogs that selectively activate type Iα, compared with type Iβ, cGMP-dependent protein kinase. *Mol. Pharmacol.* **42**, 103–108.

Shen, S. S. (1989). Protein kinase C and regulation of the Na^+–H^+ antiporter during fertilization of the sea urchin egg. *In* "Mechanisms of Egg Activation" (R. Nuccitelli, G. N. Cherr, and W. N. Clark, Jr., eds.), pp. 173–199. Plenum Press, New York.

Shen, S. S., and Buck, W. R. (1990). A synthetic peptide of the pseudosubstrate domain of protein kinase C blocks cytoplasmic alkalinization during activation of the sea urchin egg. *Dev. Biol.* **140**, 272–280.

Shen, S. S., and Buck, W. R. (1993). Sources of calcium in sea urchin eggs during the fertilization response. *Dev. Biol.* **157**, 157–169.

Shen, S. S., and Burgart, L. J. (1985). Intracellular sodium activity in the sea urchin egg during fertilization. *J. Cell Biol.* **101**, 420–426.

Shen, S. S., and Steinhardt, R. A. (1978). Direct measurement of intracellular pH during metabolic depression of the sea urchin egg. *Nature* **272**, 253–254.

Shen, S. S., and Steinhardt, R. A. (1984). Time and voltage windows for reversing the electrical block to fertilization. *Proc. Natl. Acad. Sci. USA* **81**, 1436–1439.

Shen, S. S., and Sui, A. L. (1989). K^+ activity and regulation of intracellular pH in the sea urchin egg during fertilization. *Exp. Cell Res.* **183**, 343–352.

Shilling, F., Mandel, G., and Jaffe, L. A. (1990). Activation by serotonin of starfish eggs expressing the rat serotonin 1c receptor. *Cell Regul.* **1**, 465–469.

Shilling, F. M., Carroll, D. J., Muslin, A. J., Escobedo, J. A., Williams, L.T., and Jaffe, L. A. (1994). Evidence for both tyrosine kinase and G-protein-coupled pathways leading to starfish egg activation. *Dev. Biol.* **162**, 590–599.

Shimomura, H., and Garbers, D. L. (1986). Differential effects of resact analogues on sperm respiration rates and cyclic nucleotide concentrations. *Biochemistry* **25**, 3405–3410.

Shimomura, O., and Johnson, F. H. (1970). Mechanisms of the quantum yield of *Cypridina* bioluminescence. *Photochem. Photobiol.* **12**, 291–295.

Shimomura, O., Johnson, F. H., and Saiga, Y. (1962). Extraction, purification and properties of aequorin, a bioluminescent protein from the luminous hydromedusan, *Aequorea*. *J. Cell. Comp. Physiol.* **59**, 223–239.

Shimomura, O., Musicki, B., and Inouye, S. (1993). Light-emitting properties of recombinant semi-synthetic aequorins and recombinant fluorescein-conjugated aequorin for measuring cellular calcium. *Cell Calcium* **14**, 373–378.

Silver, R. A., Whitaker, M., and Bolsover, S. R. (1992). Intracellular ion imaging using fluorescent dyes: Artifacts and limits to resolution. *Pflügers Arch.* **420**, 595–602.

Sorrentino, V., and Volpe, P. (1993). Ryanodine receptors: How many, where and why? *Trends Pharmacol. Sci.* **14**, 98–103.

States, D. I., Walseth, T. F., and Lee, H. C. (1992). Similarities in amino acid sequences of *Aplysia* ADP-ribosyl cyclase and human lymphocyte antigen CD38. *Trends Pharmacol. Sci.* **17**, 495.

Steinhardt, R. A., and Epel, D. (1974). Activation of sea urchin eggs by a calcium ionophore. *Proc. Natl. Acad. Sci. USA* **71**, 1915–1919.

Steinhardt, R. A., Lundin, L., and Mazia, D. (1971). Bioelectric responses of the echinoderm egg to fertilization. *Proc. Natl. Acad. Sci. USA* **68**, 2426–2430.

Steinhardt, R. A., Epel, D., Carroll, E. J., Jr., and Yanagimachi, R. (1974). Is calcium ionophore a universal activator for unfertilized eggs? *Nature* **252**, 41–43.

Steinhardt, R. A., Zucker, R. S., and Schatten, G. (1977). Intracellular calcium release at fertilization in the sea urchin egg. *Dev. Biol.* **58**, 185–196.

Stice, S. L., and Robl, J. M. (1990). Activation of mammalian oocytes by a factor obtained from rabbit sperm. *Mol. Reprod. Dev.* **25**, 272–280.

Stith, B. I., Goalstone, M., Silva, S., and Jaynes, C. (1993). Inositol 1,4,5-trisphosphate mass changes from fertilization through first cleavage in *Xenopus laevis*. *Mol. Biol. Cell* **4**, 435–443.

Stith, B. J., Espinoza, R., Roberts, D., and Smart, T. (1994). Sperm increase inositol 1,4,5-trisphosphate mass in *Xenopus laevis* eggs preincubated with calcium buffers or heparin. *Dev. Biol.* **165**, 206–215.

Stricker, S. A., Centonze, V. E., Paddock, S. W., and Schatten, G. (1992). Confocal microscopy of fertilization-induced calcium dynamics in sea urchin eggs. *Dev. Biol.* **149**, 370–380.

Sui, A. L., and Shen, S. S. (1986). Intracellular free magnesium concentration in the sea urchin egg during fertilization. *Dev. Biol.* **114**, 208–213.

Swann, K. (1990). Cytosolic sperm factor stimulates repetitive calcium increases and mimics fertilization in hamster eggs. *Development* **110**, 1295–1302.

Swann, K. (1991). Thimerosal causes calcium oscillations and sensitizes calcium-induced calcium release in unfertilized hamster eggs. *FEBS Lett.* **278**, 175–178.

Swann, K. (1992). Different triggers for calcium oscillations in mouse eggs involve a ryanodine-sensitive calcium store. *Biochem. J.* **287**, 79–84.

Swann, K., and Whitaker, M. (1986). The part played by inositol trisphosphate and calcium in the propagation of the fertilization wave in sea urchin eggs. *J. Cell Biol.* **103**, 2333–2342.

Swann, K., Ciapa, B., and Whitaker, M. (1987). Cellular messengers and sea urchin activation. *In* "Molecular Biology of Invertebrate Development" (J. D. O'Connor, ed.), pp. 45–69. Liss, New York.

Swann, K., McCulloh, D. H., McDougall, A., Chambers, E. L., and Whitaker, M. (1992). Sperm-induced currents at fertilization in sea urchin eggs injected with EGTA and neomycin. *Dev. Biol.* **151**, 552–563.

Takasawa, S., Nata, K., Yonekura, H., and Okamoto, H. (1993). Cyclic ADP-ribose in insulin secretion from pancreatic β cells. *Science* **259**, 370–373.

Terasaki, M., and Jaffe, L. A. (1991). Organization of the sea urchin egg endoplasmic reticulum and its organization at fertilization. *J. Cell. Biol.* **114**, 929–940.

Terasaki, M., and Sardet, C. (1991). Demonstration of calcium uptake and release by sea urchin egg cortical endoplasmic reticulum. *J. Cell Biol.* **115**, 1031–1037.

Thastrup, O., Cullen, P. I., Drøbak, B. K., Hanley, M. R., and Dawson, A. P. (1990). Thapsigargin, a tumor promoter, discharges intracellular Ca^{2+} stores by specific inhibition of the endoplasmic reticulum Ca^{2+}-ATPase. *Proc. Natl. Acad. Sci. USA* **87**, 2466–2470.

Tsien, R. Y., and Poenie, M. (1986). Fluorescence ratio imaging: A new window into intracellular ionic signaling. *Trends Biochem. Sci.* **11**, 450–455.

Turner, P. R., Sheetz, M. P., and Jaffe, L. A. (1984). Fertilization increases the polyphosphoinositide content of sea urchin eggs. *Nature* **310**, 414–415.

Turner, P. R., Jaffe, L. A., and Fein, A. (1986). Regulation of cortical vesicle exocytosis in sea urchin eggs by inositol 1,4,5-trisphosphate and GTP-binding protein. *J. Cell Biol.* **102**, 70–76.

Turner, P. R., Jaffe, L. A., and Primakoff, P. (1987). A cholera toxin-sensitive G-protein stimulates exocytosis in sea urchin eggs. *Dev. Biol.* **120**, 577–583.

Vacquier, V. D. (1975). The isolation of intact cortical granules from sea urchin eggs: Calcium ion triggers granule discharge. *Dev. Biol.* **43**, 62–74.

Vacquier, V. D., and Moy, G. W. (1977). Isolation of bindin: The protein responsible for adhesion of sperm to sea urchin egg. *Proc. Natl. Acad. Sci. USA* **74**, 2456–2460.

Vacquier, V. D., Carner, K. R., and Stout, C. D. (1990). Species-specific sequences of abalone lysin, the sperm protein that creates a hole in the egg envelope. *Proc. Natl. Acad. Sci. USA* **87**, 5792–5796.

Walseth, T. F., and Lee, N. C. (1993). Synthesis and characterization of cyclic ADP-ribose induced Ca^{2+} release. *Biochim. Biophys. Acta* **1178**, 235–242.

Walseth, T. F., Aarhus, R., Zeleznikar, R. J., Jr., and Lee, H. C. (1991). Determination of endogeneous levels of cyclic ADP-ribose in rat tissues. *Biochim. Biophys. Acta* **1094**, 113–120.

Walseth, T. F., Aarhus, R., Kerr, J. A., and Lee, H. C. (1993). Identification of cyclic ADP-ribose-binding proteins by photoaffinity labeling. *J. Biol. Chem.* **268**, 26686–26891.

Walton, P. D., Airey, J. A., Sutko, J. L., Beck, C. F., Mignery, G. A., Südhof, T. C., Deerinck, T. J., and Ellisman, M. H. (1991). Ryanodine and inositol trisphosphate receptors coexist in avian cerebellar Purkinje neurons. *J. Cell Biol.* **113**, 1145–1157.

Ward, G. E., Garbers, D. L., and Vacquier, V. D. (1985). Effects of extracellular egg factors on sperm guanylate cyclase. *Science* **227**, 768–770.

Weiss, A. (1993). T-cell antigen receptor signal transduction: A tale of tails and cytoplasmic protein-tyrosine kinases. *Cell* **73**, 209–212.

Whalley, T., and Whitaker, M. (1988). Guanine nucleotide activation of phosphoinositidase C at fertilization in sea urchin eggs. *J. Physiol. (London)* **406**, 126P.

Whalley, T., McDougall, A., Crossley, I., Swann, K., and Whitaker, M. (1992). Internal calcium release and activation of sea urchin eggs by cGMP are independent of the phosphoinositide signaling pathway. *Mol. Biol. Cell* **3**, 373–383.

Whitaker, M. J., and Aitchison, M. (1985). Calcium-dependent polyphosphoinositide hydrolysis is associated with exocytosis in vitro. *FEBS Lett.* **182,** 119–124.
Whitaker, M. J., and Crossley, I. (1990). How does sperm activate a sea urchin egg? *NATO ASI Series* **H45,** 433–443.
Whitaker, M. J., and Irvine, R. F. (1984). Microinjection of inositol trisphosphate activates sea urchin eggs. *Nature* **312,** 636–639.
Whitaker, M. J., and Steinhardt, R. A. (1985). Ionic signaling in the sea urchin egg at fertilization. *In* "Biology of Fertilization" (C. B. Metz and A. Monroy, eds.), Vol. 3, pp. 167–221. Academic Press, San Diego.
Whitaker, M., and Swann, K. (1993). Lighting the fuse at fertilization. *Development* **117,** 1–12.
Whitaker, M., Swann, K., and Crossley, I. (1989). What happens during the latent period at fertilization? *In* "Mechanisms of Egg Activation" (R. Nuccitelli, G. N. Cherr, and W. N. Clark, Jr., eds.), pp. 157–171. Plenum Press, New York.
White, A. M., Watson, S. P., and Galione, A. (1993). Cyclic ADP-ribose-induced Ca^{2+} release from rat brain microsomes. *FEBS Lett.* **318,** 259–263.
Williams, C. J., Schultz, R. M., and Kopf, G. S. (1992). Role of G proteins in mouse egg activation: Stimulatory effects of acetylcholine on the ZP2-to-$ZP2_f$ conversion and pronuclear formation in eggs expressing a functional m1 muscarinic receptor. *Dev. Biol.* **151,** 288–296.
Worley, P. F., Baraban, J. M., Supattone, S., Wilson, W. S., and Snyder, S. N. (1987). Characterization of inositol trisphosphate receptor binding in brain. *J. Biol. Chem.* **262,** 12132–12136.
Yim, D. L., Opresko, L. K., Wiley, H. S., and Nuccitelli, R. (1994). Highly polarized EGF receptor tyrosine kinase activity initiates egg activation in *Xenopus*. *Dev. Biol.* **162,** 41–55.
Yoshimoto, Y., and Hiramoto, Y. (1991). Observation of intracellular Ca^{2+} with aequorin luminescence. *Int. Rev. Cytology* **129,** 45–73.
Zocchi, E., Franco, L., Guida, L., Benatti, U., Bargellesi, A., Malavasi, F., Lee, H. C., and DeFlora, A. (1993). A single protein immunologically identified as CD38 displays NAD^+ glycohydrolase, ADP-ribosyl cyclase and cyclic ADP-ribose hydrolase activities at the outer surface of human erythrocytes. *Biochem. Biophys. Res. Commun.* **196,** 1459–1465.
Zucker, R. S., and Steinhardt, R. A. (1978). Prevention of the cortical reaction in fertilized sea urchin eggs by injection of calcium-chelating ligands. *Biochim. Biophys. Acta* **541,** 459–466.
Zucker, R. S., Steinhardt, R. A., and Winkler, M. M. (1978). Intracellular calcium release and the mechanisms of parthenogenetic activation of the sea urchin egg. *Dev. Biol.* **65,** 285–295.

4
Regulation of Oocyte Growth and Maturation in Fish

Yoshitaka Nagahama, Michiyasu Yoshikuni, Masakane Yamashita, Toshinobu Tokumoto, and Yoshinao Katsu
Laboratory of Reproductive Biology
Department of Developmental Biology
National Institute for Basic Biology
Okazaki 444, Japan

I. Introduction
II. Gonadotropin
III. Structure of Follicles
IV. Oocyte Growth
    A. Endocrine Regulation of Oocyte Growth
    B. Mechanisms of Follicular Production of Estradiol-17β
V. Oocyte Maturation
    A. Gonadotropin—Primary Mediator of Oocyte Maturation
    D. Maturation-Inducing Hormone—Secondary Mediator of Oocyte Maturation
    C. Maturation-Promoting Factor—Tertiary Mediator of Oocyte Maturation
VI. Summary
References

I. Introduction

Oogenesis in fish, as in other vertebrates, can be divided into two parts: the prolonged growth phase, which is characterized by the enlargement of oocytes, and the maturation phase, during which fully grown oocytes become fertilizable. The endocrine regulation of these two phases primarily involves the actions of the pituitary gonadotropins. In most cases, however, gonadotropin action on oogenesis is not direct, but is mediated through the follicular production of ovarian steroid hormones, which in turn mediate various stages of oogenesis including oocyte growth and maturation. Over the past several years, a series of studies in our laboratory using several species of teleosts as experimental animals has provided new information about the endocrine regulation of oocyte growth and maturation, the interactions of gonadotropins with ovarian somatic cell types, and the mechanisms of their action. This chapter briefly reviews the endocrine regulation of oocyte growth and maturation, and describes the data on the mechanisms by which gonadotropin exerts its action on the production of two major naturally occurring steroid hormones, estradiol-17β and 17α, 20β-dihydroxy-4-pregnen-3-one (17α,20β-DP). Finally, our recent studies on the characterization and mechanisms of activation and inactivation of fish maturation-promoting factor (MPF), the key cell cycle molecule during the G_2/M transition in both meiosis and mitosis, are described.

II. Gonadotropin

Researchers have established that in teleosts, as in other vertebrates, pituitary gonadotropins are the major hormones that stimulate various ovarian activities. Although a number of biochemical studies have been conducted to purify fish gonadotropins, the number and identity of fish gonadotropins has been controversial (Burzawa-Gerard, 1982; Fontain and Dufour, 1987; Idler and Ng, 1983). Recently, two types of gonadotropins, designated GTH-I and GTH-II, have been isolated from several species of fish. In chum salmon (*Oncorhynchus keta*) and coho salmon (*Oncorhynchus kisutch*) these two gonadotropins are distinctly different from each other in chemical characteristics, and are structurally homologous to tetrapod follicle-stimulating hormone (FSH) and luteinizing hormone (LH) (Suzuki *et al.*, 1988; Kawauchi *et al.*, 1989; Swanson, 1991; Swanson *et al.*, 1991). Each gonadotropin consists of α and β subunits; the β subunits have only about 31% amino acid sequence identity. In both species, GTH-I and GTH-II exhibit similar *in vitro* steroidogenic potencies. Blood and pituitary levels of these two gonadotropins vary significantly during reproductive development. GTH-I is the predominant gonadotropin in the plasma and pituitary of vitellogenic females, whereas GTH-II is the predominant gonadotropin at the time of final oocyte maturation (Swanson, 1991). In this article, the term gonadotropin refers to the GTH-II ("maturational") gonadotropin.

III. Structure of Follicles

The female reproductive system of fish, unlike that of mammals, is highly variable, reflecting the wide range of reproductive patterns including oviparity, ovoviviparity, and viviparity. The ovary consists of oogonia, oocytes and their surrounding follicle cells, supporting tissue or stroma, and vascular and nervous tissue. Most teleosts are cyclical breeders, and the ovary varies greatly in appearance at different times in the cycle. At least three ovarian types have been classified according to the pattern of oocyte development (Wallace and Selman, 1981; Nagahama, 1983). The synchronous ovary contains oocytes all at the same stage of development; this type is found in teleosts that spawn only once and then die, such as anadromous *Oncorhynchus* species or catadromous eels. The group synchronous ovary consists of at least two populations of oocytes at different

4. Oocyte Growth and Maturation in Fish

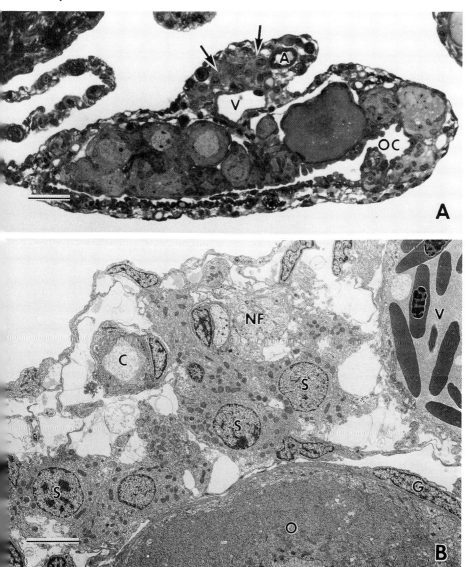

Fig. 1 Light and electron micrographs of ovaries of tilapia, *O. niloticus*, 50 days after hatching. (A) Light micrograph (1-μm section) showing autocytes and clusters (arrows) of steroid-producing cells. A, Artery; OC, ovarian cavity; V, vein. Bar: 25 μm. (B) Electron micrograph showing clusters of steroid-producing cells (S). A layer of granulosa cells (G) encloses an oocyte in the perinucleus stage. C, Capillary; O, oocyte; NF, nerve fiber; V, vein. Bar: 5 μm.

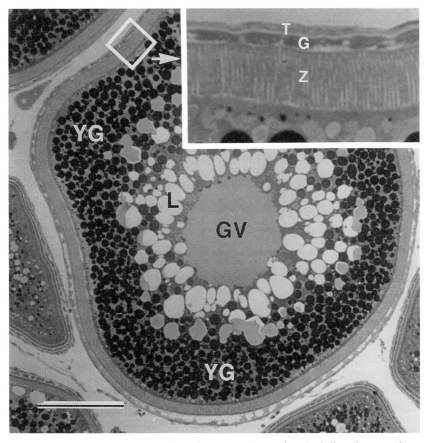

Fig. 2 Light micrograph (1-μm section) of *Pagrus major* ovaries. A vitellogenic oocyte is surrounded by two follicle cell layers, the thecal cell layer (T) and the granulosa cell layer (G). GV, Germinal vesicle; L, lipid droplet; YG, yolk granule; Z, zona radiata. Bar: 100 μm.

developmental stages; teleosts with this type of ovary, such as rainbow trout (*Oncorhynchus mykiss*) generally spawn once a year and have a relatively short breeding season. The asynchronous ovary contains oocytes at all stages of development; this type occurs in those species that spawn many times during a prolonged breeding season [e.g., medaka (*Oryzias latipes*) and goldfish (*Carassius auratus*)].

Each oocyte during its development becomes surrounded by a layer of follicle cells (Nakamura *et al.*, 1993) (Fig. 1). As the oocytes grow, follicle cells multiply and form a continuous follicular layer called the granulosa cell layer. Simultaneously, the surrounding stromal connective tissue elements become organized to form the distinct outer layer of the follicular envelope called the thecal cell layer

4. Oocyte Growth and Maturation in Fish

(Fig. 2). Therefore, vitellogenic and fully grown postvitellogenic oocytes are surrounded by two major cell layers, an outer thecal cell layer and an inner granulosa cell layer, that are separated by a distinct basement membrane. The thecal cell layer contains fibroblasts, capillaries, collagen fibers, and large cells designated special thecal cells. The special thecal cells possess features that characterize steroidogenic cells in general, that is, mitochondria with tubular cristae and a tubular agranular endoplasmic reticulum (Nagahama, 1983). Recently, we raised a polyclonal antibody against a synthetic peptide corresponding to an amino acid sequence that is completely conserved between mammalian and rainbow trout 3β-hydroxysteroid dehydrogenase (3β-HSD), the enzyme responsible for oxidation and isomerization of Δ^5-3β-hydroxysteroid precursors into Δ^4-3-ketosteroids (T. Kobayashi and Y. Nagahama, unpublished results). Immunocytochemical studies using this antibody reveal that in tilapia (*Oreochromis niloticus*), ovarian follicle 3β-HSD is located in special thecal cells but not in granulosa cells (M. Nakamura and Y. Nagahama, unpublished results). Histochemical studies have indicated that steroidogenic enzyme activities are localized in the granulosa cells of some teleosts. However, these histochemical observations of granulosa cells do not seem to be supported by ultrastructural studies. Because these cells contain features suggestive of protein synthesis but do not contain organelles associated with steroid-producing cells (Nagahama, 1983; Dodd, 1987).

One of the major features of salmonid ovaries is that the follicles they contain are large (about 3–5 mm in diameter) and develop synchronously, an enormous

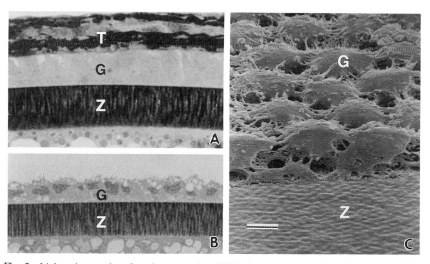

Fig. 3 Light micrographs of an intact ovarian follicle layer (A) and granulosa cell layer (B). Scanning electron micrograph (C) of a granulosa cell layer preparation, consisting of purely granulosa cells. Z, zona radiata. Bar: 5 μm.

advantage for biochemical studies since a large number of follicles at the same stage of development can be obtained easily. The large size of the follicles has also facilitated the development of a simple dissection technique to separate the ovarian follicle of salmonids into two layers (thecal cell layer and granulosa cell layer), making it possible to elucidate the relative contributions of these layers and of gonadotropin in the production of estradiol-17β and 17α,20β-DP in amago salmon (*Oncorhynchus rhodurus*) and rainbow trout (Fig. 3).

IV. Oocyte Growth

After oogonia undergo proliferation by mitotic divisions, they become oocytes and enter a period of growth. During growth, the increase in oocyte size is very considerable. For example, in salmonids a young oocyte may be about 50 μm in diameter and the fully developed egg is between 3.0 mm and 5.0 mm in diameter. Simultaneously with growth, the oocyte nucleus goes through the early stages of prophase but is arrested in the diplotene stage of the first meiotic division. The increase in oocyte size is mainly due to the accumulation of yolk proteins. In teleosts, as in most other vertebrates, the site of synthesis of the precursor protein of yolk, vitellogenin, is the liver. Vitellogenin is then transported via the circulatory system to the ovary, where it is taken up into the developing oocytes by receptor-mediated endocytosis.

A. Endocrine Regulation of Oocyte Growth

1. Vitellogenesis

Investigations on the endocrine regulation of vitellogenesis in teleosts have been reviewed elsewhere (Ng and Idler, 1983; Wallace, 1985; Mommsen and Walsh, 1988; Ho, 1991; Specker and Sullivan, 1994). Estrogen, in most cases estradiol-17β, produced by the ovary under the influence of gonadotropins (probably GTH-I in salmonid fishes) is introduced into the vascular system and stimulates the hepatic synthesis and secretion of vitellogenin (Fig. 4). Researchers have demonstrated that injection with pituitary extracts or various preparations of gonadotropins induces estradiol-17β synthesis and subsequent vitellogenesis. Plasma levels of estradiol-17β and vitellogenin are positively correlated, and increase in parallel. Exogenous estrogen can induce vitellogenin synthesis in the liver and its appearance in the plasma. This chapter does not review the biochemistry and molecular biology of vitellogenesis in fishes, nor the estrogen induction of vitellogenin synthesis in the liver. For details of these events, consult reviews by Mommsen and Walsh (1988), Lazier and MacKay (1993), and Specker and Sullivan (1994).

4. Oocyte Growth and Maturation in Fish 109

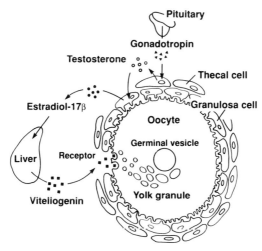

Fig. 4 Hormonal regulation of vitellogenesis in teleosts (see text for details).

2. Vitellogenin Uptake

Fish oocytes accumulate large quantities of protein yolk during vitellogenesis to meet the nutritional needs of the eventual embryo. This yolk is derived from the proteolytic cleavage of vitellogenin. Vitellogenin is selectively taken up from the bloodstream by developing oocytes (Wallace, 1985; Tyler, 1991); ultrastructural evidence shows that it is incorporated into the oocyte by micropinocytosis (Selman and Wallace, 1982). Studies using *in vitro* culture systems have provided evidence to indicate that vitellogenin uptake by fish oocytes is mediated by receptor-coupled endocytosis (Kanungo *et al.*, 1990; Tyler *et al.*, 1990; Tyler, 1991). A vitellogenin receptor was characterized in coho salmon. Scatchard analyses revealed a single class of binding sites with a very high affinity for vitellogenin (Stifani *et al.*, 1990). Binding for vitellogenin was also characterized in isolated oocyte membranes of tilapia (*O. niloticus*) (Chan *et al.*, 1991). Saturation and Scatchard analyses reveal only a single class of binding site. The number and affinity of these bindings greatly increase from the previtellogenic to the vitellogenic stage and remain unchanged until the preovulatory stage, at which time the affinity of binding is also highest. Similarly, proteins with an affinity for vitellogenin have been isolated from vitellogenic follicles of rainbow trout (Lancaster and Tyler, 1991; LeMenn and Nunez Rodriguez, 1991). Further studies on the expression and regulation of these receptor proteins will provide new insight into the mechanism controlling vitellogenin uptake into oocytes.

Little is known about hormonal involvement in vitellogenin uptake by oocytes of teleosts. Several hormones have been implicated in regulating oocyte growth, including gonadotropin, thyroxine, tri-iodothyronine, insulin, and growth hormone, although the role of each of these has yet to be determined fully (Tyler, 1991). *In vitro* studies showed that GTH-I, but not GTH-II, stimulated uptake of vitellogenin into ovarian follicles of intact vitellogenic rainbow trout, as well as *in vitro* incorporation of vitellogenin by partially denuded follicles (Tyler *et al.*, 1990). We have developed an *in vitro* assay system for vitellogenin uptake using follicle-free oocytes of rainbow trout (Shibata *et al.*, 1993). Using this *in vitro* incubation system, we have shown that insulin and thyroxine significantly increase vitellogenin uptake. In contrast, neither GTH-I nor GTH-II from chum salmon is effective. When using the oocytes that matured *in vivo*, only $1/15$ of vitellogenin is incorporated. Insulin also has no effect on these mature oocytes. We hypothesized from these findings that the previously observed stimulation of vitellogenin uptake into intact follicles or partially denuded oocytes of rainbow trout was most likely mediated by the actions of gonadotropins on the ovarian follicles.

3. Egg Membrane Formation

Another possible function of estrogen during oocyte growth has been demonstrated in medaka. One of the most striking features observed during oocyte growth in teleosts is the formation of a thick, highly differentiated zone (egg membrane, vitelline membrane, zona radiata, zona pellucida) between granulosa cell layers and oocytes. Although the exact mechanism by which the egg membrane is formed in teleost oocytes is still unknown, findings by Hamazaki *et al.* (1987a,b,1989) are of great interest. These authors reported that the major glycoprotein constituent of the inner layer of the medaka egg membrane is synthesized in the liver under the influence of estradiol-17β. A similar mechanism also functions in the formation of egg membranes in other teleost species (Hyllner *et al.*, 1991).

B. Mechanisms of Follicular Production of Estradiol-17β

1. Two-Cell-Type Model

Investigators have reported that in a number of fish species, estradiol-17β levels in the plasma increase during vitellogenesis and rapidly decrease prior to oocyte maturation (Kagawa *et al.*, 1983). This seasonal pattern is reflected in the ability of the ovarian follicle to produce estradiol-17β in response to gonadotropins. Using various follicular preparations obtained from vitellogenic amago salmon and rainbow trout, we examined the effects of partially purified salmon gonadotropins on estradiol-17β production. Gonadotropins stimulate estradiol-17β production by intact follicles and thecal and granulosa cell layer coculture preparations, but not by isolated thecal and granulosa cell layers. These results indi-

4. Oocyte Growth and Maturation in Fish

cate that both cell layers are necessary for gonadotropin-stimulated estradiol-17β production (Kagawa et al., 1982). Experiments examining the effects of conditioned media from incubates of one follicle layer on steroidogenesis by the other layer reveal that the thecal cell layer produces a steroid precursor that is metabolized to estradiol-17β in the granulosa cell layer. We further identified testosterone as the steroid precursor produced by thecal cell layers, but not by granulosa cell layers, in response to gonadotropins (Adachi et al., 1990). Gonadotropins greatly stimulate testosterone production by thecal cell layers. Estradiol-17β production is stimulated when granulosa cell layers are incubated with exogenous testosterone. Relatively small amounts of estradiol-17β are also produced by thecal cell layers incubated with testosterone, but this production can be attributed to contamination of thecal cell layer preparations with granulosa cells.

Considering all these data, a two-cell-type model in the production of follicular estradiol-17β in salmonids has been proposed (Kagawa et al., 1985; Nagahama, 1987b; Fig. 5). In this model, the thecal cell layer, under the influence of gonadotropin, secretes aromatizable androgen (mainly testosterone) that is converted to estradiol-17β by the granulosa cell layer. Restricted distribution of testosterone and estradiol-17β has been confirmed immunohistochemically in vitellogenic ovarian follicles of rainbow trout (Schulz, 1986). We have also found that the thecal cell layer from amago salmon and the granulosa cell layer from rainbow trout can produce the same effect that has been reported using combinations of thecal and granulosa cell layers from the same species. The reciprocal use of amago salmon granulosa and rainbow trout thecal cell layers is also effective. This finding implies that there may be little species specificity in each of these cell layers among salmonids. This two-cell-type model is the first of its kind to be reported in lower vertebrates, and is of evolutionary interest considering the situation in mammals and birds. A two-cell-type model similar to that seen in teleosts has been demonstrated in mammals (Dorrington and Armstrong, 1979), but in this case two kinds of gonadotropins, FSH and LH, are required, each of which acts on a separate follicular cell type. The thecal internal cells, under the influence of LH, produce androgens that are transferred to granulosa cells where, in the presence of FSH, they are converted to estrogens. In some larger mammals such as monkeys, the thecal internal tissue is considered an additional source of estrogens. Under gonadotropin stimulation in the domestic hen, progesterone is produced by the granulosa cells, most of it diffusing to the thecal cells where it is converted to estradiol-17β (Huang and Nalvandov, 1979). This process is in sharp contrast to the events in salmonids and mammals, and thus is of evolutionary interest. However, the two-cell-type model described for salmonid ovarian follicles does not seem to be valid for *Fundulus heteroclitus* (Petrino et al., 1989) and medaka (Onitake and Iwamatsu, 1986) ovarian follicles. In these species, in which steroidogenic thecal cells are not evident in the thecal cell layer, the granulosa cells are the major sites of steroid synthesis and

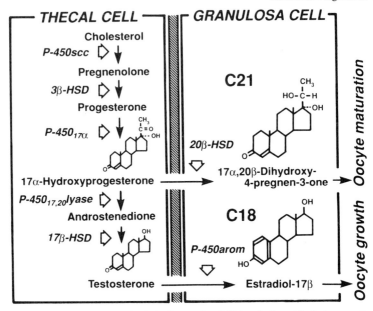

Fig. 5 Pathway of steroid biosynthesis in the ovarian follicle of salmonids during oocyte growth and maturation, showing the relative contribution of thecal and granulosa cell layers in the production of estradiol-17β and 17α,20β-dihydroxy-4-pregnen-3-one (17α,20β-DP). P-450scc, Cholesterol side-chain cleavage cytochrome P-450; 3β-HSD, 3β-hydroxysteroid dehydrogenase-isomerase; P-450$_{17\alpha}$, P-450 17α-hydroxylase; P-450$_{17,20}$lyase, P-450 17,20-lyase; 17β-HSD, 17β-hydroxysteroid dehydrogenase; P-450arom, aromatase cytochrome P-450; 20β-HSD, 20β-hydroxysteroid dehydrogenase.

the production of estradiol-17β does not require the involvement of two cell types.

Gonadotropin action on the thecal cell layer to stimulate testosterone production is mediated through a receptor-coupled adenylate cyclase–cAMP system (Nagahama, 1987a; Kanamori and Nagahama, 1988a,b; Kanamori et al., 1988). Other intracellular signaling molecules including calcium, protein kinase, and arachidonic acid, are also associated with gonadotropin-induced testosterone production by intact preovulatory follicles of goldfish (Van Der Kraak 1990, 1991; Van Der Kraak and Chang, 1990). The site of gonadotropin action on the thecal cell layer remains unclear, but our *in vitro* data using steroid synthesis inhibitors suggest that its major site is on the pathway from cholesterol to pregnenolone.

The mechanism of induction or activation of aromatase in salmonid granulosa cells is still unknown. In certain mammals, one of the major roles of gonadotropin action on follicular estradiol-17β secretion is known to be the stimu-

4. Oocyte Growth and Maturation in Fish

lation of aromatase activity; the most fundamental regulator is FSH (Dorrington and Armstrong, 1979). However, very few studies have been conducted in nonmammalian species. As discussed earlier, isolated salmonid granulosa cells have no capacity to produce testosterone and estradiol-17β from endogenous substrates in response to gonadotropins. Furthermore, there is no significant difference in estradiol-17β production by granulosa cell layers incubated with testosterone in the presence of gonadotropins at any stage of follicle development (Young et al., 1983a). In addition, pituitary homogenates from vitellogenic amago salmon are unable to induce aromatase activity (H. Kagawa, unpublished results). Nevertheless, aromatase activity in granulosa cells increases during vitellogenesis to reach a peak in late vitellogenesis, and therefore rapidly declines in the postvitellogenic period. Thus whether gonadotropins enhance aromatase activity in salmonid granulosa cells is still unknown.

We have previously shown that in medaka vitellogenic follicles, aromatase activity is markedly enhanced by pregnant mare serum gonadotropin (PMSG) via an adenylate cyclase–cAMP system. Furthermore, the PMSG-induced aromatase activation is blocked completely by actinomycin D and cycloheximide, suggesting that this action of PMSG is dependent on both transcriptional and translational processes (Nagahama et al., 1991).

2. Developmental Changes in the Capacity of Ovarian Follicles to Produce Estradiol-17β in Response to Gonadotropin

In vitro production of testosterone by the isolated thecal cell layer preparations obtained each month during vitellogenesis and oocyte maturation has revealed that the capacity of the thecal cell layer to produce testosterone in response to salmon gonadotropins gradually increases during the course of vitellogenic growth and peaks during the postvitellogenic period. This capacity of thecal cell layers is maintained by the period of oocyte maturation and ovulation (Fig. 6). An identical seasonal pattern of stimulation of testosterone production is observed when thecal cell layers are incubated with forskolin and agents known to raise intracellular cAMP levels.

The foregoing findings indicate the presence of aromatase enzymes in granulosa cells of salmonid vitellogenic follicles. Aromatase activity increases during the period of vitellogenesis to reach a peak in late vitellogenesis, and then declines rapidly in association with the ability of the oocyte to mature in response to gonadotropins (Young et al., 1983a). These findings, compared with our observations of the enhanced capacity of the thecal cell layer to produce testosterone in the postvitellogenic period, provide strong evidence that the decrease in the capacity of postvitellogenic follicles to produce estradiol-17β is due, in part, to the decrease in aromatase activity during the course of oocyte development.

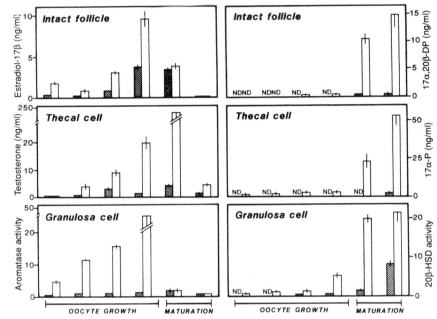

Fig. 6 Developmental changes in steroidogenic capacity of intact follicles, thecal cell layers, and granulosa cell layers in amago salmon during oocyte growth (left to right: follicles collected in June, July, August, and September) and maturation (in October, postvitellogenic immature follicle, *left,* and postovulatory follicles, *right,* were collected). Intact follicles and thecal layers were incubated in Ringer alone (shaded bars) or with chum salmon gonadotropin (SGA, 1μg/ml; open bars) for 18 hr at 15°C. Aromatase activity was assessed by incubating granulosa cells with Ringer with (*open bars*) or without testosterone (100 ng/ml; *shaded bars*). 20β-Hydroxysteroid dehydrogenase (20β-HSD) activity was assessed by incubating granulosa layers with 17α-hydroxyprogesterone (17α-P) in the presence (*open bars*) or absence (*shaded bars*) of 1 μg/ml SGA. Levels of steroid hormones in media were measured by radioimmunoassay. 17α-P, 17α-Hydroxyprogesterone; 17α,20β-DP, 17α,20β-dihydroxy-4-pregnen-3-one.

3. Cloning of Genes Encoding Various Steroidogenic Enzymes Necessary for Estradiol-17β Synthesis

To understand more fully the molecular mechanisms by which gonadotropins or other hormones regulate estradiol-17β biosynthesis, cDNA clones encoding several steroidogenic enzymes necessary for the production of estradiol-17β were isolated from rainbow trout and medaka ovarian cDNA libraries. Four steroidogenic enzymes involved in the production of estradiol-17β were selected to be cloned: cholesterol side-chain cleavage cytochrome P-450 (P-450scc), 3β-hydroxysteroid dehydrogenase-isomerase (3β-HSD), 17α-hydroxylase/17,20-lyase cytochrome P-450 (P-450c17), and aromatase cytochrome P-450 (P-450arom). The cDNA inserts were confirmed to encode each steroidogenic

4. Oocyte Growth and Maturation in Fish 115

enzyme by introduction into nonsteroidogenic COS-1 monkey kidney tumor cells.

a. P-450scc. The first and rate-limiting step in steroidogenesis is the conversion of cholesterol to pregnenolone. This reaction is catalyzed by P-450scc. A cDNA clone encoding P-450scc was isolated from a rainbow trout ovarian follicle cDNA library (Takahashi et al., 1993). The cDNA contains an open reading frame of 1542 nucleotides encoding a protein of 514 amino acids. The predicted amino acid sequence of trout P-450scc shows 40% homology with that of human and 46% homology with that of rat, bovine, and pig. The cDNA only hybridizes to a single 1.8-kb RNA transcript. The 1.8-kb transcripts increase in quantity toward the end of vitellogenesis.

b. 3β-HSD. A cDNA clone encoding 3β-HSD, the enzyme responsible for the oxidation and isomerization of Δ^5-3β-hydroxysteroid precursors into Δ^4-3-ketosteroids, was isolated from a cDNA library of rainbow trout ovarian thecal cells (Sakai et al., 1993). The cDNA contains an open reading frame of 1122 nucleotides encoding a protein of 374 amino acid residues. Comparison of the deduced amino acid sequence of trout 3β-HSD with that of mammalian 3β-HSD reveals at least five conserved regions, including a putative pyridine-nucleotide binding domain. The trout 3β-HSD expressed in COS-1 cells shows a unique enzymatic 3β-HSD activity. Dehydroepiandrosterone is a more favorable substrate of trout 3β-HSD than is 17α-hydroxypregnenolone. Interestingly, trout 3β-HSD expressed in COS-1 cells exhibited minimal ability to convert pregnenolone to progesterone. This activity profile is distinct from that of rat and human forms of 3β-HSD that catalyze the conversion of pregnenolone to progesterone at rates similar to the conversion of other Δ^5-3β-hydroxysteroids to Δ^4-3-ketosteroids. Southern hybridization analysis of trout genomic DNA with the cDNA suggests the presence of a single gene encoding 3β-HSD in rainbow trout, showing a total genomic size of less than 4 kb. The cDNA only hybridizes to a single 1.4-kb transcript isolated from rainbow trout ovaries. The 1.4-kb transcripts are barely detected in early vitellogenic follicles but increase markedly in ovaries during oocyte maturation.

c. P-450c17. A full-length cDNA encoding P-450c17, the enzyme involved in the 17α-hydroxylation of both pregnenolone and progesterone as well as in their conversion to C_{19} steroids, was cloned from a rainbow trout ovarian thecal cell layer cDNA library (Sakai et al., 1992). The cDNA contains an open reading frame encoding a protein of 514 amino acid residues. The amino acid sequence of trout P-450c17 shows a much greater homology with chicken P-450c17 than with that of human, bovine, and rat. The trout P-450c17 expressed in COS-1 cells shows both 17α-hydroxylase and 17,20-lyase activities, a finding similar to that reported for mammalian P-450c17, which mediates both 17α-hydroxylase and 17,20-lyase activities in the synthesis of steroid hormones (Nakajin and Hall, 1981; Zuber et al., 1986). The cDNA only hybridizes to a single 2.4-kb transcript isolated from rainbow trout ovaries. The 2.4-kb transcripts increase in number toward the end of vitellogenesis. As discussed earlier, a dramatic switch in salmonid steroidogenesis from testosterone to 17α-hydroxyprogesterone occurs only in thecal cell layers immediately prior to oocyte maturation. The formation of testosterone requires the activities of both P-450 17α-hydroxylase and 17,20-lyase, whereas 17α-hydroxyprogesterone synthesis requires only 17α-hydroxylase. Therefore, investigating the molecular mechanisms responsible for the differential regulation of these two enzymatic activities of P-450c17 in thecal cell layers immediately prior to oocyte maturation is of great interest.

d. P-450arom. P-450arom is responsible for catalyzing the conversion of C_{19} steroids to C_{18} steroids. A full-length cDNA encoding P-450arom was cloned from the rainbow trout ovary (Tanaka et al., 1992a). The cDNA insert of rainbow trout aromatase was sequenced and found to contain an open reading frame predicted to encode a protein of 522 amino acid residues. The deduced polypeptide is 52% homologous with aromatase of human, mouse, and rat, and 53% homologous with that of chicken. The cDNA hybridizes to a single 2.6-kb RNA transcript present in the trout ovary only during vitellogenesis. The 2.6-kb P-450arom RNA transcripts are present in the ovary during active vitellogenesis, but are absent in the ovary in the stage of oocyte maturation or in the ovary containing postovulatory follicles (Tanaka et al., 1992a; Fig. 7). These results are consistent with the rapid decrease in aromatase activity in the granulosa cell layers during the postovulatory period (Young et al., 1983a). Thus we conclude that the ability of the granulosa cells to produce estradiol-17β is regulated by the amount of the 2.6-kb RNA transcripts present (Tanaka et al., 1992a).

We also isolated the structural gene encoding P-450arom for the first time from a nonmammalian vertebrate, medaka, using rainbow trout P-450arom cDNA as a probe (M. Tanaka, S. Fukada, and Y. Nagahama, unpublished results). The medaka P-450 arom gene is composed of nine exons, but spans only 2.6 kb; it is much smaller than the human P-450arom gene (at least 70 kb) as a result of extremely small introns (medaka, 73–213 bp; human, 1.3–10 kb). The splicing junctions are located at exactly the same positions as in the human P-450arom gene. The deduced amino acid sequence is 51–52% identical to those of mammals and chicken, and 75% identical to the rainbow trout amino acid sequence. Genomic Southern blots reveal the presence of a single medaka gene. S1 nuclease mapping and primer extension indicate two major transcription initiation sites 60 and 61 bp upstream from a putative initiation codon. The promoter region of the medaka P-450arom gene also contains potential Ad4 binding protein (Ad4BP) sites (Morohashi et al., 1993) and estrogen responsive element (ERE) half sites. These results suggest that the basic structural organization of P-450arom genes and their regulatory mechanisms of expression are well conserved throughout the vertebrates.

4. Oocyte Growth and Maturation in Fish

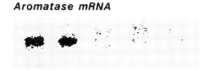

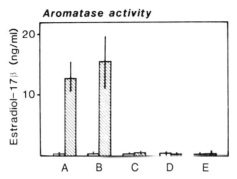

Fig. 7 Changes in aromatase activity and mRNA levels in rainbow trout during oocyte growth and maturation. Aromatase activity was assessed by incubating intact follicles in Ringer with (*open bars*) or without (*shaded bars*) 100ng/ml testosterone. Levels of estradiol-17β were measured by radioimmunoassay. Poly(A)+ RNAs were extracted from the same ovaries used for determining aromatase activity. RNA transcripts (2.6 kb) are present in ovaries during vitellogenesis. (A) Early vitellogenic stage; (B) late vitellogenic stage; (C) migrated nucleus stage; (D) mature stage; (E) postovulatory stage.

V. Oocyte Maturation

After the oocyte is fully grown during vitellogenesis, it becomes ready for the next phase of oogenesis—the resumption of meiosis, or final oocyte maturation, which is accompanied by several maturational processes in the nucleus and cytoplasm of the oocyte. Fully grown fish oocytes are arrested in late G_2 of meiosis I and must progress to the second meiotic metaphase before fertilization is possible. The oocyte nucleus (germinal vesicle, GV) of fish immature oocytes is generally inconspicuous because of the opaque cytoplasm. The GV of this state is generally located centrally or halfway between the center and the oocyte periphery. The first visible event associated with final oocyte maturation is the migration of the GV to the animal pole where the micropyle is situated; at this stage, the GV becomes visible under the dissecting microscope. Although the process(es) that allows GV migration to the animal pole remains unknown, evidence suggests the possible role of cytoskeletal components in the regulation of GV migration (Habibi and Lessman, 1986). Confocal immunofluorescence microscope studies using an anti-tubulin antibody have revealed that microtubules distribute in yolk-free regions of cytoplasm throughout fully grown immature oocytes of goldfish, forming a well-distributed microtubule network. Significant reorganization of cytoplasmic microtubule distribution occurs during the early phase of hormonally induced oocyte maturation, suggesting an important role for microtubules in the migration of the GV. After the completion of GV migration, the membrane of the GV breaks down (GV breakdown, GVBD) and its contents become intermingled with the surrounding cytoplasm. After dissolution of the nuclear membrane, the condensed chromosomes align on the first metaphase spindle, complete meiosis I, and realign on the second spindle, where they remain until the mature egg is fertilized. In some species, particularly marine teleosts, an enormous volume increase occurs concomitant with oocyte maturation due to rapid water uptake (Fig. 8).

A. Gonadotropin—Primary Mediator of Oocyte Maturation

The primary hormone that triggers oocyte maturation in fish, as in other vertebrates, is gonadotropin (probably GTH-II in salmonid fish) released from the pituitary gland (Nagahama, 1987c; Fig. 9). A distinct preovulatory gonadotropin surge occurs in several teleosts, including goldfish (Stacey *et al.*, 1979; Kobayashi *et al.*, 1987) and carp (*Cyprinus carpio*) (Santos *et al.*, 1986). In a number of teleosts, females can be induced to mature and ovulate by injection of a variety of gonadotropin preparations. Similarly, follicle-enclosed, fully grown postvitellogenic oocytes of several teleosts undergo GVBD *in vitro* when they are incubated with a number of gonadotropin preparations. However, denuded oocytes are incapable of responding to gonadotropins. Cyanoketone, a specific inhibitor of 3β-HSD, completely abolishes the maturational effects of gonadotropin and pregnenolone, but not of Δ^4 steroids such as progesterone or 17α,20β-DP (Young *et al.*, 1982; Iwamatsu and Onitake, 1983). These findings

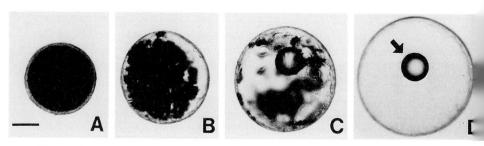

Fig. 8 17α,20β-Dihydroxy-4-pregnen-3-one-induced oocyte maturation of *Pagrus major*. Oocytes were incubated with the steroid (0.1 μg/ml) for (A) 0, (B) 8, (C) 10, and (D) 12 hr. Arrow indicates an oil drop. Bar: 200 μm.

4. Oocyte Growth and Maturation in Fish

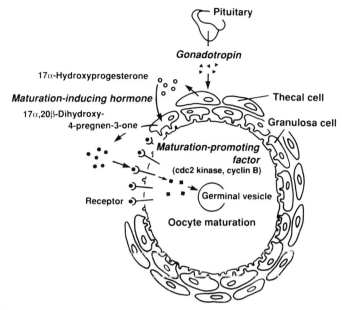

Fig. 9 Hormonal regulation of oocyte maturation in teleosts. Three major mediators—gonadotropin, maturation-inducing hormone, and maturation-promoting factor—are involved.

indicate that the action of gonadotropin in inducing oocyte maturation is dependent on the synthesis of a second Δ^4-steroidal mediator of meiotic maturation by ovarian follicles (Wasserman and Smith, 1978; Masui and Clarke, 1979; Nagahama, 1987b; Jalabert et al., 1991; Fig. 9).

B. Maturation-Inducing Hormone—Secondary Mediator of Oocyte Maturation

In a number of teleost species, C_{21} steroids have been shown to be potent initiators of GVBD *in vitro* and to be present at high levels in plasma of fish undergoing final oocyte maturation (Jalabert, 1976; Fostier et al., 1983; Goetz, 1983; Nagahama, 1987b; Scott and Canario, 1987; Jalabert et al., 1991). Among the C_{21} steroids, however, only two were identified as the naturally occurring maturation-inducing hormone (MIH) in fish: 17α,20β-dihydroxy-4-pregnen-3-one (17α,20β-DP) in amago salmon (Nagahama and Adachi, 1985) and 17α,20β,21-trihydroxy-4-pregnen-3-one (20β-dihydro-11-deoxycortisol, 20β-S) in the Atlantic croaker (*Micropogonias undulatus*) and the spotted seatrout (*Cynoscion nebulosus*) (Trant et al., 1986; Trant and Thomas, 1988,1989a,b; Thomas and Trant, 1989; see subsequent discussion). Testosterone, as well as other C_{19} steroids, induces

GVBD only at high concentrations. Estradiol-17β and other C_{18} steroids are generally not effective inducers of oocyte maturation in fish.

Defolliculated fully grown oocytes of amago salmon undergo GVBD when incubated in media in which postvitellogenic follicles or follicle layers isolated from postvitellogenic follicles have previously been incubated with a partially purified chum salmon gonadotropin. The MIH of amago salmon was isolated from these incubation media. Of the fractions obtained by reversed-phase high performance liquid chromatography (HPLC), only one fraction that had a retention time coinciding exactly with a 17α,20β-DP standard exhibited maturation-inducing activity as assessed by a homologous *in vitro* GVBD assay. The purity and final characterization of the residue in this fraction were further confirmed by a comparison with authentic 17α,20β-DP using thin layer chromatography (TLC) and mass spectroscopy (Nagahama and Adachi, 1985). 17α,20β-DP levels in the plasma are low in vitellogenic amago salmon but are strikingly elevated in mature and ovulated females. The increase in plasma 17α,20β-DP levels correlates well with a dramatic rise in plasma gonadotropin levels (Young *et al.*, 1983b). Of the nine pregnene derivatives tested, 17α,20β-DP is the most effective inducer of GVBD in four species of teleosts including amago salmon (Nagahama *et al.*, 1983). Collectively, these results indicate that 17α,20β-DP is the major naturally occurring MIH in amago salmon. Other investigations also suggest that 17α,20β-DP functions as the MIH in several other species of salmonids (Jalabert, 1976; Fostier *et al.*, 1983; Goetz, 1983; Nagahama, 1987a; Scott and Canario, 1987). Other studies have provided convincing evidence for the important role of 17α,20β-DP in inducing oocyte maturation in three non-salmonid teleosts (*Clarias batrachus*, Haider and Rao, 1992; *F. heteroclitus*, Petrino *et al.*, 1993; *O. latipes*, Fukada *et al.*, 1994).

More recently, the MIH of Atlantic croaker was isolated from media previously incubated with human chorionic gonadotropin (HCG) and pregnenolone for 8 hr (Trant *et al.*, 1986). Steroids were extracted from the medium and fractionated by HPLC and TLC. Fractions were bioassayed for their potency to induce GVBD in Atlantic croaker oocytes *in vitro*. 20β-S was identified as the predominant steroid product and the major MIH produced by the ovary of Atlantic croaker *in vitro*. The involvement of 20β-S in the induction of oocyte maturation was also demonstrated in spotted seatrout, a species closely related to Atlantic croaker (Thomas and Trant, 1989). However, 20β-S does not appear to be involved in the induction of oocyte maturation in salmonids, since there is no evidence for the presence of significant amounts of this steroid in the blood of female salmonids undergoing maturation or ovulation (Scott and Canario, 1987).

1. Two-Cell-Type Model for 17α,20β-DP Production

The identification of the MIH in salmonids as 17α,20β-DP permitted a study of the role of the follicle layer in the production of this hormone. 17α,20β-DP production by intact follicles at different stages of development shows that the capacity of

4. Oocyte Growth and Maturation in Fish

the follicles to respond to salmon gonadotropins by synthesizing this steroid is acquired immediately prior to the natural maturation period (Young et al., 1983b). Using various follicular preparations obtained from postvitellogenic amago salmon and rainbow trout, we examined the effects of salmon gonadotropins on 17α,20β-DP production. Salmon gonadotropins stimulate 17α,20β-DP production by intact follicles and cocultures of thecal and granulosa cell layers, but not by isolated thecal or granulosa cell layers alone, indicating that both cell layers are necessary for gonadotropin-stimulated 17α,20β-DP production (Young et al., 1986). Experiments examining the effects of conditioned media from incubates of one follicle layer on steroidogenesis by the other layer reveal that the thecal cell layer produces a steroid precursor that is metabolized to 17α,20β-DP in the granulosa cell layer. We further identified 17α-hydroxyprogesterone as the steroid precursor produced by thecal cell layers in response to gonadotropins. Salmon gonadotropins greatly stimulate 17α-hydroxyprogesterone production by thecal cell layers but not by granulosa cell layers. Levels of 17α-hydroxyprogesterone in media from intact follicles and coculture incubations peak at 12 hr and rapidly decrease concomitant with a rapid rise in 17α,20β-DP. Incubation of granulosa cell layers with exogenous 17α-hydroxyprogesterone results in elevated 17α,20β-DP levels, indicating the presence of 20β-hydroxysteroid dehydrogenase (20β-HSD), the key enzyme involved in the conversion of 17α-hydroxyprogesterone to 17α,20β-DP. Based on the results of these *in vitro* incubation studies, a two-cell-type model has been proposed for the first time in any vertebrate for the follicular production of MIH (Young et al., 1986; for review, see Nagahama, 1987a). In this model, the thecal cell layer produces 17α-hydroxyprogesterone, which traverses the basal lamina and is converted to 17α,20β-DP by the granulosa cell layer, where gonadotropins act to enhance the activity of 20β-HSD (see Fig. 5).

In the two-cell-type model for the production of 17α,20β-DP just described, gonadotropin has at least two sites of action: the thecal cell layer and the granulosa cell layer. This situation suggests the existence of gonadotropin receptors in these cell layers. Specific gonadotropin receptors have been demonstrated in both thecal and granulosa cell layers from amago salmon postvitellogenic follicles (Kanamori and Nagahama, 1988a). Scatchard analysis shows that both cell layers exhibit essentially the same characteristics of binding to labeled gonadotropins. The values of the dissociation constants (K_d) are in the range of 0.2–0.8 nM. These results suggest that amago salmon preovulatory follicles contain a single population of gonadotropin receptors in both thecal and granulosa cell layers. More recently, Yan et al. (1992) have provided evidence suggesting the existence of at least two types of gonadotropin receptor, designated type I and type II receptors, in coho salmon. The type I receptor binds both GTH-I and GTH-II, but with higher affinity for GTH-I, whereas the type II receptor binds GTH-II specifically and may have only limited interaction with GTH-I. The type I receptor exists in both thecal and granulosa cell layers, whereas the type II receptor exists only in granulosa cell layers.

The stimulatory effects of gonadotropins on 17α-hydroxyprogesterone production by thecal cell layers are all mimicked by forskolin, an adenylate cyclase activator, and cAMP (Kanamori and Nagahama, 1988b). Furthermore salmon gonadotropins and forskolin stimulate cAMP formation in thecal cell layers. A significant increase in 17α-hydroxyprogesterone production by thecal cell layers in response to salmon gonadotropins and forskolin can be detected within 1 hr after the onset of incubation. These findings are consistent with the view that gonadotropin acts on thecal cell layers to stimulate 17α-hydroxyprogesterone production through a mechanism involving receptor-mediated activation of adenylate cyclase and formation of cAMP. Gonadotropin appears to promote formation of one or more labile proteins required for the delivery of cholesterol to the mitochondrial cytochrome P-450 system in thecal cell layers (Y. Nagahama, unpublished results). Furthermore, gonadotropin- and cAMP-induced 17α-hydroxyprogesterone production by the thecal cell layer is abolished by cycloheximide and puromycin, but not by actinomycin D and cordycepin, suggesting that stimulatory effects of gonadotropin in the thecal cell layer require the synthesis of a new protein (see Nagahama, 1987a).

Unlike thecal cell layers, granulosa cell layers lack the side-chain cleavage enzyme systems. Granulosa cells obtained from fully grown follicles respond to salmon gonadotropins by enhancing 20β-HSD activity when exogenous 17α-hydroxyprogesterone is added to the incubation medium. The action of gonadotropin on 20β-HSD enhancement in granulosa cells can be mimicked by forskolin, by dbcAMP (but not dbcGMP), or by two phosphodiesterase inhibitors (Nagahama et al., 1985a). Furthermore, salmon gonadotropins and forskolin cause a rapid accumulation of cAMP with maximum levels at 30–60 min. These findings suggest that gonadotropin enhances the activity of granulosa cell 20β-HSD through a cAMP-dependent step. In addition to cAMP, calcium appears to play an important role in the gonadotropin regulation of steroidogenesis in amago salmon granulosa cells (Y. Nagahama, unpublished results). Calmodulin inhibitors such as trifluoroperazine (TFP), N-(6-aminohexyl)-1-naphthalenesulfonamide hydrochloride (W5), and N-(6-aminohexyl)-5-chloro-1-naphthalenesulfonamide (W7) prevent salmon gonadotropin-, cAMP-, and forskolin-stimulated 20β-HSD activation in granulosa cell layers in a dose-dependent manner.

When protein synthesis inhibitors (cycloheximide and puromycin) and RNA synthesis inhibitors (actinomycin D, cordycepin, and α-amanitin) are added to the incubation medium containing 17α-hydroxyprogesterone, stimulatory effects of salmon gonadotropins and dbcAMP on the production of 17α,20β-DP are significantly inhibited. Furthermore, the stimulatory effects of salmon gonadotropins and cAMP on 17α,20β-DP production by granulosa cell layers require at least 12 hr to be manifested. These results are consistent with the suggestion that one of the major actions of gonadotropin and cAMP on 20β-HSD activation in amago salmon granulosa cells is the enhancement of new RNA and protein synthesis (Nagahama et al., 1985b; Nagahama, 1987a). Our time course

4. Oocyte Growth and Maturation in Fish

studies further suggest that *de novo* synthesis of 20β-HSD *in vitro* in response to salmon gonadotropins and cAMP occurs, and consists of gene transcriptional events within the first 6 hr of exposure to gonadotropin and cAMP and translational events 6–9 hr after the exposure to gonadotropin and cAMP. Thus, these results suggest that gonadotropin causes the *de novo* synthesis of 20β-HSD in the amago salmon granulosa cell through a mechanism dependent on RNA synthesis. The induction of 20β-HSD activity by gonadotropin in amago salmon granulosa cells is a good example of differentiated functions expressed by target cells in response to peptide hormone stimulation.

2. Developmental Changes in the Capacity of Ovarian Follicles to Produce 17α,20β-DP in Response to Gonadotropin

Immediately prior to oocyte maturation, intact follicles develop the potential to respond to gonadotropin by secreting 17α,20β-DP. However, granulosa cell layers isolated from vitellogenic follicles about 2 mo prior to oocyte maturation also have the ability to respond to gonadotropin by enhancing the conversion of 17α-hydroxyprogesterone to 17α,20β-DP, although his ability is much lower than that detected in granulosa cell layers from fully grown postvitellogenic follicles. In contrast, only the thecal cell layers that are isolated from preovulatory follicles immediately prior to or during the maturation period possess the capacity to produce 17α-hydroxyprogesterone in response to gonadotropin. Thus, the acute increase in 17α,20β-DP production in intact follicles during oocyte maturation is largely dependent on the secretion of 17α-hydroxyprogesterone by thecal cell layers (Kanamori *et al.*, 1988).

In contrast to the constancy of the affinity of ovarian gonadotropin receptors in both follicular layers, progressive increases occur in the number of gonadotropin receptors in both thecal and granulosa cell layers during follicular development. The enhanced responsiveness of thecal (17α-hydroxyprogesterone production) and granulosa (20β-HSD activity) cell layers to gonadotropin may be related, in part, to the increased number of gonadotropin receptors (Kanamori and Nagahama, 1988a).

In addition to the cell surface receptor system, post-cAMP mechanisms appear to be involved in regulating the steroidogenic action of gonadotropin. Cyclic AMP is believed to be the intracellular mediator of gonadotropin-induced steroidogenesis in ovarian cells. If the increase in steroid production in mature follicle layers compared to developing follicles is due solely to an increase in the number of cell surface receptors, then both immature and mature follicle layers should have similar steroidogenic responses if the receptor–adenylate cyclase step is bypassed. This possibility was tested using exogenous cAMP and agents that raise intracellular cAMP levels (Kanamori and Nagahama, 1988b). The effects of these agents paralleled gonadotropin-induced steroidogenic responses of thecal and granulosa cell layers from both vitellogenic and postvitellogenic follicles. Furthermore, regardless of the stage of follicular development, gonadotropin treatment stimulated cAMP formation in both thecal and granulosa cell layers (Kanamori and Nagahama, 1988b). Clearly, mechanisms subsequent to cAMP generation must differentiate before the 20β-HSD system becomes gonadotropin sensitive. A similar argument can be made for the mechanisms governing 17α-hydroxyprogesterone secretion until the time of oocyte maturation. Changes in the properties of cAMP-dependent protein kinases, or changes in their substrates, might occur in the process of the steroidogenic shift in both thecal and granulosa cell layers.

3. Changes in mRNA Levels of Various Steroidogenic Enzymes in Rainbow Trout Ovarian Follicles during Oocyte Maturation

A distinct steroidogenic shift from estradiol-17β to 17α,20β-DP occurs in ovarian follicles of salmonid fish immediately prior to or during oocyte maturation (see Fig. 6). This shift may depend, in part, on the activities of the particular steroidogenic enzymes involved in the biosynthesis of these two steroids. Therefore, to understand the mechanisms of control responsible for the shift, it is important to examine the changes in the levels of steroidogenic enzymes and their transcripts during oocyte maturation.

As described earlier, we have already isolated cDNA clones encoding most of the steroidogenic enzymes responsible for the production of estradiol-17β and 17α,20β-DP from ovarian cDNA libraries of rainbow trout and medaka. However, genes encoding two hydroxysteroid dehydrogenases, 17β-HSD and 20β-HSD, have not yet been isolated. Unfortunately, unlike most of the ovarian steroidogenic enzymes, 20β-HSD cDNA has not been cloned in any animal species. Therefore, as a first step, we isolated a cDNA encoding pig testis 20β-HSD, since purified preparations of 20β-HSD are available only from pig (Nakajin et al., 1988). Using synthetic oligonucleotides deduced from the partially determined amino acid sequences, we have for the first time isolated and cloned cDNA encoding 20β-HSD from a pig testis cDNA library (Tanaka et al., 1992b). The cDNA contains an open reading frame predicted to encode 289 amino acid residues. Surprisingly, it has 85% amino acid homology with human carbonyl reductase. We are now using this cDNA clone to isolate cDNAs encoding rainbow trout 20β-HSD from a rainbow trout ovarian cDNA library.

As mentioned in the preceding section, the mRNA levels of P-450scc, 3β-HSD, and P-450c17 are barely detected in follicles during the midvitellogenic stage, and are abundant in follicles during the postvitellogenic stage and oocyte maturation (Sakai et al., 1992,1993; Takahashi et al., 1993). Our preliminary results indicate that forskolin-induced 17α,20β-DP production is accompanied by a dramatic decrease in P-450arom mRNA levels in granulosa cells isolated from postvitellogenic follicles. A 2- to 3-fold increase in P-450scc and 3β-HSD mRNAs and a slight decrease in P-450c17 mRNA are also observed during forskolin-induced 17α,20β-DP production (M. Tanaka and Y. Nagahama, unpublished results). Our preliminary Northern hybridization analysis using the pig 20β-HSD cDNA as a probe has revealed that 20β-HSD mRNA transcripts are low in granulosa cells prior to oocyte maturation, but markedly increase during naturally occurring and gonadotropin-induced oocyte maturation.

4. Surface Site of MIH Action

a. MIH Receptors. Previous *in vitro* studies in fish have shown that the mechanism of action involved in the steroid stimulation of GVBD has special characteristics not typical of the classical steroid mechanism of action. Steroid-induced oocyte maturation is prevented by puromycin, a translational inhibitor, but not by actinomycin D, a transcriptional inhibitor, indicating that the steroid mechanism of action in final maturation is nongenomic (Dettlaff and Skoblina, 1969; Jalabert, 1976; DeManno and Goetz, 1987). Investigators have also reported that cAMP, activators of adenylyl cyclase, and phosphodiesterase inhibitors also inhibit steroid-stimulated maturation *in vitro* in several species (Goetz and Hennessy, 1984; Jalabert and Finet, 1986; DeManno and Goetz, 1987; Finet *et al.*, 1988). These results suggest that an intracellular second messenger system such as a cAMP cascade is involved in the mechanism of action of MIH-induced oocyte maturation *in vitro*. We have also demonstrated that 17α,20β-DP is ineffective in inducing oocyte maturation when microinjected into the immature oocytes of goldfish, but that external application of the steroid is effective (Nagahama, 1987a). Collectively, these *in vitro* results suggest that the site of MIH action in inducing meiotic maturation in fish oocytes is at or near the oocyte surface. Researchers have shown that, in amphibians, progesterone apparently acts via a receptor on the oocyte surface plasma membrane and not through cytoplasmic or nuclear receptors (Maller, 1985).

More direct evidence for the existence of MIH receptors in oocyte plasma membranes has been obtained by binding studies using labeled MIH. We have identified and characterized specific binding of [^{3}H]17α,20β-DP to plasma membranes prepared from defolliculated oocytes of rainbow trout (Yoshikuni *et al.*, 1993). Binding is rapid and reaches equilibrium in 30 min. 17α,20β-DP strongly inhibits [^{3}H]17α,20β-DP binding in a competitive manner. Scatchard analysis reveals two different binding sites: a high affinity binding site with a K_d of 18 nM and a B_{max} of 0.2 pmol/mg protein and a low affinity binding site with a K_d of 0.5 μM and a B_{max} of 1 pmol/mg protein. Maneckjee *et al.* (1989) also reported, without full characterization, 17α,20β-DP binding activity in the zona radiata–oocyte membrane complex of rainbow trout and brook trout (*Salvelinus fontinalis*). We have also described a specific receptor for 17α,20β-DP in the oocyte cortices of the Japanese flounder (*Paralichthys olivaceus*) (M. Yoshikuni *et al.*, unpublished results). This 17α,20β-DP binding increases, reaching an equilibrium within 1 hr. Scatchard analysis reveals a single binding site with a K_d of 63 nM and a B$_{max}$ of 25 fmol/cortex. The 17α,20β-DP receptor is present in oocyte cortices from both postvitellogenic oocytes (500–600 μm in diameter) and ovulated eggs, but not in those from either mid vitellogenic oocytes (425–500 μm) or early vitellogenic oocytes (<425 μm).

A specific binding site for 20β-S exists in plasma membranes from spotted seatrout ovaries (Patino and Thomas, 1990b). The association of 20β-S to saturable binding sites is extremely swift, reaching a maximum after 5 min of incubation of 0°C. The 20β-S receptor has a high affinity (K_d = 1 nM), that is several orders of magnitude higher than that of progesterone receptors described for amphibian oocytes (Kostellow et al., 1982; Sadler and Maller, 1982). In agreement with this high affinity, the apparent in vitro ED$_{50}$ for 20β-S (approximately 0.1–1 nM) in the seatrout GVBD bioassay (Thomas and Trant, 1989) appears to be several orders of magnitude lower than the reported ED$_{50}$ for progesterone (100 nM) in the Xenopus bioassay (Sadler and Maller, 1982). The concentrations of 20β-S receptors in ovarian plasma membranes are three times higher in seatrout undergoing oocyte maturation than in vitellogenic females. This increase in receptor concentrations in fish collected during their natural spawning cycle is similar to that previously observed in seatrout undergoing final maturation following luteinizing hormone releasing hormone (LHRH) injections (Patino and Thomas, 1990c). These observations suggest that changes in the concentration of MIH receptors in the ovaries are of physiological importance during natural oocyte maturation.

b. Development of Oocyte Maturational Competence.

Studies with several teleost species such as kisu (*Sillago japonica*; Kobayashi et al., 1988), dragonet (*Repomucenus beniteguri*; Zhu et al., 1989), Atlantic croaker (Patino and Thomas, 1990a), and the Japanese flounder (M. Yoshikuni and Y. Nagahama, unpublished results) have demonstrated two distinct stages of final oocyte maturation: an early MIH-insensitive phase in which oocytes can mature *in vitro* with gonadotropin but not with MIH alone followed by an MIH-sensitive stage. The MIH-insensitive oocytes develop the ability to respond to MIH if they are previously primed with gonadotropin. Thomas and Patino (1991) examined the relationship between 20β-S receptor concentrations and the development of maturational competence using an ovarian incubation system. Treatment of spotted seatrout ovarian follicles with gonadotropin causes a twofold increase in 20β-S receptor concentrations; concomitantly, fully grown oocytes acquire the ability to mature in response to 20β-S. In contrast, 20β-S treatment does not induce an increase in receptor concentrations or the development of oocyte maturational competence. Similarly, *in vitro* treatment of Japanese flounder follicles with HCG also caused a threefold increase in 17α,20β-DP receptor concentrations. Moreover, this HCG-induced elevation in receptor concentrations is accompanied by the appearance of maturational competence in the follicle-enclosed oocytes (M. Yoshikuni et al., unpublished results). Collectively, these results suggest that a gonadotropin-induced increase in MIH receptor concentrations is responsible for the development of oocyte maturational competence in seatrout and in Japanese flounder, and is not mediated by increases in the production of MIH.

C. Maturation-Promoting Factor—Tertiary Mediator of Oocyte Maturation

The existence of $17\alpha,20\beta$-DP receptors at the oocyte plasma membrane suggests that there is a cytoplasmic factor that mediates the action of $17\alpha,20\beta$-DP. This factor, named maturation-promoting factor or metaphase-promoting factor (MPF), was first demonstrated in amphibian oocytes. In this study, cytoplasm taken from progesterone (a proposed MIH in amphibians)-treated oocytes induced GVBD and all subsequent maturational events when injected into recipient oocytes in the absence of progesterone (Masui and Markert, 1971; Smith and Ecker, 1971). Since then, MPF has been demonstrated in the maturing oocytes of a number of species.

1. Existence of MPF in Fish Eggs

This existence of MPF activity in fish oocytes was first demonstrated by the induction of GVBD in starfish (*Asterias pectinifera*) oocytes by microinjection of cytoplasm from mature unfertilized goldfish oocytes (Kishimoto, 1988). More recently MPF activity was extracted from goldfish oocytes matured by HCG treatment *in vivo* and naturally, and was injected into immature *Xenopus* oocytes (Yamashita *et al.*, 1992a). After centrifugation of mature goldfish oocytes at 100,000 g for 1 hr, the five major layers from centripetal to centrifugal pole were obtained: supernatant, green layer, yolk, chorion, and cortical alveoli. MPF activity can be detected in the supernatant and the green layer containing cellular organelles such as mitochondria and endoplasmic reticulum. MPF activity extracted from goldfish oocytes is also effective when injected into immature goldfish oocytes under conditions of inhibited protein synthesis. The MPF-injected oocytes undergo meiotic maturation much more rapidly than oocytes induced to mature *in vitro* by incubation with $17\alpha,20\beta$-DP. GVBD usually occurs at the center of the MPF-injected oocytes, because the movement of the GV to the animal pole does not take place. In contrast, GV migration always occurs in the oocytes matured naturally, *in vivo* by HCG, or *in vitro* by $17\alpha,20\beta$-DP.

MPF activity is also detected in extracts from oocytes matured *in vitro* by $17\alpha,20\beta$-DP, but is not present in extracts from either immature oocytes or activated eggs. MPF transfers were carried out between oocytes of goldfish and the frog *Xenopus laevis* (Yamashita *et al.*, 1992a; M. Yamashita, unpublished results). MPF from mature oocytes of either source induced maturation in oocytes of the other species. The induction of GVBD in *Xenopus* oocytes was, in fact, used as a test for MPF activity during goldfish MPF purification (see subsequent discussion). We further demonstrated that microinjection of extracts of pachytene microsporocytes of a higher plant (lily, *Lilium longiflorum*) induces GVBD and chromosome condensation in *Xenopus* oocytes (Yamaguchi *et al.*, 1991). These findings further support the notion that MPF is a common initiator of metaphase in a wide variety of meiotic cells. Other studies on MPF in diverse cells further established the generality of MPF as a metaphase-promoting factor in mitotic as well as meiotic cells (Sunkara *et al.*, 1979; Kishimoto *et al.*, 1982, 1984).

Changes in MPF activity during the 17α,20β-DP-induced maturation of goldfish oocytes were assessed by measuring histone H1 kinase activity (Fig. 10). Although a weak H1 kinase activity is detectable in immature GV-stage goldfish oocytes, it increases before GVBD and reaches a peak at metaphase I. Activity then decreases during anaphase and telophase I, but is found to be at maximal levels again in metaphase II oocytes (Yamashita *et al.*, 1992b). The H1 kinase activity decreases immediately after egg activation (fertilization).

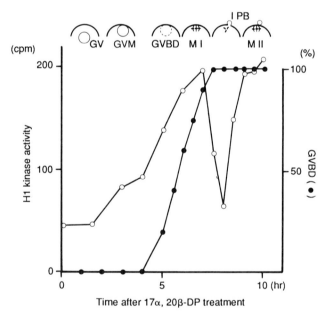

Fig. 10 Oscillation of MPF activity during goldfish oocyte maturation induced by 17α,20β-dihydroxy-4-pregnen-3-one (17α,20β-DP). Maturation-promoting factor (MPF) activity was assessed by measuring histone H1 kinase activity (○) using histone H1 as a substrate. Schematic drawings on the figure indicate the maturational stage of the oocytes. Germinal vesicle breakdown (GVBD, ●) is also shown. GV, Germinal vesicle; GVM, germinal vesicle migration; MI, first meiotic metaphase; MII, second meiotic metaphase; IPB, first polar body.

2. Purification and Characterization of Fish MPF and Histone H1 Kinase

Despite the biological importance of MPF, little progress was made in its chemical analysis until Lohka et al. (1988) were able to purify MPF from mature oocytes of *Xenopus* as a 200-kDa complex containing 32- and 45-kDa proteins. Later, MPF was also purified from mature oocytes of starfish (Labbé et al., 1989a,b). MPF of *Xenopus* and starfish consists of two components (Gautier et al., 1988,1990; Lohka et al., 1988; Labbé et al., 1989a,b). A catalytic subunit of MPF is the homolog of the serine/threonine protein kinase encoded by the fission yeast (*Schizosaccharomyces pombe*) $cdc2^+$ gene (cdc2 kinase; Simanis and Nurse, 1986). This kinase can use histone H1 as an exogenous substrate (Brizuela et al., 1989). A regulatory subunit of MPF is cyclin (Gautier et al., 1990), which was first discovered, independently of MPF, in the early embryos of marine invertebrates (Rosenthal et al., 1980; Evans et al., 1983).

3. Purification of Carp MPF

MPF purification was carried out using carp (*C. carpio*), a species closely related to goldfish. Carp is an excellent source of MPF since a large number of eggs arrested in a metaphase II, 300–1000 ml/fish, can be obtained easily by an *in vivo* injection of HCG into gravid females. MPF was purified from the 100,000 g supernatant of crushed, naturally spawned carp oocytes using four chromatography columns, including a p13^{suc1}-affinity Sepharose column (Yamashita et al., 1992a). MPF activity was assayed by injecting the sample into fully grown immature *Xenopus* oocytes in the presence of cycloheximide, a protein synthesis inhibitor. H1 kinase activity was also determined using histone H1 and a synthetic peptide (SP-peptide, KKAAKSPKKAKK) as exogenous substrates (Yamashita et al., 1992b). MPF activity comigrates with histone H1 kinase activity throughout purification. Analyses by SDS-PAGE showed that the most active fraction after Mono S contained four proteins with molecular masses of 33, 34, 46, and 48 kDa. When the active fractions after Mono S were applied to Superose 12, MPF and kinase activities coeluted as a single peak with an apparent molecular mass of about 100 kDa, indicating that these four proteins form complex(es) of about 100 kDa in their native form.

The 34-, 46-, and 48-kDa proteins correspond well to MPF and H1 kinase activities, but the 33-kDa protein peak seems to be different from the peaks of MPF and H1 kinase activities. The 46- and 48-kDa proteins are labeled when [γ-^{32}P]ATP is applied in the fractions, indicating that these proteins are among the endogenous substrates for the kinase. The final preparation of MPF was purified over 1,000-fold, with a recovery of about 1%. The final preparation was also purified 5,000-fold with a recovery of 5% when histone H1 was used for the kinase assay, and 10,000-fold with a recovery of 7% when SP-peptide was used.

4. Characterization of Carp MPF

Researchers have suggested, from the comparison of the components of purified carp MPF with those of *Xenopus* MPF, that among the four proteins found in carp MPF, the 34-kDa protein and the 46- and 48-kDa proteins correspond to cdc2 kinase and cyclin B of *Xenopus* MPF, respectively. To characterize these four proteins fully, we isolated cDNA clones encoding fish homologs of MPF-related proteins, including cdc2 kinase, cdk2 kinase, cyclin A, and cyclin B, from cDNA libraries of goldfish oocytes.

a. cdc2 Kinase. The isolated goldfish cdc2 kinase cDNA clone has an insert of 1284 bp containing a poly(A)+ tail with an open reading frame encoding 302 amino acids. This clone encodes a PSTA*V*R sequence instead of a PSTA*I*R sequence motif (EGVPSTA/REISLLKE), a hallmark of cdc2, cdk2, and cdk3 (Kajiura *et al.*, 1993). The predicted molecular weight of the protein encoded by this gene is 34,499. This clone has higher homology with cdc2 (95% for *Xenopus*, 85% for human, and 84% for mouse) than cdk2 (67% for goldfish, 66% for *Xenopus*, and 67% for human) at the amino acid sequence level. An anti-goldfish cdc2 kinase monoclonal antibody was raised against a peptide (CPYFD-DLDKSTLPASNLKI) that corresponds to the C-terminal sequence of goldfish cdc2 kinase cDNA, with an additional cysteine in the N terminus (Kajiura *et al.*, 1993). A monoclonal antibody against the PSTAIR sequence of cdc2 kinase was also raised (Yamashita *et al.*, 1991). This antibody recognizes 31-kDa–35-kDa proteins by immunoblotting in all species examined to date (Fig. 11). The proteins recognized by the anti-PSTAIR antibody are probably either cdc2 kinase itself or proteins that are highly homologous to cdc2 kinase in the given species since, in all species studied to date, all proteins are precipitated with p13^{suc1}, the fission yeast *suc1*+ gene product that binds cdc2 kinase with high specificity.

b. cdk2 Kinase. A cdk2 kinase cDNA clone was isolated from a cDNA library constructed from mature goldfish oocytes (Hirai *et al.*, 1992b). The isolated clone contains an open reading frame encoding 298 amino acids. This gene encodes a putative protein kinase containing the PSTAIR sequence and four tryptophan residues conserved in the cdc2 family. The predicted molecular weight of the protein encoded by this gene is 33,998. The amino acid sequence of this clone has significant homology (64–69%) with the cdc2 of other species, but much higher homology is found with the cdk2 of *Xenopus* (88%) and human (90%). A monoclonal antibody against the C-terminal sequence of goldfish cdk2 was raised.

c. Cyclin A. A cyclin A cDNA clone was isolated from a cDNA library constructed from immature goldfish oocytes. The isolated clone contains an open reading frame encoding 391 amino acids. The amino acid sequence has signifi-

4. Oocyte Growth and Maturation in Fish   131

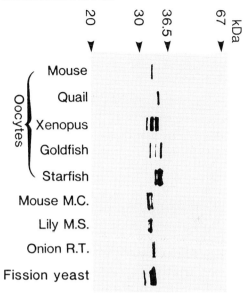

Fig. 11 Immunoblotting with a monoclonal antibody against an anti-PSTAIR. Mouse M. C., mouse myeloma cell; Lily M. S., lily microsporocyte; Onion R. T., onion root tip cell.

cant homology with cyclin A of other species (43–65%) rather than with cyclin B (35–37%). A monoclonal antibody against *Escherichia coli*-produced goldfish cyclin A was raised (Katsu *et al.*, 1993).

d. Cyclin B. Two different cyclin B clones were isolated from a cDNA library constructed from mature goldfish oocytes (Hirai *et al.*, 1992a). Sequence comparison reveals that these two clones are highly homologous (85%) and are found to be similar to *Xenopus* cyclin B1. One clone encodes 397 amino acids covering the entire coding region. The predicted molecular weight of the protein encoded by this gene is 44,763. The goldfish cyclin B contains a "cyclin box," a hallmark of cyclins (Pines, 1991), and resembles cyclin B more closely than cyclin A. At the amino acid level, the goldfish cyclin B cDNA shows homology of 66% with *Xenopus* cyclin B1 and 50% with *Xenopus* cyclin B2, whereas it shows 33% homology with *Xenopus* cyclin A. An anti-cyclin B monoclonal antibody was raised against *E. coli*-produced full-length goldfish cyclin B (Katsu *et al.*, 1993).

e. MPF Characterization. The monoclonal antibodies just described were used to characterize the purified carp MPF, which contains 33-, 34-, 46-, and 48-kDa proteins. Both the 33- and the 34-kDa protein are recognized by the anti-PSTAIR antibody, indicating that they are the cyclin-dependent kinases. The anti-cdc2 C-terminal antibody reacts with the 34-kDa but not the 33-kDa protein. The latter is recognized by the anti-cdk2 C-terminal antibody (Hirai et al., 1992b). The 46- and 48-kDa proteins are recognized by the anti-full-length cyclin B antibody, but not by an anti-cyclin A antibody (Yamashita et al., 1992a; Katsu et al., 1993). From these findings, we conclude that carp MPF is a complex of cdc2 kinase (34 kDa) and cyclin B (46 and 48 kDa; Fig. 11).

5. Changes in cdc2 Kinase and Cyclin B Protein Levels during Oocyte Maturation

Researchers have reported that in immature oocytes of *Xenopus* and starfish, cdc2 kinase forms a complex with cyclin B as pre-MPF. In contrast, there is no detectable cyclin B in immature goldfish oocytes, suggesting that cdc2 kinase in immature goldfish oocytes is monomeric. This suggestion was confirmed by the consecutive fractionation of immature and mature oocyte extracts eluted from gel filtration columns, followed by immunoblotting with anti-cdc2 C-terminal and anti-cyclin B antibodies. When immature oocyte extracts were applied on the gel filtration column, cdc2 kinase was eluted as a single peak at the monomeric position around 35 kDa. However, cdc2 kinase in mature oocyte extracts eluted as two peaks at around 100 kDa and 35 kDa. The 34-kDa cdc2 kinase was present only in the first peak at 100 kDa, coexisting with cyclin B. The vast majority of the 35-kDa cdc2 kinase in mature oocytes migrated at the monomeric position. These results demonstrate that most, if not all, cdc2 kinase (inactive 35-kDa form) in immature oocytes is monomeric and that, when oocytes mature, a part of cdc2 kinase forms a complex with cyclin B and is activated (active 34-kDa form). A minor fraction of the 35-kDa cdc2 kinase is also detectable at 100 kDa; therefore not all cdc2 kinase that binds to cyclin B is activated. Collectively, these data suggest that inactive 35-kDa cdc2 kinase binds to cyclin B first and is activated subsequently; this activation is associated with an electrophoretic mobility shift from 35 to 34 kDa.

To investigate further the cdc2 kinase and cyclin B protein levels during oocyte maturation, oocyte extracts were precipitated at various times after the addition of 17α,20β-DP with either p13^{suc1} or anti-cyclin B antibody, and were immunoblotted with anti-cdc2 kinase and anti-cyclin B antibodies. As described earlier, immature oocytes contain the 35-kDa inactive cdc2 kinase but not cyclin B and mature oocytes contain both the 35-kDa inactive and the 34-kDa active cdc2 kinases, as well as cyclin B. The appearance of the 34-kDa active cdc2 kinase coincides with the appearance of cyclin B just before GVBD. Anti-cyclin B immunoblots of p13^{suc1} precipitates and anti-cdc2 kinase immunoblots of anti-cyclin B immunoprecipitates show that the binding of cdc2 kinase and cyclin B coincides with the appearance of cyclin B and the 34-kDa active cdc2 kinase. The cyclin B that appears during oocyte maturation can be labeled with [^{35}S]methionine (Hirai et al., 1992a), demonstrating *de novo* synthesis during

4. Oocyte Growth and Maturation in Fish   133

oocyte maturation. On the other hand, anti-cyclin B immunoprecipitates from mature oocyte extracts sometimes contain the 35-kDa inactive cdc2 kinase (Hirai et al., 1992a). This 35-kDa cdc2 kinase, as well as the 34-kDa form, can bind to cyclin B in a cell-free system (M. Yamashita et al., unpublished results). Therefore, the 35-kDa inactive cdc2 kinase most likely binds to *de novo* synthesized cyclin B at first, and then is rapidly converted into the 34-kDa active form.

Using the anti-cyclin A antibody, we also examined changes in cyclin A protein levels in goldfish oocytes during 17α,20β-DP-induced meiotic maturation. In contrast to cyclin B, cyclin A is already present in immature oocytes, exhibiting no measurable differences during maturation (Y. Katsu et al., unpublished results). This situation differs from that in *Xenopus*, in which cyclin A is absent but cyclin B is present in immature oocytes.

6. Mechanism of MPF Activation

The preceding results indicate that the appearance of cyclin B is required and is sufficient for inducing oocyte maturation in goldfish. To confirm that the appearance of cyclin B is sufficient for inducing oocyte maturation, *E. coli*-produced full-length goldfish cyclin B protein was injected into immature oocytes. Even under conditions of inhibited protein synthesis, injected cyclin B induces oocyte maturation within 1 hr after injection in a dose-dependent manner. Injection of 1 ng cyclin B fully induces GVBD in the recipient oocytes. The concentration of cyclin B within the injected oocyte is estimated to be 2 µg/ml, which is about equal to the cyclin B concentration in mature oocytes.

MPF activation can also be induced when cyclin B protein is introduced into immature oocyte extracts (which contain the 35-kDa inactive cdc2 kinase but not cyclin B), occurring in an almost "all or nothing" manner. The threshold concentration of cyclin B for inducing the activation is around 2 µg/ml, equivalent to that required to induce oocyte maturation by injection. These results demonstrate that the presence of 2 µg/ml cyclin B, corresponding to the concentration in mature oocytes, is sufficient for inducing oocyte maturation.

Introduction of *E. coli*-produced cyclin B into immature oocyte extracts also induces the activation of cdc2 kinase concurrent with the change in apparent molecular weight from 35 to 34 kDa, as found in oocytes matured with 17α,20 β-DP. Phosphoamino acid analysis shows that threonine phosphorylation of the 34-kDa cdc2 kinase and serine phosphorylation of cyclin B are associated with activation. The same phosphorylation is also found in oocytes matured using 17α,20β-DP. Cyclin B-induced cdc2 activation is not observed when threonine phosphorylation of cdc2 kinase and serine phosphorylation of cyclin B are inhibited by protein kinase inhibitors, although the binding of the 35-kDa cdc2 kinase to cyclin B occurs even in the presence of the inhibitors. On the other hand, cdc2 kinase can be activated by mutant cyclins that undergo no serine phosphorylation during activation. These results suggest that threonine phosphorylation of cdc2 kinase, but not serine phosphorylation of cyclin B, is required for cdc2 kinase activation (M. Yamashita et al., unpublished results).

To determine the site of threonine phosphorylation on cdc2 kinase that is responsible for cdc2 kinase activation, peptide map analyses were carried out using ^{32}P-labeled 34-kDa cdc2 kinase digested with trypsin and V8 protease. A peptide (VYTHE) that is phosphorylated at a single threonine was also synthesized. VYTHE is a sequence that is predicted to occur after digestion of goldfish cdc2 kinase with trypsin and V8 protease. After digestion, only a single ^{32}P-labeled peptide fragment was obtained. The chromatographic mobility of this fragment corresponds exactly to that of the synthetic VYTHE. These results indicate that the site of threonine phosphorylation on active cdc2 kinase of goldfish is residue Thr-161. To confirm that phosphorylation at Thr-161 is required for cdc2 kinase activity, threonine at position 161 was replaced with alanine by site-directed mutagenesis of the goldfish cdc2 kinase. No mobility shift from 35 to 34 kDa occurs when the mutant protein is added to goldfish immature oocyte extracts in the presence of a bacterially expressed goldfish cyclin B. Furthermore, a double mutant, T14AY15F cdc2, was also constructed in which Thr-14 was changed to alanine and Tyr-15 was changed to phenylalanine. In this mutant, as in the wild-type cdc2 kinase, the mobility shift from 35 to 34 kDa accompanies cdc2 kinase activation. Collectively, these findings indicate that the shift of cdc2 kinase from the inactive 35-kDa form to the active 34-kDa form is absolutely dependent on Thr-161 phosphorylation (Fig. 12).

We have isolated a cDNA clone encoding a goldfish homolog of p40^{MO15}, the catalytic subunit of a protein kinase that has been shown to activate cdc2 kinase through phosphorylation of Thr-161, from a goldfish oocyte cDNA library (Onoe et al., 1993). This clone contains an open reading frame encoding a protein of 344 amino acid residues. The overall amino acid sequence of the goldfish p40^{MO15} homolog exhibits 83%, 42%, and 43% identity with Xenopus p40^{MO15} (Shuttleworth et al., 1990), goldfish cdc2 kinase (Kajiura et al., 1993), and goldfish cdk2 kinase (Hirai et al., 1992b), respectively. Northern and Western blot analyses reveal that both p40^{MO15} mRNA and protein are already present in goldfish immature oocytes and do not appear to exhibit any measurable changes during hormonally induced maturation (Onoe et al., 1993; S. Onoe, unpublished results). Thus, p40^{MO15} may be regulated by subunit association and/or protein phosphorylation. The substrate preference of p40^{MO15} also remains to be determined.

In Xenopus oocytes, dephosphorylation of threonine (Thr-14) and tyrosine (Tyr-15) is required for cdc2 kinase activation (Maller, 1991). However, it is doubtful that Tyr-15 dephosphorylation of cdc2 kinase is a prerequisite for its activation during goldfish oocyte maturation, since the activation is not inhibited by vanadate, a protein phosphatase inhibitor commonly used for inhibiting the cdc25 activity that dephosphorylates Tyr-15 of cdc2 kinase, thereby inhibiting its

4. Oocyte Growth and Maturation in Fish 135

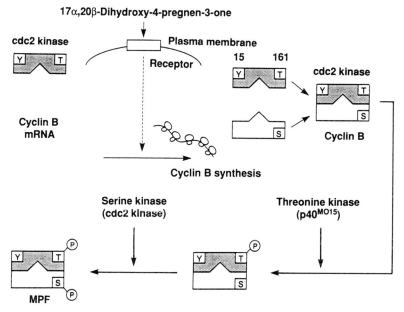

Fig. 12. Current model of the formation and activation of maturation-promoting factor (MPF) during fish oocyte maturation. P, Phosphorylation; S, serine; T, threonine; Y, tyrosine (see text for details).

activation. Furthermore, anti-phosphotyrosine antibodies do not cross-react with the 35-kDa cdc2 kinase nor with the 34-kDa form that binds to cyclin B. Also, an antibody that recognizes the dephosphorylated form of Thr-14 cross reacts with both the 34- and the 35-kDa cdc2 kinase. Therefore, the 35-kDa cdc2 kinase found in goldfish immature oocytes may already be dephoshorylated on Thr-14 and Tyr-15, and threonine phosphorylation (Thr-161) may be the only step required for activation. These findings strongly suggest that $17\alpha,20\beta$-DP induces oocytes to synthesize cyclin B, which in turn activates pre-existing 35-kDa cdc2 kinase through Thr-161 phosphorylation, producing the 34-kDa active cdc2 kinase (see Fig. 11). These mechanisms of MPF activation in fish apparently differ from those in *Xenopus* and starfish, in which cyclin B is present in immature oocytes and forms a complex with cdc2 kinase (pre-MPF).

7. Mechanism of MPF Inactivation

At present, less is known about exit from mitosis than about entry into mitosis. In *Xenopus* oocytes, MPF activity has been reported to decrease rapidly after fertilization (egg activation) through a mechanism that involves the degradation of the cyclin B subunit (Murray *et al.*, 1989; Nurse, 1990). In goldfish oocytes, MPF activity also declines within 3 min after egg activation, coinciding with the disappearance of cyclin B proteins (T. Tokumoto, unpublished results). Cyclin B degradation must be a highly selective process, since few other proteins are degraded only at this time. However, the mechanisms of cyclin degradation are poorly understood. Researchers have suggested that the ubiquitin pathway may be involved in cyclin B proteolysis (Glotzer et al., 1991). Since active proteasome (26S proteasome) is thought to be involved in the ATP/ubiquitin-dependent proteolysis of target proteins (Hershko and Ciechanover, 1982), proteasomes may be involved in the degradation of cyclin B. A newly developed purification procedure involving five steps of chromatography was used to purify 26S proteasome from goldfish ovarian homogenates. The purified proteasome has chymotrypsin-, trypsin-, and V8 protease-like activities. The enzyme exhibits two bands on native PAGE. Electrophoresis and Western blot analyses show that the enzyme consists of at least 15 protein components ranging in molecular mass from 35.5 to 140 kDa, as well as the multiple subunits of the 20S proteasome ranging in molecular mass from 23.5 to 31.5 kDa. The molecular mass of the 26S proteasome is estimated to be 1200 kDa. The availability of E.coli-produced goldfish cyclin B and purified goldfish 26S proteasome has made it possible for the first time to investigate the role of proteasomes in the regulation of cyclin B degradation. Researchers found that purified 26S proteasome can digest the wild-type cyclin B (cyclin B Δ0) in vitro, producing an intermediate cyclin B (42 kDa). In contrast, cyclin B mutants lacking the first 42, 68, and 96 N-terminal amino acids (cyclin B Δ42, Δ68, and Δ96, respectively) are not digested by the proteasome, suggesting that the N-terminal amino acids are necessary for cyclin B degradation. Amino acid sequence analysis of the 42-kDa intermediate cyclin B reveals that the 26S proteasome cleaves the C-terminal peptide bond of Lys-57. An intermediate cyclin B similar to that found in vitro appears transiently in activated goldfish eggs immediately before the complete disappearance of cyclin B (Fig. 13). Finally, goldfish cyclin B degradation in a Xenopus cycling

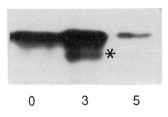

0   3   5

Minutes after egg activation

Fig. 13 Degradation of cyclin B after egg activation in goldfish. Egg activation was induced by immersing goldfish ovulated eggs in water. At the indicated time (min), activated eggs were homogenized in a buffer. Proteins were separated by SDS-PAGE and analyzed by Western blotting with anti-goldfish cyclin B antibody. Note that an intermediate cyclin B protein (asterisk) is seen 3 min after egg activation.

extract (Murray et al., 1989) is prevented when 26S proteasome is depleted from the extract with an anti-goldfish proteasome antibody. This is the first evidence for the crucial role of 26S proteasome in cyclin B degradation. The first cut of cyclin B by 26S proteasome may be crucial to cyclin B recognition by the ubiquitin-conjugating system. An understanding of this recognition step will be required for a complete understanding of the mechanisms regulating cyclin B degradation that lead to the completion of mitosis and meiosis.

VI. Summary

This chapter has briefly reviewed the current status of investigations on the hormonal regulation of oocyte growth and maturation in fish (see Figs. 4 and 9). Pituitary gonadotropins are of primary importance in triggering these processes in fish oocytes. In both cases, however, the actions of gonadotropins are not direct, but are mediated by the follicular production of steroidal mediators, estradiol-17β (oocyte growth) and 17α,20β-DP or 20β-S (oocyte maturation). Investigators have established that both estradiol-17β and 17α,20β-DP are biosynthesized by salmonid ovarian follicles via an interaction of two cell layers, the thecal and granulosa cell layers (two-cell-type model). The granulosa cell layers are the sites of production of these two steroidal mediators, but their production depends on the provision of precursor steroids by the thecal cell layers. A distinct steroidogenic shift from estradiol-17β to 17α,20β-DP, occurring in salmonid ovarian follicles immediately prior to oocyte maturation, is a prerequisite for the growing oocytes to enter the maturation stage, and requires a complex and integrated network of gene regulation involving cell specificity, hormonal regulation, and developmental patterning. The cDNAs for most of the steroidogenic enzymes responsible for estradiol-17β and 17α,20β-DP biosynthesis have been cloned from rainbow trout ovaries. Our next task is to determine how gonadotropin and other factors act on ovarian follicle cells to turn the expression of these specific genes on and off at specific times during oocyte growth and maturation. Increasing evidence now suggests that a variety of neuromodulatory, autocrine, and paracrine factors may also be involved in the regulation of steroidogenesis in fish ovarian follicles. Molecular biological technologies should be applied to identify these substances.

Of considerable interest is the finding that MIH, unlike most steroid hormones, acts on its receptors at the surface of oocytes. Further studies of the association of the MIH–MIH receptor complex with a G_i protein, probably resulting in the inactivation of adenylate cyclase, should lead to a discovery of a new mechanism of steroid hormone action. The early steps following MIH action involve the formation of the major cytoplasmic mediator of MIH, MPF. Fish MPF, like that of *Xenopus* and starfish, consists of two components: cdc2 kinase and cyclin B. Nevertheless, the mechanism of MIH-induced MPF activation in fish oocytes differs from that in *Xenopus* and starfish because the appearance of cyclin B protein is a crucial step for 17α,20β-DP-induced oocyte maturation in fish. Further work should focus on determining the mechanism of *de novo* synthesis of cyclin B and Thr-161 phosphorylation, both of which are induced by 17α,20β-DP. A recent finding highlighting the crucial role of the 26S proteasome in cyclin B degradation in goldfish oocytes generates new perspectives for understanding the mechanism of MPF inactivation. We believe that ongoing studies using fish oocytes as experimental models hold much promise for understanding the basis of hormonal regulation of oogenesis.

Acknowledgments

The authors thank the many wonderful colleagues who have contributed to the work described herein, especially G. Young, M. Nakamura, H. Kagawa, S. Adachi, M. Tanaka, N. Sakai, T. Hirai, M. Matsuyama, M. Takahashi, H. Kajiura, and S. Fukada. A special thanks to N. Nagahama and K. Noda for their assistance in the preparation of this manuscript. This work was supported in part by a grant-in-aid from the Ministry of Education, Science, and Culture in Japan (02102010 to Y. N.).

References

Adachi, S., Kajiura, H., Kagawa, H., and Nagahama, Y. (1990). Identification of aromatizable androgens secreted by isolated thecal cell layers from vitellogenic ovarian follicles of a teleost, amago salmon (*Oncorhynchus rhodurus*). *Biomed. Res.* **11**, 359–363.

Brizuela, L., Draetta, G., and Beach, D. (1989). Activation of human cdc2 protein as a histone H1 kinase is associated with complex formation with the p62 subunit. *Proc. Natl. Acad. Sci. USA* **86**, 4362–4366.

Burzawa-Gerard, E. (1982). Chemical data on the pituitary gonadotropins and their implication to evolution. *Can. J. Fish Aquat. Sci.* **39**, 80–91.

Chan, S. L., Tan, C. H., Pang, K. M., and Lam, T. M. (1991). Vitellogenin purification and development of assay for vitellogenin receptor in oocyte membranes of the tilapia (*Oreochromis niloticus*, Linnaeus 1766). *J. Exp. Zool.* **257**, 96–109.

DeManno, D. A., and Goetz, F. W. (1987). Steroid-induced final maturation in brook trout (*Salvelinus fontinalis*) oocytes in vitro: The effects of forskolin and phosphodiesterase inhibitors. *Biol. Reprod.* **36**, 1321–1332.

Dettlaff, T. A., and Skoblina, M. N. (1969). The role of germinal vesicle in the process of oocyte maturation in Anura and Acipenseridae. *Ann. Embryol. Morphog. Suppl.* **1**, 133–151.

Dodd, J. M. (1987). The ovary. *In* "Vertebrate Endocrinology: Fundamentals and Biochemical Implications" (P. K. T. Pang and M. P. Schreibman, eds.), pp. 351–397. Academic Press, New York.

Dorrington, J. H., and Armstrong, D. T. (1979). Effects of FSH on gonadal function. *Rec. Prog. Horm. Res.* **35**, 301–332.

Evans, T., Rosenthal, E. T., Youngblom, J., Distel, D., and Hunt, T. (1983). Cyclin: A protein specified by maternal mRNA in sea urchin eggs that is destroyed at each cleavage division. *Cell* **33**, 389–396.

Finet, B., Jalabert, B., and Garg, S. K. (1988). Effect of defolliculation and 17α-hydroxy,20β-dihydroxyprogesterone on cyclic AMP level in full-grown oocytes of the rainbow trout *Salmo gairdneri*. *Gamete Res.* **19**, 241–252.

Fontain, Y. A., and Dufour, S. (1987). Current status of LH-FSH-like gonadotropin in fish. *In* "Reproductive Physiology of Fish" (D. R. Idler, L. W. Crim, and J. W. Walsh, eds.), pp. 48–56. Memorial University Press, St. Johns.

Fostier, A., Jalabert, B., Billard, R., Breton, B., and Zohar, Y. (1983). The gonadal steroidogenesis. *In* "Fish Physiology" (W. S. Hoar, D. J. Randall, and E. M. Donaldson, eds.), Vol. IX A, pp. 277–372. Academic Press, New York.

Fukada, S., Sakai, N., Adachi, S., and Nagahama, Y. (1994). Steroidogenesis in the ovarian follicle of medaka (*Oryzias latipes*, a daily spawner) during oocyte maturation. *Dev. Growth Diff.* **36**, 81–88.

Gautier, L., Norbury, S., Lohka, M., Nurse, P., and Maller, J. (1988). Purified maturation-promoting factor contains the product of a *Xenopus* homolog of the fission yeast cell cycle control gene cdc2$^+$. *Cell* **54**, 433–439.

Gautier, L., Minshull, J., Lohka, M., Glotzer, M., Hunt, T., and Maller, J. L. (1990). Cyclin is a component of maturation-promoting factor from *Xenopus*. *Cell* **60**, 487–494.

Glotzer, M. N., Murray, A. W., and Kirshner, M. W. (1991). Cyclin is degraded by the ubiquitin pathway. *Nature* **349**, 132–138.

Goetz, F. W. (1983). Hormonal control of oocyte final maturation and ovulation in fishes. *In* "Fish Physiology" (W. S. Hoar, D. J. Randall, and E. M. Donaldson, eds.), Vol. IX A, pp. 117–170. Academic Press, New York.

Goetz, F. W., and Hennessy, T. (1984). The *in vitro* effects of phosphodiesterase inhibitors on 17α,20β-dihydroxy-4-pregnen 3 one-induced germinal vesicle breakdown in brook trout *Salvelinus fontinalis* oocytes. *Comp. Biochem. Physiol.* **77A**, 785–786.

Habibi, H., and Lessman, C. A. (1986). Effect of demecolcine (Colcemid) on goldfish oocyte meiosis *in vitro*. *Gamete Res.* **13**, 103–114.

Haider, S., and Rao, N. V. (1992). Oocyte maturation in *Clarias batrachus*. III. Purification and characterization of maturation-inducing steroid. *Fish Physiol. Biochem.* **9**, 505–512.

Hamazaki, T. S., Iuchi, I., and Yamagami, K. (1987a). Production of a "spawning female-specific substance" in hepatic cells and its accumulation in the ascites of the estrogen-treated adult fish, *Oryzias latipes*. *J. Exp. Zool.* **241**, 325–332.

Hamazaki, T. S., Iuchi, I., and Yamagami, K. (1987b). Isolation and partial characterization of a "spawning female specific substance" in the teleost, *Oryzias latipes*. *J. Exp. Zool.* **242**, 343–349.

Hamazaki, T. S., Nagahama, Y., Iuchi, I., and Yamagami, K. (1989). A glycoprotein from the liver constitutes the inner layer of the egg envelope (zona pellucida) of the fish, *Oryzias latipes*. *Develop. Biol.* **133**, 101–110.

Hershko, A., and Ciechanover, A. (1982). Mechanisms of intracellular protein breakdown. *Annu. Rev. Biochem.* **51**, 335–364.

Hirai, T., Yamashita, M., Yoshikuni, M., Lou, Y.-H., and Nagahama, Y. (1992a). Cyclin B in fish oocytes: Its cDNA and amino acid sequences, appearance during maturation, and induction of p34^{cdc2} activation. *Mol. Reprod. Dev.* **33**, 131–140.

Hirai, T., Yamashita, M., Yoshikuni, M., Tokumoto, T., Kajiura, H., Sakai, N., and Nagahama, Y. (1992b). Isolation and characterization of goldfish cdk2, a cognate variant of the cell cycle regulator cdc2. *Dev. Biol.* **152**, 113–120.

Ho, S. M. (1991). Vitellogenesis. *In* "Vertebrate Endocrinology: Fundamentals and Biomedical Implications" (P. K. T. Pang and M. P. Schreibman, eds.), Vol. 4, Part A, pp. 23–90. Academic Press, New York.

Huang, E. S., and Nalvandov, A. V. (1979). Synthesis of sex steroids by cellular components of chicken. *Biol. Reprod.* **20**, 454–461.

Hyllner, S. J., Oppen-Berntsen, D. O., Helvik, J. V., Walther, B. T., and Haux, C. (1991). Oestradiol-17β induces the major vitelline envelope proteins in both sexes in teleosts. *J. Endocrinol.* **131**, 229–236.

Idler, D. R., and Ng, T. B. (1983). Teleost gonadotropins: Isolation, biochemistry and function. In "Fish Physiology" (W. S. Hoar, D. J. Randall, and E. M. Donaldson, eds.), Vol. IX A, pp. 187–221. Academic Press, New York.

Iwamatsu, T., and Onitake, K. (1983). On the effects of cyanoketone on gonadotropin and steroid-induced *in vitro* maturation of *Oryzias latipes*. *Gen. Comp. Endocrinol.* **52**, 418–425.

Jalabert, B. (1976). *In vitro* oocyte maturation and ovulation in rainbow trout (*Salmo gairdneri*), northern pike (*Esox lucius*), and goldfish (*Carassius auratus*). *J. Fish Res. Board Can.* **33**, 974–988.

Jalabert, B., and Finet, B. (1986). Regulation of oocyte maturation in rainbow trout, *Salmo gairdneri*: Role of cyclic AMP in the mechanism of action of the maturation inducing steroid (MIS), 17α-hydroxy-20β-dihydroxyprogesterone. *Fish Physiol. Biochem.* **2**, 65–74.

Jalabert, B., Fostier, A., Breton, B., and Weil, C. (1991). Oocyte maturation in vertebrates. In "Vertebrate Endocrinology: Fundamentals and Biomedical Implications" (P. K. T. Pang and M. P. Schreibman, eds.), Vol. 4, Part A, pp. 23–90. Academic Press, New York.

Kagawa, H., Young, G., Adachi, S., and Nagahama, Y. (1982). Estradiol-17β production in amago salmon (*Oncorhynchus rhodurus*) ovarian follicles: Role of the thecal and granulosa cells. *Gen. Comp. Endocrinol.* **47**, 440–448.

Kagawa, H., Young, G., and Nagahama, Y. (1983). Relationship between seasonal plasma estradiol-17β and testosterone levels and *in vitro* production by ovarian follicles of amago salmon (*Oncorhynchus rhodurus*). *Biol. Reprod.* **29**, 301–309.

Kagawa, H., Young, G., Adachi, S., and Nagahama, Y. (1985). Estrogen synthesis in the teleost ovarian follicle: The two-cell type model in salmonids. In "Salmonid Reproduction" (R. N. Iwamoto and S. Sower, eds.), pp. 20–25. Washington Sea Grant Program, University of Washington, Seattle.

Kajiura, H., Yamashita, M., Katsu, Y., and Nagahama, Y. (1993). Isolation and characterization of goldfish cdc2, a catalytic component of maturation-promoting factor. *Dev. Growth Diff.* **35**, 647–654.

Kanamori, A., and Nagahama, Y. (1988a). Developmental changes in the properties of gonadotropin receptors in the ovarian follicles of amago salmon (*Oncorhynchus rhodurus*) to chum salmon gonadotropin during oogenesis. *Gen. Comp. Endocrinol.* **72**, 25–38.

Kanamori, A., and Nagahama, Y. (1988b). Involvement of 3′,5′-cyclic adenosine monophosphate in the control of follicular steroidogenesis of amago salmon (*Oncorhynchus rhodurus*). *Gen. Comp. Endocrinol.* **72**, 39–53.

Kanamori, A., Adachi, S., and Nagahama, Y. (1988). Devclopmental changes in steroidogenic responses of ovarian follicles of amago salmon (*Oncorhynchus rhodurus*) to chum salmon gonadotropin during oogenesis. *Gen. Comp. Endocrinol.* **72**, 13–24.

Kanungo, J., Patrino, T. R., and Wallace, R. A. (1990). Oogenesis in *Fundulus heteroclitus*. VI. Establishment and verification of conditions for vitellogenin incorporation by oocytes *in vitro*. *J. Exp. Zool.* **254**, 313–321.

Katsu, Y., Yamashita, M., Kajiura, H., and Nagahama, Y. (1993). Behavior of the components of maturation-promoting factor, cdc2 kinase and cyclin B, during oocyte maturation of goldfish. *Dev. Biol.* **160**, 99–107.

Kawauchi, H., Suzuki, K., Itoh, H., Swanson, P., Naito, N., Nagahama, Y., Nozaki, M., Nakai, Y., and Itoh, S. (1989). The duality of teleost gonadotropins. *Fish Physiol. Biochem.* **7**, 29–38.

Kishimoto, T. (1988). Regulation of metaphase by a maturation-promoting factor. *Dev. Growth Diff.* **30**, 105–115.

Kishimoto, T., Kuriyama, R., Kondo, H., and Kanatani, H. (1982). Generality of the action of various maturation-promoting factors. *Exp. Cell Res.* **137,** 121–126.

Kishimoto, T., Yamazaki, K., Kato, Y., Koide, S., and Kanatani, H. (1984). Induction of starfish oocyte maturation and maturation promoting factor of mouse and surf clam oocytes. *J. Exp. Zool.* **231,** 293–295.

Kobayashi, M., Aida, K., and Hanyu, I. (1987). Hormone changes during ovulation and effects of steroid hormones on plasma gonadotropin levels and ovulation in goldfish. *Gen. Comp. Endocrinol.* **67,** 24–32.

Kobayashi, M., Aida, K., Furukawa, K., Law, Y. K., Moriwaki, T., and Hanyu, I. (1988). Development of sensitivity to maturation-inducing steroids in the oocytes of the daily spawning teleost, the kisu *Sillago japonica*. *Gen. Comp. Endocrinol.* **72,** 264–271.

Kostellow, A. B., Weinstein, S. P., and Morrill, G. A. (1982). Specific binding of progesterone to the cell surface and its role in the meiotic divisions in *Rana* oocytes. *Biochim. Biophys. Acta* **720,** 356–363.

Labbé, J. C., Picard, A., Peaucellier, G., Cavadore, J. C., Nurse, P., and Dorée, M. (1989a). Purification of MPF from starfish: Identification as the H1 histone kinase p34^{cdc2} and a possible mechanism of its periodic activation. *Cell* **57,** 253–263.

Labbé, J.-C., Capony, J.-P., Caput, D., Cavadore, J.-C., Derancourt, J., Kaghad, M., Lelias, L.-M., Picard, M., and Dorée, M. (1989b). MPF from starfish oocytes at first meiotic metaphase is a heterodimer containing one molecule of cdc2 and one molecule of cyclin B. *EMBO J.* **8,** 3053–3058.

Lancaster, P. M., and Tyler, C. R. (1991). Isolation of a membrane bound protein with an affinity for vitellogenin from rainbow trout (*Oncorhynchus mykiss*) follicles. In "Proc. IVth Int. Symp. Reproductive Physiology of Fish" (A. P. Scott, J. P. Sumpter, D. E. Kime, and M. S. Rolfe, eds.), p. 128. FishSymp **91,** Sheffield, England.

Lazier, C. B., and MacKay, M. E. (1993). Vitellogenin gene expression in teleost fish. In "Biochemistry and Molecular Biology of Fishes" (P. W. Hochachka and T. P. Mommsen, eds.), Vol. 2. Elsevier Biomedical Publishers, New York.

LeMenn, F., and Nunez Rodriguez, J. (1991). Receptors mediate endocytosis of VTG in fish follicle. In "Proc. IVth Int. Symp. Reproductive Physiology of Fish" (A. P. Scott, J. P. Sumpter, D. E. Kime, and M. S. Rolfe, eds.), pp. 300–302. FishSymp **91,** Sheffield, England.

Lohka, M. L., Hayes, M. K., and Maller, J. L. (1988). Purification of maturation-promoting factor, an intracellular regulator of early mitotic events. *Proc. Natl. Acad. Sci. USA* **85,** 3009–3013.

Maller, J. L. (1985). Regulation of amphibian oocyte maturation. *Cell Diff.* **16,** 211–221.

Maller, J. L. (1991). Mitotic control. *Curr. Opin. Cell Biol.* **3,** 269–275.

Maneckjee, A., Weisbert, M., and Idler, D. R. (1989). The presence of 17α,20β-dihydroxy-4-pregnen-3-one receptor activity in the ovary of the brook trout, *Salvelinus fontinalis*, during terminal stages of oocyte maturation. *Fish Physiol. Biochem.* **6,** 19–38.

Masui, Y. and Clarke, H. J. (1979). Oocyte Maturaion. *Int. Rev. Cytol.* **57,** 185–282.

Masui, Y., and Markert, C. L. (1971). Cytoplasmic control of nuclear behavior during meiotic maturation of frog oocytes. *J. Exp. Zool.* **177,** 129–146.

Mommsen, T. P., and Walsh, P. J. (1988). Vitellogenesis and oocyte assembly. In "Fish Physiology" (W. S. Hoar and D. J. Randall, eds.), Vol. IX A, pp. 347–406. Academic Press, New York.

Morohashi, K., Honda, S., Inomata, Y., Handa, H., and Omura, T. (1993). A common transacting factor, Ad4BP, to the promoters of steroidogenic P-450s. *J. Biol. Chem.* **267,** 17913–17917.

Murray, A. W., Solomon, M. J., and Kirshner, M. W. (1989). The role of cyclin synthesis and degradation in the control of maturation promoting factor activity. *Nature* **339,** 280–286.

Nagahama, Y. (1983). The functional morphology of teleost gonads. In "Fish Physiology" (W. S. Hoar, D. J. Randall, and E. M. Donaldson, eds.), Vol. IX A, pp. 223–275. Academic Press, New York.

Nagahama, Y. (1987a). 17α,20β-Dihydroxy-4-pregnen-3-one: A teleost maturation-inducing hormone. *Dev. Growth Diff.* **29**, 1–12.

Nagahama, Y. (1987b). Gonadotropin action on gametogenesis and steroidogenesis in teleost gonads. *Zool. Sci.* **4**, 209–222.

Nagahama, Y. (1987c). Endocrine control of oocyte maturation. In "Hormones and Reproduction in Fishes, Amphibians, and Reptiles" (D. O. Norris and R. E. Jones, eds.), pp. 171–202. Plenum Press, New York.

Nagahama, Y., and Adachi, S. (1985). Identification of maturation-inducing steroid in a teleost, the amago salmon (*Oncorhynchus rhodurus*). *Dev. Biol.* **109**, 428–435.

Nagahama, Y., Hirose, K., Young, G., Adachi, S., Suzuki, K., and Tamaoki, B. (1983). Relative *in vitro* effectiveness of 17α,20β-dihydroxy-4-pregnen-3-one and other pregnene derivatives on germinal vesicle breakdown in oocytes of ayu (*Plecoglossus altivelis*), amago salmon (*Oncorhynchus rhodurus*), rainbow trout (*Salmo gairdneri*) and goldfish (*Carassius auratus*). *Gen. Comp. Endocrinol.* **51**, 15–23.

Nagahama, Y., Kagawa, H., and Young, G. (1985a). Stimulation of 17α,20β-dihydroxy-4-pregnen-3-one production in the granulosa cells of amago salmon, *Oncorhynchus rhodurus*, by cyclic nucleotides. *J. Exp. Zool.* **236**, 371–375.

Nagahama, Y., Young, G., and Adachi, S. (1985b). Effect of actinomycin D and cycloheximide on gonadotropin-induced 17α,20β-dihydroxy-4-pregnen-3-one production by intact follicles and granulosa cells of the amago salmon, *Oncorhynchus rhodurus*. *Dev. Growth Diff.* **27**, 213–221.

Nagahama, Y., Matsuhisa, A., Iwamatsu, T., Sakai, N., and Fukada, S. (1991). A mechanism for the action of pregnant mare serum gonadotropin on aromatase activity in the ovarian follicle of the medaka, *Oryzias latipes*. *J. Exp. Zool.* **259**, 53–58.

Nakajin, S., and Hall, P. F. (1981). Microsomal cytochrome P-450 from neonatal pig testis. Purification and properties of a C_{21} steroid side-chain cleavage system (17α-hydroxylase-$C_{17,20}$ lyase). *J. Biol. Chem.* **256**, 3871–3876.

Nakajin, S., Ohno, S., and Shinoda, M. (1988). 20β-Hydroxysteroid dehydrogenase of neonatal pig testis: Purification and some properties. *J. Biochem.* **104**, 565–569.

Nakamura, M., Specker, J. L., and Nagahama, Y. (1993). Ultrastructural analysis of the developing follicle during early vitellogenesis in tilapia, *Oreochromis niloticus*, with special reference to the steroid-producing cells. *Cell & Tissue Res.* **272**, 33–39.

Ng, T. B., and Idler, D. R. (1983). Yolk formation and differentiation in teleost fishes. In "Fish Physiology" (W. S. Hoar, D. J. Randall, and E. M. Donaldson, eds.), Vol. IX A, pp. 373–404. Academic Press, New York.

Nurse, P. (1990). Universal control mechanism regulating onset of M-phase. *Nature* **344**, 503–508.

Onitake, K., and Iwamatsu, T. (1986). Immunocytochemical demonstration of steroid hormones in the granulosa cells of the medaka, *Oryzias latipes*. *J. Exp. Zool.* **239**, 97–103.

Onoe, S., Yamashita, M., Kajiura, H., Katsu, Y., Jianquao, J., and Nagahama, Y. (1993). A fish homolog of the cdc2-related protein p40[MO15]: Its cDNA cloning and expression in oocytes. *Biomed. Res.* **14**, 441–444.

Patino, R., and Thomas, P. (1990a). Induction of maturation of Atlantic croaker oocytes by 17α,20β,21-trihydroxy-4-pregnen-3-one *in vitro*: Consideration of some biological and experimental variables. *J. Exp. Zool.* **255**, 97–109.

Patino, R., and Thomas, P. (1990b). Characterization of membrane receptor activity for 17α,20β,21-trihydroxy-4-pregnen-3-one in ovaries of spotted seatrout (*Cynoscion nebulosus*). *Gen. Comp. Endocrinol.* **78**, 204–217.

Patino, R., and Thomas, P. (1990c). Effects of gonadotropin on ovarian intrafollicular processes during the development of oocyte maturational competence in a teleost, the Atlantic croaker:

4. Oocyte Growth and Maturation in Fish

Evidence for two distinct stages of gonadotropic control of final oocyte maturation. *Biol. Reprod.* **43**, 818–827.

Petrino, T. R., Greeley, M. S., Jr., Selman, K., Lin, Y.-W. P., and Wallace, R. A. (1989). Steroidogenesis in *Fundulus heteroclitus*. II. Production of 17α-hydroxy,20β-dihydroprogesterone, testosterone, and 17β-estradiol by various components of the ovarian follicle. *Gen. Comp. Endocrinol.* **76**, 230–240.

Petrino, T. R., Lin, Y.-W. P., Netherton, J. C., Powell, D. H., and Wallace, R. A. (1993). Steroidogenesis in *Fundulus heteroclitus*. V.: Purification, characterization, and metabolism of 17α,20β-dihydroxy-4-pregnen-3-one by intact follicles and its role in oocyte maturation. *Gen. Comp. Endocrinol.* **92**, 1–15.

Pines, J. (1991). Cyclins: Wheels within wheels. *Cell Growth Diff.* **2**, 305–310.

Rosenthal, E. T., Hunt, T., and Ruderman, J. V. (1980). Selective translation of mRNA controls the pattern of protein synthesis during early development of the surf clam, *Spisula solidissima*. *Cell* **20**, 487–494.

Sadler, S. E., and Maller, J. L. (1982). Identification of a steroid receptor on the surface of *Xenopus* oocytes by photoaffinity labelling. *J. Biol. Chem.* **257**, 355–361.

Sakai, N., Tanaka, M., Adachi, S., Miller, W. L., and Nagahama, Y. (1992). Rainbow trout cytochrome P-450$_{c17}$ (17α-hydroxylase/17,20-lyase): cDNA cloning, enzymatic properties and temporal pattern of ovarian p450$_{c17}$mRNA expression during oogenesis. *FEBS Lett.* **301**, 6–64.

Sakai, N., Tanaka, M., Takahashi, M., Adachi, S., and Nagahama, Y. (1993). Isolation and expression of rainbow trout (*Oncorhynchus mykiss*) ovarian cDNA encoding 3β-hydroxysteroid dehydrogenase/Δ$^{4-5}$-isomerase. *Fish Physiol. Biochem.* **11**, 273–279.

Santos, A. J. G., Furukawa, K., Kobayashi, M., Banno, K., Aida, K., and Hanyu, I. (1986). Plasma gonadotropin and steroid hormone profiles during ovulation in the carp *Cyprinus carpio*. *Nippon Suisan Gakkaishi* **52**, 1159–1166.

Schulz, R. (1986). Immunohistological localization of 17β-estradiol and testosterone in the ovary of the rainbow trout (*Salmo gairdneri* Richardson) during the preovulatory period. *Cell Tiss. Res.* **245**, 629–633.

Scott, A. P., and Canario, A. V. M. (1987). Status of oocyte maturation-inducing steroids in teleosts. In "Reproductive Physiology of Fish" (D. R. Idler, L. W. Crim, and J. M. Walsh, eds.), pp. 224–234. Memorial University Press, St. Johns.

Selman, K., and Wallace, R. A. (1982). Oocyte growth in the sheephead minnow: Uptake of exogenous proteins by vitellogenic oocytes. *Tiss. Cell* **14**, 555–571.

Shibata, N., Yoshikuni, M., and Nagahama, Y. (1993). Vitellogenin incorporation into oocytes of rainbow trout, *Oncorhynchus mykiss*, in vitro: Effect of hormones on denuded oocytes. *Dev. Growth Diff.* **35**, 115–121.

Shuttleworth, J., Godfrey, R., and Colmona, A. (1990). p40^{MO15}, a *cdc2*-related protein kinase involved in negative regulation of meiotic maturation of *Xenopus* oocytes. *EMBO J.* **9**, 3233–3240.

Simanis, V., and Nurse, P. (1986). The cell cycle control gene $cdc2^+$ of fission yeast encodes a protein kinase potentially regulated by phosphorylation. *Cell* **45**, 261–268.

Smith, L. D., and Ecker, R. E. (1971). The interaction of steroids with *Rana pipiens* oocytes in the induction of maturation. *Dev. Biol.* **25**, 233–247.

Specker, J. L., and Sullivan, C. V. (1994). Vitellogenesis in fishes: Status and perspective. In "Perspectives in Comparative Endocrinology" (K. G. Davey, R. E. Peter, and S. S. Tobe, eds.), pp. 304–315. National Research Council of Canada, Ottawa.

Stacey, N. E., Cook, A. F., and Peter, R. E. (1979). Ovulatory surge of gonadotropin in the goldfish, *Carassius auratus*. *Gen. Comp. Endocrinol.* **37**, 246–249.

Stifani, S., Le Menn, M., Nunez Rodriguez, J., and Schneider, W. (1990). Regulation of oogenesis: The piscine receptor of vitellogenin. *Biochim. Biophys. Acta* **1045**, 271–279.

Sunkara, P. S., Wright, D. A., and Rao, P. N. (1979). Mitotic factors from mammalian cells induce germinal vesicle breakdown and chromosome condensation in amphibian oocytes. *Proc. Natl. Acad. Sci. USA* **85**, 8747–8750.
Suzuki, K., Kawauchi, H., and Nagahama, Y. (1988). Isolation and characterization of two distinct gonadotropins from chum salmon pituitary glands. *Gen. Comp. Endocrinol.* **71**, 292–301.
Swanson, P. (1991). Salmon gonadotropins: Reconciling old and new ideas. *In* "Proc. IVth Int. Symp. Reproductive Physiology of Fish" (A. P. Scott, J. P. Sumpter, D. E. Kime, and M. S. Rolfe, eds.), pp. 2–7. FishSymp **91**, Sheffield.
Swanson, P., Suzuki, K., Kawauchi, H., and Dickhoff, W. W. (1991). Isolation and characterization of two coho salmon gonadotropins, GTH I and GTH II. *Biol. Reprod.* **44**, 29–38.
Takahashi, M., Tanaka, M., Sakai, N., Adachi, S., Miller, W. L., and Nagahama, Y. (1993). Rainbow trout ovarian cholesterol side-chain cleavage cytochrome P450 (P450scc): cDNA cloning and mRNA expression during oogenesis. *FEBS Lett.* **319**, 45–48.
Tanaka, M., Telecky, T. M., Fukada, S., Adachi, S., Chen, S., and Nagahama, Y. (1992a). Cloning and sequence analysis of the cDNA encoding P-450 aromatase (P450arom) from a rainbow trout (*Oncorhynchus mykiss*) ovary; relationship between the amount of P450arom mRNA and the production of eostradiol-17β in the ovary. *J. Mol. Endocrinol.* **8**, 53–61.
Tanaka, M., Ohno, S., Adachi, S., Nakajin, S., Shinoda, N., and Nagahama, Y. (1992b). Pig testicular 20β-hydroxysteroid dehydrogenase exhibits carbonyl reductase-like structure and activity: cDNA cloning of pig testicular 20β-hydroxysteroid dehydrogenase. *J. Biol. Chem.* **267**, 13451–13455.
Thomas, P., and Patino, R. (1991). Changes in 17α,20β,21-trihyroxy-4-pregnen-3-one membrane receptor concentrations in ovaries of spotted seatrout during final oocyte maturation. *In* "Proc. IVth Int. Symp. Reproductive Physiology of Fish" (A. P. Scott, J. P. Sumpter, D. E. Kime, and M. S. Rolfe, eds.), pp. 122–124. FishSymp **91**, Sheffield.
Thomas, P., and Trant, J. K. (1989). Evidence that 17α,20β,21-trihydroxy-4-pregnen-3-one is a maturation-inducing steroid in spotted seatrout. *Fish Physiol. Biochem.* **7**, 185–191.
Trant, J. M., and Thomas, P. (1988). Structure-activity relationships of steroids in inducing germinal vesicle breakdown of Atlantic croaker oocytes *in vitro*. *Gen Comp. Endocrinol.* **71**, 307–317.
Trant, J. M., and Thomas, P. (1989a). Isolation of a novel maturation-inducing steroid produced *in vitro* by ovaries of Atlantic croaker. *Gen. Comp. Endocrinol.* **75**, 397–404.
Trant, J. M., and Thomas, P. (1989b). Changes in ovarian steroidogenesis *in vitro* associated with final maturation of Atlantic croaker oocytes. *Gen. Comp. Endocrinol.* **75**, 405–412.
Trant, J. M., Thomas, P., and Shackleton, C. H. L. (1986). Identification of 17α,20β,21-trihydroxy-4-pregnen-3-one as the major ovarian steroid produced by the teleost *Micropogonias undulatus* during final oocyte maturation. *Steroids* **47**, 89–99.
Tyler, C. R. (1991). Vitellogenesis in salmonids. *In* "Proc. IVth Int. Symp. Reproductive Physiology of Fish" (A. P. Scott, J. P. Sumpter, D. E. Kime, and M. S. Rolfe, eds.), pp. 295–299. FishSymp **91**, Sheffield.
Tyler, C. R., Sumpter, J. P., and Bromage, N. R. (1990). An *in vitro* culture system for studying vitellogenin uptake into ovarian follicles of the rainbow trout, *Salmo gairdneri*. *J. Exp. Zool.* **255**, 216–231.
Van Der Kraak, G. (1990). The influence of calcium ionophore and activators of protein kinase C on steroid production by preovulatory ovarian follicles of the goldfish. *Biol. Reprod.* **42**, 231–238.
Van Der Kraak, G. (1991). Role of calcium in the control of steroidogenesis in preovulatory ovarian follicles of the goldfish. *Gen. Comp. Endocrinol.* **81**, 268–275.
Van Der Kraak, G., and Chang, J. P. (1990). Arachidonic acid stimulates steroidogenesis in goldfish preovulatory ovarian follicles. *Gen. Comp. Endocrinol.* **77**, 221–228.

Wallace, R. A. (1985). Vitellogenesis and oocyte growth in nonmammalian vertebrates. *Dev. Biol.* **1,** 127–177.
Wallace, R. A., and Selman, K. (1981). Cellular and dynamic aspects of oocyte growth in teleosts. *Am. Zool.* **21,** 325–343.
Wasserman, W. J., and Smith, L. D. (1978). Oocyte maturation in nonmammalian vertebrates. *In* "The Vertebrate Ovary" (R. E. Jones, ed.), pp. 443–468. Plenum Press, New York.
Yamaguchi, A., Yamashita, M., Yoshikuni, M., Hotta, Y., Nurse, P., and Nagahama, Y. (1991). The involvement in meiotic metaphase of H1 histone kinase, p34^{cdc2} homologues and maturation-promoting factor (MPF) in lily (*Lilium longiflorum*) microsporocytes. *Dev. Growth Diff.* **33,** 625–632.
Yamashita, M., Yoshikuni, M., Hirai, T., Fukada, S., and Nagahama, Y. (1991). A monoclonal antibody against the PSTAIR sequence of p34^{cdc2}, catalytic subunit of maturation-promoting factor and key regulator of the cell cycle. *Dev. Growth Diff.* **33,** 617–624.
Yamashita, M., Fukada, S., Yoshikuni, M., Bulet, P., Hirai, T., Yamaguchi, A., Lou, Y.-H., Zhao, Z., and Nagahama, Y. (1992a). Purification and characterization of maturation-promoting factor in fish. *Dev. Biol.* **149,** 8–15.
Yamashita, M., Fukada, S., Yoshikuni, M., Bulet, P., Hirai, T., Yamaguchi, A., Yasuda, H., Ohba, Y., and Nagahama, Y. (1992b). M-phase specific histone H1 kinase in fish oocytes: Purification, components and biochemical properties. *Eur. J. Biochem.* **205,** 537–543.
Yan, L., Swanson, P., and Dickhoff, W. W. (1992). A two-receptor model for salmon gonadotropins (GTH I and GTH II). *Biol. Reprod.* **47,** 418–427.
Yoshikuni, M., Shibata, N., and Nagahama, Y. (1993). Specific binding of [^{3}H]17α,20β-dihydroxy-4-pregnen-3-one to oocyte cortices of rainbow trout (*Oncorhynchus mykiss*). *Fish Physiol. Biochem.* **11,** 15–24.
Young, G., Kagawa, H., and Nagahama, Y. (1982). Oocyte maturation in the amago salmon (*Oncorhynchus rhodurus*): *In vitro* effects of salmon gonadotropin, steroids, and cyanoketone (an inhibitor of 3β-hydroxy-Δ^4-steroid dehydrogenase). *J. Exp. Zool.* **224,** 265–275.
Young, G., Kagawa, H., and Nagahama, Y. (1983a). Evidence for a decrease in aromatase activity in the ovarian granulosa cells of amago salmon (*Oncorhynchus rhodurus*) associated with final oocyte maturation. *Biol. Reprod.* **29,** 310–315.
Young, G., Crim, L. W., Kagawa, H., Kambegawa, A., and Nagahama, Y. (1983b). Plasma 17α,20β-dihydroxy-4-pregnen-3-one levels during sexual maturation of amago salmon (*Oncorhynchus rhodurus*): Correlation with plasma gonadotropin and *in vitro* production by ovarian follicles. *Gen. Comp. Endocrinol.* **51,** 96–106.
Young, G., Adachi, S., and Nagahama, Y. (1986). Role of ovarian thecal and granulosa layers in gonadotropin-induced synthesis of a salmonid maturation-inducing substance (17α,20β-dihydroxy-4-pregnen-3-one). *Dev. Biol.* **118,** 1–8.
Zhu, Y., Aida, K., Furukawa, K., and Hanyu, I. (1989). Development of sensitivity to maturation-inducing steroids and gonadotropins in the oocytes of the tobinumeri-dragonet, *Repomucenus beniteguri*, Callionymidae (Teleostei). *Gen Comp. Endocrinol.* **76,** 250–260.
Zuber, M. X., Maliyakal, E. J., Okamura, T., Simpson, E. R., and Waterman, M. R. (1986). Bovine adrenocortical cytochrome P-450$_{17α}$. *J. Biol. Chem.* **261,** 2475–2482.

5
Nuclear Transplantation in Mammalian Eggs and Embryos

Fang Zhen Sun[1] and Robert M. Moor
Development and Differentiation Laboratory
Babraham Institute
Babraham, Cambridge CB2 4AT, England

I. Introduction
II. Historical Background
III. Nuclear Transplantation Procedures
    A. Preparation of Donor Cells
    B. Preparation of Recipient Cytoplasts
    C. Procedure
IV. Early Events in Cells Reconstituted by Nuclear Transplantation
    A. Nuclear Remodeling
    B. Regulation of Gene Activity
V. Biological Factors Influencing Development of Cells Reconstituted by Nuclear Transplantation
    A. Developmental Stage of Donor Cells
    B. Recipient Cytoplasm
    C. Cell Cycle Regulation
    D. Initiation of Egg Activation Program
VI. Serial Nuclear Transplantation
VII. Epigenetic Modifications of Genomic Totipotency
VIII. Summary and Prospects
References

I. Introduction

At fertilization, a sperm fuses with an egg to restore diploidy and to initiate development. In the course of embryonic development, many different types of cells are generated from fertilized eggs. Some are specialized as muscle, others as neurons, others as blood cells, and so on. A fundamental biological question to be answered by developmental biologists is whether irreversible changes in the genome occur during mammalian embryonic development and cell differentiation. Nuclear transplantation, which refers to the replacement of the nucleus of one cell with that of

---

[1]Current address: Institute of Developmental Biology, Chinese Academy of Sciences, Beijing, PRC.

another, has been recognized as the most powerful experimental approach to answering this question (Gurdon, 1986; DiBerardino, 1987). Clearly, if researchers can prove that a fully differentiated mammalian cell nucleus is capable of directing full development to term after transplantation into a suitable recipient cytoplast they will have generated the most reliable evidence that all the genetic information required for development is still present in the nucleus after cell differentiation.

The technique of nuclear transplantation in mammals developed rapidly from the beginning of the 1980s because many of the limitations, such as inadequate culture techniques and the difficulty of carrying out micromanipulation procedures on small-sized eggs and embryos, were largely overcome. The initial attempts to introduce nuclei into mammalian embryos utilized the Sendai virus to fuse somatic cells with oocytes or embryos (Graham, 1969; Baranska and Koprowski, 1970; Lin et al., 1973). Although these early studies demonstrated the possibility of exploiting viruses for nuclear transplantation in mammals, they failed to produce reconstituted embryos that could develop beyond the first few cleavage divisions. Microsurgery is a more direct method of introducing donor nuclei into recipient eggs or zygotes; the first microsurgical nuclear transplantation experiment in mammals was reported in the rabbit by Bromhall (1975). He reported that the transfer of radioactively labeled nuclei from morula cells into unfertilized and parthenogenetically activated mature oocytes resulted in the development of a few embryos; however, because of the lack of chromosomal analysis and donor-specific markers, it was impossible to determine whether these embryos were derived solely from the transplanted nuclei, or whether they derived from the native egg nucleus with only limited participation by the injected nuclei in the very early cleavage divisions. Evidence that foreign nuclei could persist in mammalian embryos was supplied when Modlinski (1978) injected 8-cell nuclei with the T6 marker chromosome into nonenucleated zygotes and demonstrated that the marker chromosome was present in the resultant tetraploid blastocysts. In another investigation, Modlinski (1981) demonstrated the ability of nuclei from inner cell mass cells (ICM), but not trophectoderm cells, to contribute to tetraploid blastocysts when injected into nonenucleated zygotes. The first offspring produced by nuclear transplantation in mammals were claimed by Illmensee and Hoppe (1981), who reported that some ICM nuclei, when transferred to enucleated 1-cell mouse zygotes, supported development to term. However, this result has been questioned because other independent groups were unable to repeat the experiments. McGrath and Solter (1983) were first to reconstitute embryos in the mouse by microsurgery and cell fusion. Using this technique, they demonstrated via pronuclear exchange that chromosomes of both paternal and maternal origin are necessary for development to term (McGrath and Solter, 1984a; Surani et al., 1984). Furthermore, Surani et al. (1986) reported inheritable changes in parental genomes after activation of the embryonic genome. In farm animals, researchers have shown that fusion of early blastomeres with enucleated eggs by electrofusion leads to the birth of live progeny in both sheep (Willadsen, 1986) and cattle (Prather et al., 1987). Since then, successful production of nuclear transplant mammals using similar

5. Nuclear Transplantation in Eggs and Embryos   149

procedures has been reported by many laboratories. Although a small number of animals can occasionally be cloned by nuclear transplantation from a single donor embryo at an early stage of development, it is widely accepted that the percentage of nuclear transplant embryos that can direct normal embryonic development is generally very low. Therefore, identifying the optimal experimental and biological conditions for promoting normal development of the nuclear-transplanted eggs and embryos has been the central objective of nuclear transplantation studies in mammals. There is little doubt that the true developmental capacity of a nucleus from an embryonic or somatic cell can be tested reliably only if the conditions for nuclear transplantation have been fully optimized.

In this chapter, we review the progress of nuclear transplantation studies in mammals, discuss the important biological factors influencing the development of eggs and embryos reconstituted by nuclear transplantation, and consider the prospects for future studies in this field.

II. Historical Background

From the late 19th century on, the theory of nuclear equivalence has been of central interest to developmental geneticists. Throughout this period, biologists have sought to determine whether the differentiated cells of the adult organism retain the same set of genes found in the fertilized eggs and whether the genes in somatic cells are fundamentally equivalent to those in the zygote nucleus. In 1938, Spemann noted that "decisive information about this question (of nuclear equivalence) may perhaps be afforded by an experiment which appears, at first sight, to be somewhat fantastical," namely, "to isolate the nuclei of the morula and introduce one of them into an egg or an egg fragment without an egg nucleus" (Spemann, 1938). In 1952, Briggs and King, working with eggs from the leopard frog *Rana pipiens*, were the first to overcome the technical problems of nuclear transplantation in eukaryotes and successfully introduced nuclei into egg cytoplasts. They showed that nuclei from cells at the blastula stage, following transplantation to the enucleated eggs, were able to direct normal development to feeding-stage larvae (Briggs and King, 1952). This pioneering achievement was followed by the discovery that nuclei transplanted from the mesoderm or endoderm of *Rana pipiens* during late gastrulation were, in comparison to blastula nuclei, unable to support normal development (King and Briggs, 1955). This study and other work with this species and axolotl embryos (Briggs and King, 1957, 1960; Briggs et al., 1964) led to the conclusion that stable nuclear changes accompany normal development and early differentiation of cells. Furthermore, King and Briggs (1956) established that the developmental restrictions in individual gastrula nuclei of *Rana* differ from each other and are stable over many nuclear divisions. The nature of these developmental restrictions was clarified by detailed chromosomal analysis, which showed that abnormalities in embryos produced by nuclear transplantation could be correlated directly with microscopically visible chromosomal breaks (Briggs et al., 1960,1964; DiBerardino and King, 1965).

The discoveries made in Rana eggs were paralleled by equally important nuclear transplantation experiments with the eggs of Xenopus laevis (Fischberg et al., 1958; Gurdon et al., 1958). Two significant results were obtained from the early experiments with Xenopus eggs. First, some nuclear-transplanted embryos developed to sexually mature individuals, demonstrating that the transplanted nuclei were totipotent rather than pluripotent, as shown by the production of larvae. Second, the results have shown that it is possible to obtain a few normal larvae and, subsequently, fertile frogs from nuclei of the larval intestine (Gurdon and Uehlinger, 1966). These results with Xenopus eggs led to a different interpretation than that drawn from earlier work with Rana. Gurdon (1962) interpreted his results to show that cell differentiation does not require or depend on irreversible changes in the genome. This interpretation was further supported by subsequent results with nuclei of differentiated cells from Xenopus (Gurdon et al., 1975; Wabl et al., 1975) as well as from Rana pipiens (DiBerardino et al., 1986; DiBerardino and Orr, 1992). Indeed, researchers have shown that nuclei from these cells support development to the feeding-stage tadpole but not to adulthood, thus demonstrating that the genetic information required to promote advanced development is present after differentiation. However, no one has yet demonstrated that these cells retain full genomic totipotency. Thus, the question of the developmental totipotency of any differentiated somatic cell remains unanswered. At present, it is still not clear why embryos derived from transplantation of adult cell nuclei do not survive for any significant period beyond metamorphosis. In principle, the inability to obtain fertile nuclear transplant frogs from the transplantation of differentiated somatic cell nuclei permits two interpretations, pointed out by DiBerardino and Orr (1992): (1) that irreversible genetic changes occur during cell differentiation and thus eliminate genomic totipotency or (2) that the experimental procedures used are not yet adequately developed to reveal the full potential of the genome.

The generally accepted conclusions relating to the developmental capacity of transplanted nuclei in amphibia are: (1) that the developmental capacity of transplanted nuclei decreases with increasing age of donor cells; (2) that the developmental abnormalities of nuclear transplant embryos are often caused by an irreversible nuclear condition; and (3) that chromosome abnormalities are the causes of reduced developmental capacity. Chromosome abnormalities are considered to be likely consequences of the inability of transplanted nuclei to accelerate their replication cycle to coincide with the exceptionally rapid pace imposed by eggs (Gurdon, 1986).

Nuclear transplantation in amphibia has also highlighted the remarkable manner in which the cytoplasm influences the morphology, the extent of DNA synthesis, and the transcriptional activity of transplanted nuclei (Gurdon et al., 1979; Gurdon, 1986). Investigators have shown that nuclear swelling is the first response after the transplantation of a somatic nucleus into an egg or oocyte (Graham et al., 1966;

Gurdon et al., 1976). Studies have demonstrated further that, although most nuclei from advanced cell types do not promote normal development of reconstituted eggs, they do in most cases undergo DNA synthesis (Graham et al., 1966; DiBerardino, 1980); changes in transcriptional activity are evident very soon after nuclear transplantation. Experiments clearly show that the transplanted nuclei change their pattern of RNA synthesis soon after transplantation to coincide exactly with that of similarly staged fertilized eggs (Gurdon and Brown, 1965; Gurdon and Woodland, 1969). Gurdon and colleagues have also been able to confirm these results with more precision by studying changes in a single gene (Gurdon et al., 1984). Furthermore, an exchange of proteins between transplanted nuclei and cytoplasm has also been demonstrated (Merriam, 1969; Gurdon, 1970; DiBerardino and Hoffner, 1975; Gurdon et al., 1976) and is considered to play an essential role in nuclear reprogramming (Gurdon et al., 1979; DiBerardino, 1980; DiBerardino et al., 1984; Gurdon, 1986). Despite extensive studies in amphibia, the precise molecular mechanisms of nuclear reprogramming remain obscure.

In addition to the work done in amphibia, nuclear transplantation in fish was first achieved successfully by Tung and colleagues (1963). Work on fish in this field has been summarized by Yan (1989).

III. Nuclear Transplantation Procedures

In principle, nuclear transplantation in mammals consists of three essential steps: (1) the preparation of donor cells, (2) the preparation of recipient cytoplasts; and (3) the introduction of a donor nucleus into its recipient cytoplast. A summary of various procedures used for nuclear transplantation in mammals is presented in Fig. 1.

A. Preparation of Donor Cells

Donor cells can be prepared from pre-implantation embryos, stem cell lines, and differentiated tissues. In experiments using nuclei from zygotes with 2- and 4-cell embryos as donors, the karyoplasts are usually obtained by microsurgery (McGrath and Solter, 1984b). When embryonic cells are used as donor cells, embryos are removed from the reproductive tract of donor animals or obtained via *in vitro* techniques of oocyte maturation, fertilization, and culture. Individual blastomeres can be isolated from embryos either by aspiration (Prather et al., 1987) or by first removing the zona pellucida and disaggregating the cells in Ca^{2+} and Mg^{2+}-free culture medium (Willadsen, 1986). When blastocysts, stem cells, or cells from differentiated tissues are used as donors, the cells are usually dissociated by trypsinization (Illmensee and Hoppe, 1981; Tsunoda et al., 1989; Collas and Robl, 1990; Tsunoda and Kato, 1993).

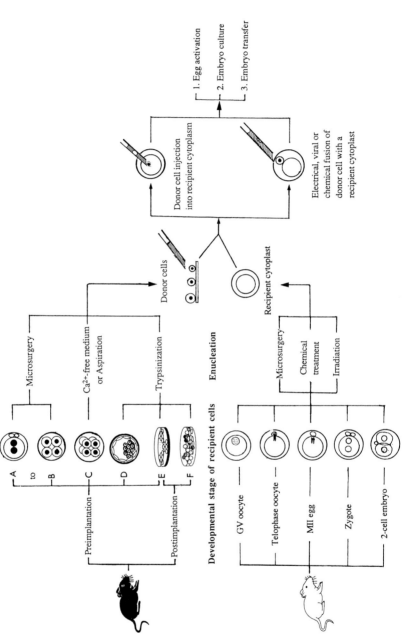

Fig. 1 Illustration of various technical procedures used for nuclear transplantation in mammals. (A) Zygote; (B) 2- to 4-cell embryos; (C) morula; (D) blastocyst; (E) stem cells (including embryonic stem cells and primordial germ cells); (F) somatic cells.

5. Nuclear Transplantation in Eggs and Embryos

B. Preparation of Recipient Cytoplasts

Several methods have been developed for enucleating zygotes, eggs, and oocytes. These include the microsurgical removal procedures reported by Illmensee and Hoppe (1981) and by McGrath and Solter (1983), the enucleation method of Willadsen (1986) on metaphase II (MII) eggs, and the process reported by Sun and Moor (1991) on germinal vesicle (GV) oocytes. Other methods, such as enucleation with chemical treatment, have also been reported (Wassarman *et al.*, 1979; Fulka and Moor, 1993). Since the methods for physical enucleation involve removing not only the chromosomes but also a certain amount of the cytoplasm, which may contain essential components for promoting nuclear reprogramming and early development, researchers have considered that enucleating the recipient eggs or zygotes by nonphysical approaches such as inactivation of the chromosomes by UV or by laser irradiation may be a better approach (Moor *et al.*, 1992).

C. Procedure

The two kinds of approaches by which a donor nucleus can be introduced successfully into a recipient cell are classified as methods of microsurgical injection (Bromhall, 1975; Illmensee and Hoppe, 1981) and cell fusion mediated by the inactivated Sendai virus (Graham, 1969; McGrath and Solter, 1983), by electrofusion (Willadsen, 1986), or by polyethylene glycol treatment (Czolowska *et al.*, 1984). Microsurgical injection of donor nuclei into recipients always causes a high percentage of cell lysis following transplantation; nuclear transfer via cell fusion (especially electrofusion and virus-mediated fusion) usually gives satisfactory results, since the rates of both fusion and survival are high (McGrath and Solter, 1983; Willadsen, 1986). Experimental conditions for electrofusion of mammalian embryos have already been well defined (Ozil and Modlinski, 1986; Sun and Moor, 1989; Robl *et al.*, 1992).

When egg cytoplasts are used as recipients, the eggs reconstituted after nuclear transfer must be activated; otherwise, they will remain at the 1-cell stage until they degenerate. Activation of the eggs can be induced by a number of artificial methods (for references, see Ozil, 1990; Sun *et al.*, 1992). However, note that to date none of the artificial activation approaches used is capable of inducing the same range of activation responses as that initiated by sperm at fertilization. Following nuclear transplantation, the reconstituted eggs and embryos are either cultured *in vitro* or transplanted into recipient animals for developmental analysis (First and Prather, 1991a).

IV. Early Events in Cells Reconstituted by Nuclear Transplantation

After insertion into recipient eggs, donor nuclei from embryonic or somatic cells will be influenced by the dominant recipient egg cytoplasm to undergo a number of morphological and functional changes. These changes are considered important for the reprogramming of the transplanted nuclei and for the normal development of embryos reconstituted by nuclear transplantation.

A. Nuclear Remodeling

Nuclear remodeling is characterized by a series of changes that occur in nuclei following transplantation into an egg or oocyte (Prather and First, 1990). An early morphological indication of nuclear remodeling in an enucleated, activated egg cytoplasm is the dispersion of nucleoli shortly after nuclear insertion (Gurdon and Brown, 1965; Szollosi et al., 1988), followed by chromatin condensation, nuclear membrane breakdown, and finally the decondensation of chromatin and the reassembly of a new nuclear membrane (Czolowska et al., 1984; Stice and Robl, 1988; Szollosi et al., 1988; Sun, 1989). A second indicator of nuclear remodeling is the dramatic swelling of the donor nuclei during the first mitotic cell cycle (Czolowska et al., 1984; Szollosi et al., 1988; Sun, 1989; Collas and Robl, 1991). Experiments have shown that the extent of nuclear swelling is critically dependent on the time interval between egg activation and nuclear insertion (Czolowska et al., 1984; Sun et al., 1991). Studies by Szollosi and colleagues (1988) have illustrated that transplanted nuclei can swell to a size similar to that of an endogenous pronucleus only when nuclei are fused to eggs in the period between metaphase II and telophase II. When nuclear transfer is delayed and occurs after telophase II, the nuclear membrane remains intact, interphase chromosomes do not condense, and the donor nuclei do not develop normal pronuclear structures. In addition, nuclear swelling does not occur when donor nuclei are transplanted into enucleated zygotes or into enucleated 2-cell-stage blastomeres (Barnes et al., 1987).

Although nuclear remodeling is of central importance in determining the fate of reconstituted embryos, little is known about the mechanisms responsible for each step in the cascade of events associated with this process. However, the remodeling process is clearly accompanied by an exchange of proteins between the nuclear and cytoplasmic compartments of reconstituted cells. This exchange process has been studied in the greatest detail in amphibian oocytes and eggs using the technique of prelabeling proteins in either the donor nucleus or the recipient cytoplasm before transplantation (Gurdon and Woodland, 1969; Gurdon et al., 1976; Leonard et al., 1982). The results show that a loss of acidic proteins from the nucleus occurs after its insertion into either an oocyte or egg cytoplasm. In parallel, both acidic and basic

5. Nuclear Transplantation in Eggs and Embryos

proteins migrate from the cytoplasm into the donor nucleus during the first 2 hr after transplantation. Attention has been focused particularly on the intracellular roles of both the protein exchange and, more specifically, the GV proteins in the immediate early period of reprogramming. We have demonstrated that nucleolar disassembly, chromatin condensation, and nuclear membrane breakdown occur readily in the absence of GV components (Sun and Moor, 1991). The requirement for cytoplasmic protein entry, or nuclear protein exit, in the regulation of these early events is still uncertain. In contrast, both protein exchange and GV components appear to be essential for the swelling of the transplanted nucleus after reconstitution (Gurdon, 1976; Balakier and Tarkowski, 1980). Thus, in amphibia, Gurdon (1976) has demonstrated that transplanted nuclei that come into contact with GV components show an exceptionally rapid rate of swelling, suggesting that GV components play an important role in this aspect of reprogramming. Similarly, Balakier and Tarkowski (1980) have shown that GV components in mouse eggs play a central role in sperm decondensation and male pronuclear formation. We conclude from the existing data that various signals are probably involved in controlling the initial disassembly and subsequent reformation and swelling of transplanted nuclei. Maturation-promoting factor (MPF) is likely to be a key component in the disassembly phase, whereas unidentified nuclear components appear to exert a crucial later role. Identifying these factors and studying their interactions with transplanted nuclei at a molecular level will be an important focus of future studies.

B. Regulation of Gene Activity

The influence of the recipient cytoplasm on gene activity in amphibian oocytes and eggs produced by nuclear transplantation has been the subject of detailed study by Gurdon and colleagues (reviewed by Gurdon, 1986). In those experiments, hybrid cells were made between *Xenopus* kidney cells and *Pleurodeles* oocytes to provide clearly identifiable nuclear and cytoplasmic protein markers of gene expression. The results showed that genes that are normally expressed in cultured *Xenopus* kidney cells but not in *Pleurodeles* oocytes became inactive when the cultured cell nuclei were injected into the recipient oocytes. Moreover, injection of nuclei from the cultured *Xenopus* cells into *Pleurodeles* oocytes resulted in the reactivation of genes that are normally expressed only in *Xenopus* oocytes but never in somatic cells. These results prove that genes that become suppressed during cell differentiation can become reactivated by the oocyte cytoplasm. The central importance of this work is that it establishes unambiguously that the oocyte cytoplasm is capable of reprogramming gene expression in differentiated nuclei in such a way that only those genes normally expressed in oocytes are expressed, and the remainder are inactivated.

In mammals, activation of the embryonic genome occurs at the 2-cell stage in mouse embryos (Bolton *et al.*, 1984); and two cell cycles later in ovine embryos (Moor and Gandolfi, 1987). After the midcleavage transition, maternal control ceases and the embryo becomes fully transcriptionally active thereafter. By examining changes in translational activity in nuclear-transplanted ovine embryos, we have analyzed the interactions between the dominant MII egg cytoplasts and the transcriptionally active nuclei derived from 32- to 64-cell embryos (Sun, 1989; Sun *et al.*, 1991). The results show that transcription in previously active nuclei is rapidly suppressed following nuclear transplantation; no new proteins are synthesized after nuclear transplantation except those originally derived from the pre-existing maternal mRNA stored in the egg cytoplasm. The period of nuclear suppression in the ovine embryos persists for at least two cell cycles. Our results show that the reactivation of the embryonic genome after nuclear transplantation occurs slightly earlier than in comparable normal embryos. Although the *trans*-acting inhibitory molecules have not been identified, our studies nevertheless demonstrate the total dominance of the egg cytoplasm in rapidly suppressing transcriptional activity in previously active nuclei. Kanka and colleagues (1991) have extended these observations to bovine eggs and have shown that the synthesis of heterogeneous nuclear RNA (hnRNA) is rapidly suppressed after transplantation of donor nuclei to cytoplasts prepared from MII eggs. These researchers further reported that the reactivation of hnRNA synthesis in reconstituted bovine embryos occurs at a time similar to that observed during normal bovine embryonic development.

V. Biological Factors Influencing Development of Cells Reconstituted by Nuclear Transplantation

Apart from the obvious biological requirement of using only cytoplasts and karyoplasts of the highest quality, four other major biological factors determine the fate of embryos produced by nuclear transplantation: (1) the developmental stage of the donor cell; (2) the biological state of the recipient cytoplasm; (3) the compatibility of the cell cycle stage of the recipient cytoplast and the donor nucleus; and (4) the initiation of the egg activation program. These factors will be discussed in relation to their effect on the future viability of reconstituted eggs and embryos.

A. Developmental Stage of Donor Cells

Studies in amphibia have shown that the developmental stage of donor nuclei has a profound effect on directing normal development of nuclear transplant embryos. The results have demonstrated that nuclei from more advanced developmental stages and from more differentiated cells always promote more abnormal development of nuclear-transplanted embryos than do nuclei from early developmental stages or undifferentiated cells (Gurdon, 1974, 1986; DiBerardino, 1987).

In mammals, the majority of studies to date have focused on the developmental capacity of early embryonic cells. In an early investigation researchers showed that

5. Nuclear Transplantation in Eggs and Embryos

more than 90% of enucleated 1-cell mouse embryos receiving pronuclei from other 1-cell embryos successfully developed to the blastocyst stage *in vitro*. In contrast, only 12.6% of embryos developed to blastocysts when nuclei from 2-cell embryos were used as donors, and no blastocyst development occurred when nuclei from 4-cell or more advanced embryos were used (McGrath and Solter, 1984b). Similar results have also been obtained by other independent research groups (Howlett *et al.*, 1987; Tsunoda *et al.*, 1987). Interestingly, the timing of the loss of nuclear totipotency is coincident with the activation of the embryonic genome at the mid-2-cell stage in mice (Bolton *et al.*, 1984). This finding has led to the suggestion that, after activation of the mouse embryonic genome, the reprogramming of transcriptionally active donor nuclei to allow normal development is "biologically impossible" (McGrath and Solter, 1984b). However, more recent investigation has revealed that this claim is not correct for donor nuclei transplanted into cytoplasts derived from late 2-cell embryos nor from MII eggs. Researchers have clearly demonstrated that when mouse 8-cell nuclei are transplanted into late 2-cell cytoplasts, the nuclei are capable of directing development to produce normal blastocysts (Robl *et al.*, 1986; Howlett *et al.*, 1987) and even live young (Tsunoda *et al.*, 1987). More recently, Cheong and colleagues (1993) have shown that both normal preimplantation and full development to term are possible when nuclei from 4- and 8-cell-stage embryos are transplanted into MII cytoplasts, provided only that the cell cycle stage of the donor nucleus is correct at transplantation. These results therefore suggest that the expression of full totipotency in transcriptionally active donor nuclei is determined by the recipient cytoplasmic environment, and that activation of the embryonic genome in the mouse is not a total obstacle to nuclear reprogramming.

Although this information is novel for the mouse, it is well established in other mammalian species that transcriptionally active embryonic cell nuclei support normal development after transplantation to MII egg cytoplasts (see Table I). In sheep (Crosby *et al.*, 1988), cattle (Camous *et al.*, 1986), and rabbits (Manes, 1977), the maternal–embryonic transition and activation of embryonic genome all occur around the 8-cell stage. However, experiments in sheep have demonstrated that some reconstituted eggs with nuclei from 16-cell (Smith and Wilmut, 1989) or 32/64-cell embryos (Sun *et al.*, 1989) and the ICM of early blastocysts (Smith and Wilmut, 1989) are able to direct normal embryo development to term following implantation to recipient mothers. In cattle, studies have shown that nuclei at the 28- and 48-cell stage of embryo development are similarly totipotent (Bondioli *et al.*, 1990). Sims and First (1993) have shown that even cultured ICM cells are capable of directing normal development to term following nuclear transplantation to MII egg cytoplasts. Likewise, Collas and Robl (1990) have demonstrated in the rabbit that nuclei derived from 16-cell embryos transplanted into MII egg cytoplasts develop into normal young. These combined results indicate clearly that totipotency is not lost after activation of the embryonic genome and that some transcriptionally active nuclei transplanted to suitable MII egg cytoplasts become fully reprogrammed and thereafter capable of directing development to term.

Table I Developmental Capacity of Transcriptionally Active Donor Nuclei after Transplantation to Recipient Cytoplasts

Species	Maternal–embryonic transition	Type of recipient cytoplast	Donor nuclear stage	Nuclear totipotency confirmed	Reference
Mouse	2-cell	Zygote	4-cell	No	McGrath and Solter (1984b); Tsunoda et al. (1987)
		Zygote	8-cell	No	McGrath and Solter (1984b); Tsunoda et al. (1987)
		Zygote	ICM	No	McGrath and Solter (1984b)
		Zygote	ICM	Yes	Illmensee and Hoppe (1981)
		2-cell embryo	4-cell	Yes	Tsunoda et al. (1987)
		2-cell embryo	8-cell	Yes	Tsunoda et al. (1987)
		MII egg	4-cell	Yes	Cheong et al. (1993)
		MII egg	8-cell	Yes	Cheong et al. (1993)
Sheep	8-cell	MII egg	16-cell	Yes	Smith and Wilmut (1989)
		MII egg	32/64-cell	Yes	Sun et al. (1989)
		MII egg	ICM	Yes	Smith and Wilmut (1989)
Cattle	8-cell	MII egg	24/48-cell	Yes	Bondioli et al. (1990)
		MII egg	Cultured ICM	Yes	Sims and First (1994)
Rabbit	8-cell	MII egg	16-cell	Yes	Collas and Robl (1990)

Although the totipotency of some transcriptionally active embryonic nuclei has been established, several reports show an inverse relationship between the developmental age of donor nuclei and the developmental frequency of embryos reconstituted by nuclear transplantation. Studies in the rabbit show that the survival of embryos reconstituted by transplantation of nuclei from 8- and 32-cell and ICM stage cells decreases gradually with the progression of embryo development; a striking drop in the potential for directing blastocyst formation is found when committed blastocyst trophectoderm cell nuclei are used as donors (Collas and Robl, 1991). Similarly, Cheong and colleagues (1993) have reported that the percentage of nuclear-transplanted mouse eggs that develops to blastocysts is reduced from 71% when 4-cell blastomeres are used as donors to 46% when nuclei are derived from 8-cell-stage embryos. At present, it is not clear whether a further reduction in developmental frequency occurs in eggs with nuclei derived from more advanced embryonic stages.

Relatively little work has been carried out to date in mammals with respect to the analysis of the developmental capacity of advanced embryonic and somatic cell nuclei in directing embryo development following nuclear transplantation. Researchers have reported that 6–20% of enucleated mouse eggs receiving male primordial germ cell (PGC) nuclei and 12–18% of eggs reconstituted with embryonic stem (ES) cell nuclei are capable of developing into normal blastocysts (Tsunoda et al., 1989; Tsunoda and Kato, 1993). In both cases, implantation sites have been observed after transplantation to maternal hosts, but fetal development has failed and totipotency has not, therefore, been proven. The developmental capacity of mouse teratocarcinoma cell nuclei in directing pre-implantation development has also been determined (Modlinski et al., 1990). These researchers reported that the majority of nuclear-transplanted eggs developed abnormally and could only undergo a few cleavages in vitro. Our laboratory has shown that the transfer of ovine embryonic stem cell nuclei to enucleated egg cytoplasts has led to both normal blastocyst formation and extended postimplantation development (Moor et al., 1992). More recently, Sims and First (1994) have reported on the establishment of pregnancies in cattle following the transplantation of nuclei from presumptive ES cells into MII egg cytoplasts. Their results provide the first evidence in any mammalian species that nuclei from ES cells support development to term. The experiments also indicated that the frequency of producing viable embryos did not decline with increasing age of the ES cell cultures from 6 to 101 days after initial ICM explantation. The combined evidence from mouse, cattle, and sheep provides a basis for future studies on the totipotency of other ES cell lines, primordial germ cells, and even differentiated somatic cells.

B. Recipient Cytoplasm

Experiments in amphibia have demonstrated that major changes in nuclear morphology, gene activity, and protein composition occur in donor nuclei following nuclear transplantation. These changes are considered essential for nuclear reprogramming in all species. Furthermore, investigators accept that the recipient cytoplasm plays an important role in determining the extent of nuclear reprogramming and, therefore, of the future viability of the reconstituted embryos. Table II represents studies on the developmental capacity of the nuclei from mouse 8-cell embryos following transplantation into recipient cytoplasts derived from zygotes, late 2-cell embryos, and eggs. The results show that when 8-cell nuclei are transplanted into enucleated zygotes, the embryos only undergo one or two cleavages. In contrast, similarly staged nuclei transplanted into cytoplasts prepared from MII eggs or 2-cell stage embryos were capable of directing normal development to blastocysts and, thereafter, to term. These results demonstrate the importance of the recipient cytoplasmic environment in determining the expression of the full developmental potential of transplanted nuclei.

At present, however, the reason cytoplasts at some developmental phases (MII eggs or 2-cell embryos) support development whereas development fails when cytoplasts are prepared from zygotes remains unknown. Thus, the identification of the molecular mechanisms responsible for full nuclear reprogramming remains an important challenge. Some evidence on the effect of the cytoplasmic environment on early morphological changes in transplanted nuclei already ex-

Table II Development of Mouse 8-Cell Nuclei after Transplantation to Recipient Cytoplasts Derived from Mouse Zygotes, 2-Cell Embryos, and MII Eggs

Type of recipient cytoplast	Number of embryos examined	Number (%) of embryos developed		Reference
		2-cell	Blastocyst	
Zygote	54	22 (44%)	0	Tsunoda et al. (1987)
2-cell embryo	139	—	49 (35%)	Tsunoda et al. (1987)
MII egg	39	36 (92%)	18 (46%)	Cheong et al. (1993)

ists. For example, morphological studies on both mouse and ovine eggs have demonstrated that after transplantation of a donor nucleus into an MII egg cytoplast, the nucleus undergoes premature chromatin condensation, decondensation, and significant nuclear swelling. On the other hand, if nuclei are introduced into egg cytoplasts a few hours after cytoplasmic activation, the donor nuclei remain in interphase and show a very limited amount of nuclear swelling during the same period of exposure to the cytoplasm (Czolowska et al., 1984; Sun, 1989; Sun et al., 1991). These studies suggest, first, that activation generates an abrupt change in the physiological state of the cytoplasm. Second, they show that this abrupt cytoplasmic change leads, within a few hours, to the suppression of those cytoplasmic factors responsible for nuclear membrane breakdown and chromatin condensation. MPF kinase is likely to be a key molecule involved in the regulation of these processes (see Norbury and Nurse, 1992). Third, the cytoplasmic changes associated with activation also determine the extent of nuclear swelling in reconstituted embryos. Using nuclear lamins as the molecular markers, researchers have extended these observations by showing that different cytoplasmic environments induce different changes in the composition of nuclear lamins following nuclear transplantation (Kubiak et al., 1991; Prather et al., 1991).

The mammalian nuclear lamina is composed of three polypeptides: the biologically related lamins A and C (also called A/C) and the distinct lamin B (Moir and Goldman, 1993). The three lamins form a network of intermediate filaments lining the inner nuclear membrane, thus forming a part of the nucleoskeleton (Aebi et al., 1986; Fisher et al., 1986). Investigators believe that the nuclear lamina plays an important role in regulating the size, shape, and assembly of the interphase nucleus (Moir and Goldman, 1993). The availability of the nuclear lamins is necessary for both the successful reformation of the interphase nucleus and the decondensation of chromatin after division (Benavente and Krohne, 1986). During mitosis, lamins A and C become fully soluble in the cytoplasm, whereas lamin B remains attached to nuclear envelope vesicles (Burke and Gerace, 1986). Researchers have demonstrated that the disassembly of the nuclear lamina proceeds because of the phosphorylation of a specific lamin motif by MPF kinase (Heald and McKeon, 1990; Ward and Kirschner, 1990). In mouse, the distribution of nuclear lamins A, B, and C in eggs and embryos has been studied (Schatten et al., 1985). Using antibodies against the lamins, investigators have shown that MII eggs have a pool of the cytoplasmic lamins A, B, and C (Houliston et al., 1988; Prather et al., 1991), which are also present in the pronuclei and nuclei of 2-cell embryos. Interestingly, lamin B remains a component of nuclei from embryos at all developmental stages, whereas lamin A/C disappears from the nuclei of embryos between the 4-cell stage and implantation. If 8-cell stage donor nuclei, which contain only lamin B, are transplanted into enucleated MII egg cytoplasts and the reconstituted eggs are activated, lamin A/C reappears in the donor nuclei during the first mitotic cell cycle. However, if similar 8-cell nuclei are inserted into pronuclear stage zygotes or enucleated zygote cytoplasts, the donor nuclei invariably fail to acquire lamin A/C (Prather et al., 1991). Clearly, these observations demonstrate that a nucleus transplanted into MII cytoplasm is remodeled to resemble a pronucleus, whereas it fails to do so when pronuclear stage cytoplasts are used as recipients. The nuclear lamina is recognized to play an important role in both morphological and functional organization of the interphase nucleus (Jackson and Cook, 1985, 1986). The inability of the zygote cytoplasm to remodel the donor nucleus presumably influences the reprogramming of developmental events, and therefore restricts the subsequent developmental capacity of the donor nuclei.

C. Cell Cycle Regulation

The classical experiments of Rao and Johnson (1970) and Johnson and Rao (1970) provide the basis for the understanding of cell cycle interactions in eggs and embryos reconstituted by nuclear transplantation. By fusing somatic cells at accurately defined stages of mitosis, these investigators established the following cell cycle hierarchy: the M phase dominates all others and induces premature chromosome condensation (PCC) on G_1, S, and G_2 chromosomes; the S phase ranks second in the hierarchy and enhances the entry of G_1-staged cells into S phase, while retarding the progression of G_2-phased cells until S phase is completed in the fusion partner. Similarly, G_2-phase cells are retarded by G_1 fusion partners until replication is completed, after which both partners enter M phase in a synchronous manner. That the cell cycle state of both the karyoplast and cytoplast is of critical importance for the development of embryos reconstituted by nuclear transplantation was initially overlooked, but is now abundantly clear. Errors in the formulation of appropriate cell cycle combinations lead to incorrect ploidy in the reconstituted embryonic cells or to irreversible structural damage to the chromosomes.

1. DNA Synthesis and Polyploidy

In amphibia, pioneering studies showed that nearly all nuclei isolated from the adult brain or liver undergo DNA synthesis within a few hours of injection into the egg cytoplasm whereas, in their normal environment, less than 1% of these nuclei replicate their DNA (Graham et al., 1966). Subsequent detailed studies by DeRoeper and colleagues (1977) refined these initial observations by establishing that G_1-staged, but not G_2-staged, HeLa cell nuclei replicate their DNA after transplantation into newly activated G_1/S-phased *Xenopus* eggs. Experiments studying DNA synthesis after nuclear transplantation in mammals have been undertaken by Barnes and colleagues (1993). These investigators have transplanted bovine embryonic cell nuclei into MII- and S-phased egg cytoplasts and have examined the influence of the recipient cell cycle stage on DNA synthesis. First, the results show that 80% of blastomeres in embryos at the 25- to 48-cell stage are in S phase at any one time. More important, when such nuclei are transplanted into MII-phased cytoplasts, a transient drop in DNA synthesis occurs during the first 3.5 hr, followed by a resumption of synthesis at 4.5 hr after transplantation and activation. In contrast, nuclei transplanted into S-phase cytoplasts show no interruption of DNA synthesis. Equally, the proportion of reconstituted embryos containing a normal chromosome complement is highest when S-phase cytoplasts are used, whereas chromosome damage is most severe when predominantly S-phase karyoplasts are transplanted into MII egg cytoplasts. This investigation supports directly the conclusions of Johnson and Rao (1970) and illustrates that the regulation of DNA synthesis in reconstituted eggs is under recipient cytoplasmic control. Interruptions to the normal pattern of DNA synthesis in S-phase nuclei induce chromosomal damage in MII egg cytoplasm and suppress subsequent developmental potential. The problems associated with the interruption of DNA synthesis in S phase are accompanied by a second series of potential problems related to the re-initiation of DNA synthesis when G_2-phase nuclei are transplanted into M-phase cytoplasts. According to the hypothesis of Blow and Laskey (1988), a cytoplasmic factor referred to as "licensing factor" interacts with the chromatin at mitosis and provides for a single round of DNA replication. Since this factor is unable to cross the nuclear membrane, no further rounds of replication can occur until the membrane is again disassembled at M phase. Since nuclear membrane breakdown is an inevitable consequence of transplanting nuclei to MII egg cytoplasts, researchers might anticipate that the licensing factor control mechanism would be disrupted, but this would be without effect on G_1 nuclei; a new round of DNA synthesis in S- and G_2-phase nuclei would be accompanied by polyploidy and a resultant failure in both chromosome constitution and development.

5. Nuclear Transplantation in Eggs and Embryos 163

2. Cell Cycle Interactions and Development

The preceding discussion on DNA replication, chromosomal damage, and polyploidy clearly shows that the cell cycle stages of the nucleus and cytoplasm are likely to have profound effects on the development of reconstituted embryos. That this is a valid assumption is shown by a number of studies. Smith and colleagues (1988) have used enucleated mouse zygotes to show that the developmental frequency of the reconstituted embryos is highest when the cell cycle stage of the donor nuclei is synchronized with that of the cytoplast. Likewise, reconstituted embryos produced by the transplantation of S-phase bovine nuclei into MII egg cytoplasts develop poorly; development is significantly enhanced when synchronized S phase to S phase transplants are made (Barnes *et al.*, 1993). These results were heavily supported by those of Cheong and colleagues (1993) working with the mouse. These investigators reported that 77.8% of reconstituted eggs developed into blastocysts when nuclei from early 2-cell-stage donors (presumptive G_1 phase) were transplanted into MII egg cytoplasts. However, not a single embryo was capable of developing into a blastocyst when nuclei at the mid-cell-cycle stage (presumptive S phase) were used. These combined results, therefore, demonstrate conclusively that cell cycle interactions after nuclear transplantation are of critical importance in determining the developmental capacity of reconstituted embryos.

3. Cell Cycle Interactions and Chromosome Defects

Two observations suggest that chromosomal defects probably account for the poor development that is observed when karyoplasts from the mid- and late cell-cycle are transplanted into MII egg cytoplasts. The first, based on chromosome behavior after transplantation of nuclei into MII egg cytoplasts, shows that the pattern of premature chromatin condensation and polar body formation differs substantially according to the stage of the donor nucleus at transplantation. When early staged (G_1)-nuclei are transplanted into MII egg cytoplasts a single chromatin clump is formed during PCC. In contrast, nuclei transplanted during the mid- or late-cell-cycle stages form two or more chromatin clumps during condensation (Cheong *et al.*, 1993). Moreover, the pattern of pronuclear and polar body formation is strikingly different in each of the different classes of reconstituted embryos. Cheong and colleagues (1993) observed that a single pronucleus with no polar body formed when the early-cell-cycle-staged nuclei were transplanted into MII cytoplasts. However, over 90% of embryos derived from the transplantation of mid- or late-staged nuclei (S or G_2 phase) not only contained a pronucleus but also extruded one or two polar bodies. Furthermore, researchers observed that the majority of these reconstituted embryos with one pronucleus and one polar body showed an abnormal chromosome constitution. The second set of observations in support of the hypothesis on cell-cycle-induced chromatin damage is that of Collas and colleagues (1992), who have examined chromatin and spindle morphology in rabbit embryos produced by nuclear transplantation. The work of this group shows convincingly that, within 2 hr of the insertion of G_1-staged nuclei to MII egg cytoplasts, normal metaphase plates and spindles are visible. In contrast, gross abnormalities are observed in all the metaphase plates formed when late S-phase nuclei were transplanted into the MII cytoplasts. In summary, these combined results suggest that G_1-staged nuclei transplanted into the MII egg cytoplasmic environment yield embryos with the least chromosomal damage and, therefore, the greatest probability of directing normal embryo development.

In conclusion, these studies have demonstrated that it is vitally important that the cell cycle stage of the donor and recipient be synchronized at the time of nuclear transplantation to direct the normal development of nuclear-transplanted eggs and embryos.

D. Initiation of Egg Activation Program

Egg activation following nuclear transplantation is another important factor in determining the developmental capacity of eggs produced by nuclear transplantation. After nuclear insertion, only those reconstituted eggs that have been activated appropriately are able to undergo further cleavage, whereas those that are not activated remain at the 1-cell stage until degeneration (Sun, 1989; F. Z. Sun, unpublished results). Using current nuclear transplantation protocols, fusion between recipient egg cytoplasts and donor cells is induced most frequently by a series of electrical pulses (Willadsen, 1986; Prather *et al.*, 1987) that also function as the trigger for causing the activation of mammalian eggs (Ozil, 1990; Fissore and Robl, 1992; Sun *et al.*, 1992). However, the problem with this procedure is that this artificial activation approach may not induce the full series of activation changes induced by sperm at fertilization.

To investigate whether mammalian eggs activated by electrostimulation undergo the full series of sperm-induced biological changes, we have compared the intracellular changes in mammalian eggs following both electrostimulation and fertilization (Sun *et al.*, 1992). The results of our experiments support the hypothesis that electrostimulation induces a form of partial activation but is unable to elicit the full range of biological changes associated with fertilization. Thus, the results in Fig. 2 show that sperm penetration causes calcium oscillations that persist for at least 4 hr whereas each electrical stimulation induces only a single transient increase that lasts for a few minutes. Clearly, these results demonstrate that the intracellular calcium oscillatory process associated with fertilization has not been reproduced by the simple electrofusion procedure currently used for nuclear transplantation studies.

Researchers now accept as a universal phenomenon the idea that sperm acti-

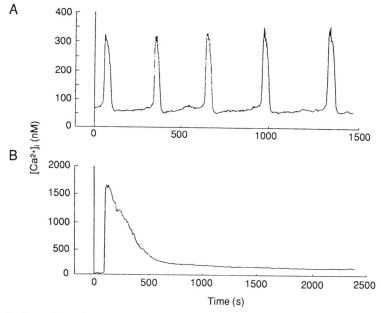

Fig. 2 Comparison of the pattern of intracellular Ca^{2+} changes in porcine eggs activated by (A) sperm or (B) electrostimulation.

vate eggs by inducing a series of transient intracellular calcium ($[Ca^{2+}]_i$) rises (Miyazaki et al., 1993; Whitaker and Swann, 1993). In addition, the early transient rises are propagated throughout the fertilizing eggs in the form of a wave (Fig. 3). The occurrence of Ca^{2+} rises is considered to be important both for the induction of cortical granule exocytosis (Miyazaki, 1990) and for subsequent cell cycle progression (Whitaker and Patel, 1990). In addition, Vitullo and Ozil (1992) have shown that the amplitude and frequency of Ca^{2+} rises induced by electrostimulation can regulate the rate and extent of meiotic resumption and the time of pronuclear formation in parthenogenetically activated mouse eggs. More recently, Collas and colleagues (1993) have reported that histone H1 kinase activity in parthenogenetically activated bovine eggs could be regulated by the number of Ca^{2+} rises induced by electrostimulation. At present, the precise biological role of calcium oscillations at fertilization in early embryogenesis remains to be fully understood. Ozil (1990) was the first to report that repetitive increases in $[Ca^{2+}]_i$ in rabbit eggs affects later developmental events such as compaction and blastocyst formation *in vitro*. The results of Ozil (1990), in conjunction with those from our laboratory (Sun et al., 1992), indicate that methods of inducing full physiological activation of eggs are vital to inducing normal development of nuclear transplanted eggs. To mimic the role of sperm in the activation process, it is important to understand how sperm activate eggs by inducing transient calcium rises. Although the precise mechanisms and pathways involved in this important biological process have not yet been identified, researchers believe that sperm induce calcium rises in eggs by one of two processes. The first hypothesis suggests that sperm act at the membrane level (sperm bind to external receptors that are linked to the egg's GTP-binding proteins, which can activate a phospholipase C that produces the intracellular-calcium-releasing second messenger inositol 1,4,5-trisphosphate). The second hypothesis postulates that fertilizing sperm deliver a cytosolic factor into egg cytoplasm during the process of sperm–egg membrane fusion (reviewed by Miyazaki *et al.*, 1993; Whitaker and Swann, 1993). This second possibility has been supported by studies in which the injection of extracts from sperm activates sea urchin, mouse, and rabbit eggs (Dale *et al.*, 1985; Stice and Robl, 1990). Moreover, a careful investigation done by Swann (1990) has shown that injecting cytosolic sperm factors into hamster eggs initiates calcium oscillations in the eggs that mimic those observed in fertilizing hamster eggs. Similarly, we have found in our experiments that a cytosolic fraction isolated from boar sperm is able to induce cortical granule exocytosis, transient calcium rises, and egg cell cycle progression (F. Z. Sun, unpublished results). By combining these observations with our previous finding that cell fusion can be induced in the absence of egg activation (Sun *et al.*, 1992), we are optimistic that a reliable egg activation procedure able to induce eggs to undergo full activation changes can be established in the near future.

VI. Serial Nuclear Transplantation

Serial nuclear transplantation is an approach that uses nuclear transplanted embryos themselves as the source of nuclei for transplantation. This technique was first used in amphibia as an experimental system designed to improve the developmental capacity of eggs produced by nuclear transplantation. In *Xenopus*, Gurdon and colleagues demonstrated that the capacity of differentiated nuclei to direct normal embryo development was significantly enhanced when nuclei from first-transfer blastulae were used for transplantation (Gurdon and Laskey, 1970; Gurdon *et al.*, 1975). In *Rana pipiens*, DiBerardino and Hoffner (1983) modified the system so somatic cell nuclei were first conditioned in meiotic oocytes for about 24 hr before egg activation; partially cleaved blastulae derived from the first transplantation served thereafter as the source of nuclei for serial transplantation to normal recipient eggs. The results of these experiments showed that erythrocyte nuclei, after sequential exposure to the cytoplasm of oocytes and eggs, can direct the formation of swimming tadpoles. In contrast, when erythrocyte nuclei were injected directly into eggs, development did not proceed beyond

Fig. 3 Pseudo-colored Ca^{2+} images showing the transient Ca^{2+} rise in a bovine egg following sperm–egg membrane fusion. The transient rise is initiated at a single site within the egg and then propagated throughout the cell in the form of a wave. Frame interval: 2.4 sec. (F. Z. Sun, J. P. Bradshaw, C. Galli, and R. M. Moor, unpublished observation).

the early gastrula stage. The authors postulated that specific molecular components in oocytes enhanced chromatin remodeling of transplanted nuclei, thus enabling the conditioned nuclei to participate more normally in the activated eggs and during subsequent development (DiBerardino, 1987).

In mammals, serial transplantation of nuclei through two or more cytoplasts is currently treated as a necessary approach for the production of a large number of cloned embryos/offspring from a single pre-implantation embryo (Westhusin et al., 1991). In theory, researchers assume that by using donor embryos that are themselves the product of nuclear transplantation, large numbers of identical cloned embryos/offspring may be generated after many generations of recloning (Westhusin et al., 1991; Stice and Keefer, 1993). In practice, 54 genetically identical morulae are the maximum number that have been produced to date from one founder embryo (Stice and Keefer, 1993). Moreover, following transfer of these embryos to recipient animals, first, second, and third generation calves have been born. The studies have also revealed that the developmental frequency of nuclear-transplanted eggs that develop to blastocyst stage varies from generation to generation and that pregnancies are lost throughout gestation from all generations of cloned embryos (Stice and Keefer, 1993). The developmental failure and variation in developmental frequency among generations may result from a number of factors, including the viability of the parent and cloned donor cells, the repeated micromanipulation and cloning treatments, and the cumulative effects of suboptimal embryo culture conditions. In addition, because both the parent and cloned donor embryos used for nuclear transplantation were at unknown stages of the cell cycle at transplantation, subsequent developmental failure probably also results from chromosomal abnormalities and/or an accumulation of transplantation-induced errors. At present, it is not clear whether serial transfer of mammalian cell nuclei through more than one egg cytoplast will enhance reprogramming of a transplanted nucleus. Determining whether the cloned mammalian embryos in each generation undergo identical paths, with respect to both reprogramming and changes in gene activity, following transplantation into recipient cytoplasts will be interesting.

VII. Epigenetic Modifications of Genomic Totipotency

Epigenetic modification plays a vital role in governing normal embryonic development in mammals (see review by Surani et al., 1993). The evidence comes from a series of elegant embryological and genetic studies in the mouse. Nuclear transplantation studies in mouse zygotes have shown that both maternal and paternal genomes are required for development to term (McGrath and Solter, 1984a; Surani et al., 1984). Hence, when both nuclear contributions to the zygote are maternal (gynogenetic embryo) or, alternatively, when both nuclear contributions are paternal (androgenetic embryo), embryonic development is aborted. Studies of a number of imbalances of parental chromosomes in the mouse, termed uniparental disomies, have refined our understanding of this requirement for both parental genomes for development (Cattanach and Beechey, 1990; Surani et al., 1993).

DNA methylation is one form of epigenetic modification (Reik et al., 1990; Chaillet, 1992; Surani et al., 1993). Studies in the mouse have shown that genome-wide changes in DNA methylation occur in the germline and during early development (Bird, 1986; Monk et al., 1987; Sanford et al., 1987; Reik et al., 1990; Kafri et al., 1992). DNA methylation is known to have profound effects on both chromatin structure and gene expression (Cedar and Razin, 1990). However, whether nuclear totipotency of embryonic cells is influenced by DNA methylation pattern changes is still not known. Nevertheless, nuclear transplantation experiments in the mouse have revealed that differences in epigenetic modifications between parental genomes are inheritable beyond the pronuclear stage and after activation of the embryonic genome. Surani and colleagues (1986) have transplanted nuclei from haploid androgenetic or gynogenetic embryos back into fertilized eggs from which either the male or female pronucleus was first removed by micromanipulation. These studies show that transplantation of paternal or maternal nuclei from early haploid pre-implantation embryos (up to the 16-cell stage) back into fertilized eggs that contain one pronucleus results in development to term, provided only that the transplanted nucleus is of the parental type opposite the remaining pronucleus. The observations suggest that reversal or removal of "imprints" that have been introduced into the parental chromosomes during gametogenesis cannot be achieved simply by exposing the donor nuclei to the egg cytoplasm. Whether or not these "imprints" are still present in nuclei derived from more advanced haploid embryos remains to be revealed.

VIII. Summary and Prospects

The transplantation of a nucleus to an appropriate cytoplast and the subsequent analysis of its embryonic development provide the most vigorous and direct tests of genomic totipotency of a cell nucleus. To date, the most advanced studies are those carried out in amphibia (Gurdon, 1986; DiBerardino, 1987). Totipotency, the capacity to direct the formation of fertile frogs, has been demonstrated for the nuclei of many kinds of embryonic cells; pluripotency, as judged by the development of heart-beating larvae, has been proved for all cell nuclei tested to date. However, whether a fully differentiated cell nucleus is capable of directing frog development to adulthood remains to be proved.

In mammals, as in frogs, researchers have demonstrated that the morphological, biochemical, and structural changes that occur in transplanted nuclei during the early phase of development are entirely under the control of the recipient cytoplasm. Nuclear swelling following transplantation into an egg cytoplasm appears to be a universal phenomenon that occurs in donor nuclei of all mammalian species studied to date; the molecules responsible for this process are,

5. Nuclear Transplantation in Eggs and Embryos					169

however, still not known. Identification of these molecules, their nature, and their interactions with the chromosomes of transplanted nuclei may provide essential information about the molecular mechanism of nuclear reprogramming. We believe these molecular studies should be a major area for future fundamental studies on nuclear transplantation in mammals.

Evidence now clearly shows that several critical biological factors influence the developmental capacity of mammalian eggs and embryos reconstituted by nuclear transplantation. The developmental age of the donor cell nuclei in determining developmental frequency has been well documented in amphibia (Gurdon, 1986; DiBerardino, 1987). Although similar systematic studies in mammals on the relationship between the developmental age of donor nuclei and the developmental frequency of the nuclear-transplanted embryos in directing normal development are still lacking, one broad generalization can be made. The combined available evidence suggests that the capacity of transplanted nuclei to direct development of reconstituted embryos is inversely related to the cleavage stage of the donor embryos from which the nuclei were derived. Likewise, in all mammalian species examined to date, activation of the embryonic genome does not constitute a total obstacle to nuclear reprogramming, and at least some transcriptionally active embryonic nuclei are totipotent following transplantation into MII egg cytoplasts.

The recipient cytoplasmic environment is another important factor in determining the development of nuclear-transplanted eggs and embryos. Studies in mammals suggest that the MII egg cytoplasm is a more suitable environment than the zygotic cytoplasm in which to reprogram transcriptionally active nuclei fully for normal development. At present, very little is known at the molecular level about why different cytoplasms have significantly different capacities for expression of the full developmental potential of transplanted nuclei. The answer to this important question depends on the identification of those molecules involved in nuclear reprogramming.

Differences in the effects of cell cycle synchronization between the karyoplast and the cytoplast on the subsequent developmental potential of reconstituted embryos reflect one probable major difference between amphibia and mammals. In amphibia, several reports suggest that the cell cycle stage of the donor nuclei at the time of nuclear transplantation has very little effect on the developmental frequency of nuclear-transplanted eggs (McAvoy et al., 1975; Ellinger, 1978; Von Beroldingen, 1981). In mammals, recent studies have demonstrated conclusively that synchronization of cell cycles between donor cells and recipient cytoplasts at the time of transplantation strongly influences the development of reconstituted eggs (Barnes et al., 1993; Cheong et al., 1993). In addition, it is evident in mammals that G_1-staged nuclei transplanted into MII egg cytoplasts yield reconstituted eggs with the least chromosomal damage and, therefore, with the greatest probability of full development to term.

Normal activation of mammalian eggs is essential for development. Evidently current egg activation protocols using electrostimulation are unable to trigger the full range of activation changes normally induced by sperm at fertilization. The induction of inadequate or incomplete activation programs using current artificial activation methods is, in our view, a further important biological component that restricts the development of nuclear-transplanted eggs following activation. Identifying the sperm factor(s) that triggers the egg activation process will likely provide us with a better physiological way of initiating the developmental programs of nuclear-transplanted eggs.

Apart from its importance in studying the genomic totipotency of cell nuclei, nuclear transplantation has also been recognized as a powerful biotechnology for the multiplication of superior farm animals. However, because the overall production efficiency of nuclear-transplanted embryos that can develop to term is still very low (reviewed by First and Prather, 1991; Yang, 1991; Wilmut and Campbell, 1992), it is unlikely that immediate commercial application of the technique in animal cloning is practical. Nevertheless, future prospects for animal cloning by nuclear transplantation are promising, especially if embryonic stem cells can be readily generated from farm animals and prove to be totipotent in directing development following transplantation. If embryonic stem cell nuclei are indeed totipotent, nuclear transplantation will also be the most powerful approach for the production of large numbers of identical transgenic animals by using stem cells with designated foreign genes as donor nuclei.

Future studies of nuclear transplantation in mammals are likely to concentrate on determining the developmental capacity of nuclei from advanced embryonic and somatic cells in directing full development to term. Determining both the effects of epigenetic modifications on nuclear totipotency and the influence of epigenetic factors on the progeny of nuclear-transplanted offsprings will also be interesting.

In conclusion, although our understanding of the factors that influence development of nuclear-transplanted eggs and embryos has been significantly advanced during the past 10 years, the fundamental question of genomic totipotency of advanced embryonic and somatic cell nuclei in the mammalian species remains unanswered.

Acknowledgments

F. Z. Sun is a Research Fellow of Wolfson College, Cambridge University. Our research was supported by the Agricultural and Food Research Council. We thank J. Fulka, Jr., X. Huang, S. M. Laurie, T. Moore, E. Notarianni, W. Reik, and K. Swann for discussion and comments on different parts of the manuscript.

References

Aebi, U., Cohn, J., Buhle, L., and Gerace, L. (1986). The nuclear lamina is a meshwork of intermediate-type filaments. *Nature* **323,** 560–564.

5. Nuclear Transplantation in Eggs and Embryos

Balakier, H., and Tarkowski, A. K. (1980). The role of germinal vesicle karyoplasm in the development of male pronucleus in the mouse. *Exp. Cell Res.* **128,** 79–84.

Baranska, W., and Koprowski, H. (1970). Fusion of unfertilized mouse eggs with somatic cells. *J. Exp. Zool.* **174,** 1–14.

Barnes, F. L., Robl, J. M., and First, N. L. (1987). Nuclear transplantation in mouse embryos: Assessment of nuclear function. *Biol. Reprod.* **36,** 1267–1274.

Barnes, F. L., Collas, P., Powell, R., King, W. A., Westhusin, M., and Shepherd, D. (1993). Influence of recipient oocyte cell cycle stage on DNA synthesis, nuclear envelope breakdown, chromatin condensation, and development in nuclear transplant bovine embryos. *Mol. Reprod. Dev.* **36,** 33–41.

Benavente, R., and Krohne, G. (1986). Involvement of nuclear lamins in postmitotic reorganization of chromatin as demonstrated by microinjection of lamin antibodies. *J. Cell Biol.* **103,** 1847–1854.

Bird, A. P. (1986). CpG-rich isoland and the function of DNA methylation. *Nature* **321,** 209–213.

Blow, J. J., and Laskey, R. A. (1988). A role for the nuclear envelope in controlling DNA replication with the cell cycle. *Nature* **332,** 546–548.

Bolton, V. N., Oades, P. J., and Johnson, M. H. (1984). The relationship between cleavage, DNA replication and gene expression in the mouse two-cell embryo. *J. Embryol. Exp. Morphol.* **190,** 139–256.

Bondioli, K. R., Westhusin, M. E., and Looney, C. R. (1990). Production of identical bovine offspring by nuclear transfer. *Theriogenology* **33,** 165–174.

Briggs, R., and King, T. J. (1952). Transplantation of living nuclei from blastula cells into enucleated frogs' eggs. *Proc. Natl. Acad. Sci. USA* **38,** 455–463.

Briggs, R., and King, T. J. (1957). Changes in the nuclei of differentiating endoderm cells as revealed by nuclear transplantation. *J. Morphol.* **100,** 269–312.

Briggs, R., and King, T. J. (1960). Nuclear transplantation studies on the early gastrula (*Rana pipiens*). *Dev. Biol.* **2,** 252–270.

Briggs, R., King, T. J., and DiBerardino, M. A. (1960). Development of nuclear-transplant embryos of known chromosome complement following parabiosis with normal embryos. *In* "Symp. Germ Cell Dev." (Inst. Intern. d'Embryologie), pp. 441–477. Fondaz A. Baselli, Pavia.

Briggs, R., Signoret, J., and Humphrey, R. R. (1964). Transplantation of nuclei of various cell types from neurulae of the mexican axolotl (*Ambystoma mexicanus*). *Dev. Biol.* **10,** 233–246.

Bromhall, J. D. (1975). Nuclear transplantation in the rabbit eggs. *Nature* **258,** 719–722.

Burke, B., and Gerace, L. (1986). A cell free system to study reassembly of the nuclear envelope at the end of mitosis. *Cell* **44,** 639–652.

Camous, S., Kopecny, V., and Flechon, J. E. (1986). Autoradiographic detection of the earliest stage of (^{3}H)-uridine incorporation. *Biol. Cell* **58,** 195–200.

Cattanach, B. M., and Beechey, C. V. (1990). Autosomal and X-chromosome imprinting. *Development (Suppl.)* 63–72.

Cedar, H., and Razin, A. (1990). DNA methylation and development. *Biochim. Biophys. Acta* **1049,** 1–8.

Chaillet, J. R. (1992). DNA methylation and genomic imprinting in the mouse. *Sem. Dev. Biol.* **3,** 99–105.

Cheong, H-T., Takahashi, Y., and Kanagawa, H. (1993). Birth of mice after transplantation of early cell-cycle stage embryonic nuclei into enucleated oocytes. *Biol. Reprod.* **48,** 958–963.

Collas, P., and Robl, J. M. (1990). Factors affecting the efficiency of nuclear transplantation in the rabbit embryo. *Biol. Reprod.* **43,** 877–884.

Collas, P., and Robl, J. M. (1991). Relationship between nuclear remodelling and development in nuclear transplant rabbit embryos. *Biol. Reprod.* **45,** 455–465.

Collas, P., Pinto-Correia, C., Ponce de Leon, F. A., and Robl, J. M. (1992). Effect of donor cell

cycle stage on chromatin and spindle morphology in nuclear transplant rabbit embryos. *Biol. Reprod.* **46**, 501–511.
Collas, P., Sullivan, E. J., and Barnes, F. L. (1993). Histone H1 kinase activity in bovine oocytes following calcium stimulation. *Mol. Reprod. Dev.* **34**, 224–231.
Crosby, I. M., Gandolfi, F., and Moor, R. M. (1988). Control of protein synthesis during early cleavage of sheep embryos. *J. Reprod. Fertil.* **82**, 769–775.
Czolowska, R., Modlinski, J. A., and Tarkowski, A. K. (1984). Behaviour of thymocyte nuclei in non-activated and activated mouse oocytes. *J. Cell Sci.* **69**, 19–34.
Dale, B., De Felice, L. J., and Ehrenstein, G. (1985). Injection of a soluble extract into sea urchin eggs triggers the cortical reaction. *Experientia* **41**, 1086–1070.
DeRoeper, A., Smith, J. A., Watt, R. A., and Barry, J. M. (1977). Chromatin dispersal and DNA synthesis in Gi and G2 HeLa cell nuclei injected into *Xenopus* eggs. *Nature* **265**, 469–470.
DiBerardino, M. A. (1980). Genetic stability and modulation of metazoan nuclei transplanted into eggs and oocytes. *Differentiation* **17**, 17–30.
DiBerardino, M. A. (1987). Genomic potential of differentiated cells analysed by nuclear transplantation. *Am. Zool.* **27**, 623–644.
DiBerardino, M. A., and Hoffner, N. J. (1975). Nucleo-cytoplasmic exchange of non-histone proteins in amphibian embryos. *Exp. Cell Res.* **94**, 235–252.
DiBerardino, M. A., and Hoffner, N. (1983). Gene reactivation in erythrocytes: nuclear transplantation in oocytes and eggs of *Rana*. *Science* **219**, 862–864.
DiBerardino, M. A., and King, T. J. (1965). Transplantation of nuclei from the frog renal adenocarcinoma. II. Chromosomal and histologic analysis of tumor nuclear-transplant embryos. *Dev. Biol.* **11**, 217–242.
DiBerardino, M. A., and Orr, N. H. (1992). Genomic potential of erythroid and leukocytic cells of *Rana pipiens* analyzed by nuclear transfer into diplotene and maturing oocytes. *Differentiation* **50**, 1–13.
DiBerardino, M. A., Hoffner, N. J., and Etkin, L. D. (1984). Activation of dormant genes in specialized cells. *Science* **224**, 946–952.
DiBerardino, M. A., Orr, N. H., and McKinnell, R. G. (1986). Feeding tadpoles cloned from *Rana* erythrocyte nuclei. *Proc. Natl. Acad. Sci. USA* **83**, 8231–8234.
Ellinger, M. S. (1978). The cell cycle and transplantation of blastula nuclei in *Bambina orientalis*. *Dev. Biol.* **65**, 81–89.
First, N. L., and Prather, R. S. (1991a). Production of embryos by oocyte cytoplast–blastomere fusion in domestic animals. *J. Reprod. Fertil. (Suppl.)* **43**, 245–254.
First, N. L., and Prather, R. S. (1991b). Genomic potential in mammals. *Differentiation* **48**, 1–8.
Fischberg, M., Gurdon, J. B., and Elsdale, T. R. (1958). Nuclear transplantation in *Xenopus laevis*. *Nature* **181**, 424.
Fisher, D. Z., Chaudhary, N., and Blobel, G. (1986). cDNA sequencing of nuclear lamins A and C reveals primary and secondary structural homology to intermediate filament proteins. *Proc. Natl. Acad. Sci. USA* **83**, 6450–6454.
Fissore, R. A., and Robl, J. M. (1992). Intracellular Ca^{2+} responses of rabbit oocytes to electrical stimulation. *Mol. Reprod. Dev.* **32**, 9–16.
Fulka, J., Jr., and Moor, R. M. (1993). Noninvasive chemical enucleation of mouse oocytes. *Mol. Reprod. Dev.* **34**, 427–430.
Graham, C. F. (1969). The fusion of cells with one- and two-cell mouse embryos. *In* "Heterospecific Genome Interaction" (V. Defendi, ed.), pp 13–35. Wistar Institute Press, Philadelphia.
Graham, C. F., Arms, K., and Gurdon, J. B. (1966). The induction of DNA synthesis by frog egg cytoplasm. *Dev. Biol.* **14**, 349–381.
Gurdon, J. B. (1962). The developmental capacity of nuclei taken from intestinal epithelium cells of feeding tadpoles. *J. Embryol. Exp. Morphol.* **10**, 622–640.

Gurdon, J. B. (1970). Nuclear transplantation and the control of gene activity in animal development. *Proc. R. Soc. London, Ser. B.* **176**, 303–314.
Gurdon, J. B. (1976). Injected nuclei in frog oocytes: Fate, enlargement, and chromatin dispersal. *J. Embryol. Exp. Morphol.* **36**, 523–540.
Gurdon, J. B. (1986). Nuclear transplantation in eggs and oocytes. *J. Cell Sci. Suppl.* **4**, 287–318.
Gurdon, J. B., and Brown, D. D. (1965). Cytoplasmic regulation of RNA synthesis and nucleolus formation in developing embryos of *Xenopus laevis*. *J. Mol. Biol.* **12**, 27–35.
Gurdon, J. B., and Laskey, R. A. (1970). The transplantation of nuclei from single cultured cells into enucleated frog's eggs. *J. Embryol. Exp. Morphol.* **24**, 227–248.
Gurdon, J. B., and Uehlinger, V. (1966). "Fertile" intestine nuclei. *Nature* **210**, 1240–1241.
Gurdon, J. B., and Woodland, H. R. (1969). The influence of the cytoplasm on the nucleus during cell differentiation, with special reference to RNA synthesis during amphibian cleavage. *Proc. R. Soc. London B* **173**, 99–111.
Gurdon, J. B., Elasdale, T. R., and Fischberg, M. (1958). Sexually mature individuals of *Xenopus laevis* from the transplantation of single somatic nuclei. *Nature* **182**, 64–65.
Gurdon, J. B., Laskey, R. A., and Reeves, O. R. (1975). The developmental capacity of nuclei transplanted from keratinized skin cells of adult frogs. *J. Embryol. Exp. Morphol.* **34**, 93–112.
Gurdon, J. B., Partington, G. A., and De Robertis, E. M. (1976). Injected nuclei in frog oocytes: RNA synthesis and protein exchange. *J. Embryol. Exp. Morphol.* **36**, 541–553.
Gurdon, J. B., Laskey, R. A., De Robertis, E. M., and Partington, G. A. (1979). Reprogramming of transplanted nuclei in amphibia. *Int. Rev. Cytol. Suppl.* **9**, 161–178.
Gurdon, J. B., Brennan, S., Fairman, S., and Mohun, T. J. (1984). Transcription of muscle-specific actin genes in early *Xenopus* development: Nuclear transplantation and cell dissociation. *Cell* **38**, 691–700.
Heald, R., and McKeon, F. (1990). Mutations of phosphorylation sites in lamin A that prevent nuclear lamina disassembly in mitosis. *Cell* **61**, 579–589.
Hoffner, N. J., and DiBerardino, M. A. (1983). Gene reactivation in erythrocytes: Nuclear transplantation in oocytes and eggs of *Rana*. *Science* **219**, 862–864.
Houliston, E., Guilly, M. N., Courvalin, J. C., and Maro, B. (1988). Expression of nuclear lamins during mouse preimplantation development. *Development* **102**, 271–278.
Howlett, S. K., Barton, S. C., and Surani, M. A. (1987). Nuclear cytoplasmic interactions following nuclear transplantation in mouse embryos. *Development* **101**, 915–923.
Illmensee, K., and Hoppe, P. C. (1981). Nuclear transplantation in *Mus musculus*: Developmental potential of nuclei from preimplantation embryos. *Cell* **23**, 9–18.
Jackson, D. A., and Cook, P. R. (1985). Transcription occurs at a nucleoskeleton. *EMBO J.* **4**, 919–925.
Jackson, D. A., and Cook, P. R. (1986). Replication occurs at a nucleoskeleton. *EMBO J.* **5**, 1403–1410.
Johnson, R. T., and Rao, P. N. (1970). Mammalian cell fusion: Induction of premature chromatin condensation in interphase nuclei. *Nature* **226**, 717–722.
Kafri, T., Ariel, M., Brandeis, M., Shemer, R., Urven, L., McCarry, J., Cedar, H., and Razin, A. (1992). Developmental pattern of gene specific DNA methylation in the mouse embryo and germ line. *Genes Dev.* **6**, 705–714.
Kanka, J., Fulka J., Jr., Fulka, J., and Peter, J. (1991). Nuclear transplantation in bovine embryos: Fine structural and autoradiographic studies. *Mol. Reprod. Dev.* **29**, 110–116.
King, T. J., and Briggs, R. (1955). Changes in the nuclei of differentiating gastrula cells, as demonstrated by nuclear transplantation. *Proc. Natl. Acad. Sci. USA* **41**, 321–325.
King, T. J., and Briggs, R. (1956). Serial transplantation of embryonic nuclei. *Cold Spring Harbor Symp. Quant. Biol.* **126**, 115–126.

King, T. J., and Briggs, R. (1956). Serial transplantation of embryonic nuclei. *Cold Spring Harbor Symp. Quant. Biol.* **126**, 115–126.

Kubiak, J. Z., Prather, R. S., Maul, G. G., and Schatten, G. (1991). Cytoplasmic modification of the nuclear lamina during pronuclear-like transformation of mouse blastomere nuclei. *Mechanisms of Development* **35**, 103–111.

Leonard, R. A., Hoffner, N. J., and DiBerardino, M. A. (1982). Induction of DNA synthesis in amphibian erythroid nuclei in *Rana* eggs following conditioning in meiotic oocytes. *Dev. Biol.* **92**, 343–355.

Lin, T. P., Florence, J., and Oh, J. O. (1973). Cell fusion induced by a virus within the zona pellucida of mouse eggs. *Nature* **242**, 47–49.

Manes, C. (1977). Nucleic acid synthesis in preimplantation rabbit embryo. III. A "dark period" immediately following fertilization and the early predominance of low molecular weight RNA synthesis. *J. Exp. Zool.* **201**, 247–258.

McAvoy, J. W., Dixon, K. E., and Marshall, J. A. (1975). Effects of differences in mitotic activity, stage of cell cycle, and degree of specialization of donor cells on nuclear transplantation in *Xenopus laevis*. *Dev. Biol.* **45**, 330–339.

McGrath, J., and Solter, D. (1983). Nuclear transplantation in the mouse embryo by microsurgery and cell fusion. *Science* **220**, 1300–1302.

McGrath, J., and Solter, D. (1984a). Completion of mouse embryogenesis requires both the maternal and paternal genomes. *Cell* **37**, 179–183.

McGrath, J., and Solter, D. (1984b). Inability of mouse blastomere nuclei transplanted to enucleated zygotes to support development in vitro. *Science* **226**, 1317–1319.

Merriam, R. W. (1969). Movement of cytoplasmic proteins into nuclei induced to enlarge and initiate DNA or RNA synthesis. *J. Cell Sci.* **5**, 333–349.

Miyazaki, S. (1990). Cell signalling at fertilization of hamster eggs. *J. Reprod. Fertil. Suppl.* **42**, 163–175.

Miyazaki, S., Shirakawa, H., and Honda, Y. (1993). Essential role of inositol 1,4,5-triphosphate receptor/Ca^{2+} release channel in Ca^{2+} waves and Ca^{2+} oscillations at fertilization of mammalian eggs. *Dev. Biol.* **158**, 62–78.

Modlinski, J. A. (1978). Transfer of embryonic nuclei to fertilized mouse eggs and development of tetraploid blastocysts. *Nature* **273**, 466–467.

Modlinski, J. A. (1981). The fate of inner cell mass and trophectoderm nuclei transplanted to fertilized mouse eggs. *Nature* **292**, 342–343.

Modlinski, J. A., Gerhauser, D., Lioi, B., Winking, H., and Illmensee, K. (1990). Nuclear transfer from teratocarcinoma cells into mouse oocytes and eggs. *Development* **108**, 337–348.

Moir, R., and Goldman, R. D. (1993). Lamin dynamics. *Curr. Opin. Cell Biol.* **5**, 408–411.

Monk, M., Boubelik, M., and Lehnert, S. (1987). Temporal and regional changes in DNA methylation in the embryonic, extraembryonic and germ cell lineages during mouse development. *Development* **99**, 371–382.

Moor, R. M., and Gandolfi, F. (1987). Molecular and cellular changes associated with maturation and early development of sheep eggs. *J. Reprod. Fertil.* **34**, 55–69.

Moor, R. M., Sun, F. Z., and Galli, C. (1992). Reconstruction of ungulate embryos by nuclear transplantation. *Anim. Reprod. Sci.* **28**, 423–431.

Norbury, C., and Nurse, P. (1992). Animal cell cycles and their control. *Annu. Rev. Biochem.* **61**, 441–470.

Ozil, J. P. (1990). The parthenogenetic development of rabbit oocytes after repetitive pulsatile electrical stimulation. *Development* **109**, 117–127.

Ozil, J. P., and Modlinski, J. A. (1986). Effects of electrical field on the fusion ratio and development of two-cell rabbit embryos. *J. Embryol. Exp. Morphol.* **96**, 211–228.

Prather, R. S., and First, N. L. (1990). Cloning embryos by nuclear transfer. *J. Reprod. Fertil. Suppl.* **41**, 125–134.

5. Nuclear Transplantation in Eggs and Embryos

Prather, R. S., Barnes, F. L., Sims, M. M., Roble, J. M., Eyestone, W. H., and First, N. L. (1987). Nuclear transplantation in the bovine embryos: assessment of donor nuclei and recipient oocyte. *Biol. Reprod.* **37,** 859–866.

Prather, R. S., Kubiak, J., Maul, G. G., First, N. L., and Schatten, G. (1991). The association of nuclear lamins A and C is regulated by the developmental stage of the mouse oocytes or embryonic cytoplasm. *J. Exp. Zool.* **257,** 110–114.

Rao, P. N., and Johnson, R. T. (1970). Mammalian cell fusion: Studies on the regulation of DNA synthesis and mitosis. *Nature* **225,** 159–164.

Reik, W., Howlett, S. K., and Surani, M. A. (1990). Imprinting by DNA methylation: From transgenes to endogenous gene sequences. *Development (Suppl.)* 99–106.

Robl, J. M., Gilligan, B., Critser, E. S., and First, N. L. (1986). Nuclear transplantation in mouse embryos: Assessment of recipient cell stage. *Biol. Reprod.* **34,** 733–739.

Robl, J. M., Callas, P., Fissore, R., and Dobrinsky, J. (1992). Electrically induced fusion and activation in nuclear transplant embryos. *In* "Guide to Electroporation and Electrofusion," pp. 535–551. Academic Press, San Diego.

Sanford, J. P., Clark, H. J., Chapman, V. M., and Rossant, J. (1987). Differences in DNA methylation during oogenesis and spermatogenesis and their persistence during early embryogenesis. *Genes Dev.* **1,** 1039–1043.

Schatten, G., Maul, G. G., Schatten, H., Chaly, N., Simerly, C., Balczon, R., and Brown, D. L. (1985). Nuclear lamins and peripheral nuclear antigens during fertilization and embryogenesis in mice and sea urchins. *Proc. Natl. Acad. Sci. USA* **82,** 4727–4731.

Sims, M., and First, N. L. (1994). Production of calves by transfer of nuclei from cultured inner cell mass cells. *Proc. Natl. Acad. Sci. USA* **91,** 6143–6147.

Smith, L. C., and Wilmut, I. (1989). Influence of nuclear and cytoplasmic activity on the development in vivo of sheep embryos after nuclear transplantation. *Biol. Reprod.* **40,** 1027–1035.

Smith, L. C., Wilmut, I., and Hunter, R. H. F. (1988). Influence of cell cycle stage at nuclear transplantation on the development in vitro of mouse embryos. *J. Reprod. Fertil.* **84,** 619–624.

Spemann, H. (1938). "Embryonic Development and Induction." Yale University Press, New Haven, Connecticut.

Stice, S. L., and Keefer, C. L. (1993). Multiple generational bovine embryo cloning. *Biol. Reprod.* **48,** 715–719.

Stice, S. L., and Robl, J. M. (1988). Nuclear reprogramming in nuclear transplant rabbit embryos. *Biol. Reprod.* **39,** 657–664.

Stice, S. L., and Robl, J. M. (1990). Activation of mammalian oocytes by a factor obtained from rabbit sperm. *Mol. Reprod. Dev.* **25,** 272–280.

Sun, F. Z. (1989). Nuclear-cytoplasmic interactions during oocyte maturation and early embryogenesis in sheep. Ph.D. Thesis. Cambridge University, Cambridge, England.

Sun, F. Z., and Moor, R. M. (1989). Factors controlling the electrofusion of murine embryonic cells. *Bioelectrochem. Bioenerg.* **21,** 149–160.

Sun, F. Z., and Moor, R. M. (1991). Nuclear-cytoplasmic interactions during ovine oocyte meiotic maturation. *Development* **111,** 171–180.

Sun, F. Z., Laurie, M. S., and Moor, R. M. (1989). Nuclear-cytoplasmic interactions in ovine reconstituted embryos. *J. Reprod. Fertil. Abstr. Ser.* **4,** 15.

Sun, F. Z., Laurie, M. S., and Moor, R. M. (1991). Cytoplasmic control of donor nuclear activity in sheep nuclear-transplanted embryos. *J. Reprod. Fertil. Suppl.* **43,** 261.

Sun, F. Z., Hoyland, J., Huang, X., Mason, W., and Moor, R. M. (1992). A comparison of intracellular changes in porcine eggs after fertilization and electrostimulation. *Development* **115,** 947–956.

Surani, M. A. H., Barton, S. C., and Norris, M. L. (1984). Development of reconstituted mouse eggs suggests imprinting of the genome during gametogenesis. *Nature* **308,** 548–550.

Surani, M. A. H., Barton, S. C., and Norris, M. L. (1986). Nuclear transplantation in the mouse: Heritable differences between parental genomes after activation of the embryonic genome. *Cell* **45,** 127–136.

Surani, M. A., Sasaki, H., Ferguson-Smith, A. C., Allen, N. D., Barton, S. C., Jones, P. A., and Reik, W. (1993). The inheritance of germline-specific epigenetic modifications during development. *Phil. Trans. R. Soc. London B* **339,** 165–172.

Swann, K. (1990). A cytosolic factor stimulates repetitive calcium increases and mimics fertilization in hamster eggs. *Development* **110,** 1295–1302.

Szollosi, D., Czolowska, R., Szollosi, M. S., and Tarkowski, A. K. (1988). Remodelling of mouse thymocyte nuclei depends on the time of their transfer into activated, homologous oocytes. *J. Cell Sci.* **91,** 19–34.

Tsunoda, Y., and Kato, Y. (1993). Nuclear transplantation of embryonic stem cells in mice. *J. Reprod. Fertil.* **98,** 537–540.

Tsunoda, Y., Yasui, T., Shioda, Y., Nakamura, M., Ochida, T., and Sugie, T. (1987). Full term development of mouse blastomere nuclei transplanted into enucleated two-cell embryos. *J. Exp. Zool.* **242,** 147–151.

Tsunoda, Y., Tokunaga, T., Imai, H., and Uchida, T. (1989). Nuclear transplantation of male primordial germ cells in the mouse. *Development* **107,** 407–411.

Tung, T. C., Wu, S. C., Tung, Y. F. Y., Yan, S. Y., Du, M., and Lu, T. Y. (1963). Nuclear transplantation in fishes. *Scientia (Peking)* **14,** 1244–1245.

Vitullo, A. D., and Ozil, J. P. (1992). Repetitive calcium stimuli drive meiotic resumption and pronuclear development during mouse oocyte activation. *Dev. Biol.* **151,** 128–136.

Von Beroldingen, C. H. (1981). The developmental potential of synchronized amphibian cell nuclei. *Dev. Biol.* **81,** 115–126.

Yan, S. Y. (1989). The nuclear-cytoplasmic interactions as revealed by nuclear transplantation in fish. *In* "Cytoplasmic Organization Systems" (G. M. Malacinski, ed.), pp. 61–81. McGraw-Hill, New York.

Yang, X. (1991). Embryo cloning by nuclear transfer in cattle and rabbit. *Embryol. Trans. Newsl.* **9,** 10–22.

Wabl, M. R., Brun, R. B., and Du Pasquier, L. (1975). Lymphocytes of *Xenopus laevis* have the gene set for promoting tadpole development. *Science* **190,** 1310–1312.

Ward, G. E., and Kirschner, M. W. (1990). Identification of cell cycle-regulated phosphorylation sites on nuclear lamin C. *Cell* **61,** 561–577.

Wassarman, P. M., Schultz, R. M., and Letourneau, G. E. (1979). Protein synthesis during meiotic maturation of mouse oocytes in vitro: Synthesis and phosphorylation of a protein localized in the germinal vesicle. *Dev. Biol.* **69,** 94–107.

Westhusin, M. E., Pryor, J. H., and Bondioli, K. R. (1991). Nuclear transfer in the bovine embryo: A comparison of 5-day, 6-day, frozen-thawed and nuclear transfer donor embryos. *Mol. Reprod. Dev.* **28,** 119–123.

Whitaker, M., and Patel, R. (1990). Calcium and cell cycle control. *Development* **108,** 525–542.

Whitaker, M., and Swann, K. (1993). Lighting the fuse at fertilization. *Development* **117,** 1–12.

Willadsen, S. M. (1986). Nuclear transplantation in sheep embryos. *Nature* **320,** 63–65.

Wilmut, I., and Campbell, K. (1992). Embryo multiplication in livestock: Present procedures and the potential for development. *In* "Embryonic Development and Manipulation in Animal Production" (A. Lauria and F. Gandolfi, eds.), pp. 135–145. Portland Press Proceedings, London.

6
Transgenic Fish in Aquaculture and Developmental Biology

Zhiyuan Gong and Choy L. Hew
Research Institute
Hospital for Sick Children, and
Departments of Clinical Biochemistry
and Biochemistry
University of Toronto
Toronto, Ontario, Canada M5G 1L5

I. Introduction
II. Fish as a Transgenic System
III. Production of Transgenic Fish: Methodology
   A. Microinjection
   B. Electroporation
   C. Electroporated Sperm as Carrier
   D. Gene Bombardment
   E. Other Potential Methods
   F. Other Considerations
IV. Transgenesis in Fish: Integration, Inheritance, and Expression
   A. Persistence and Integration of the Exogenously Introduced DNA
   B. Mosaicism and Inheritance of the Foreign DNA
   C. Expression of the Transgene
V. Application of Transgenic Fish in Aquaculture
   A. Growth Hormone Gene Transfer
   B. Antifreeze Protein Gene Transfer
   C. Other Potential Genes for Gene Transfer
   D. Ecological Concerns
VI. Application of Transgenic Fish in Developmental Biology
   A. Fish as a Model in Developmental Biology
   B. Transgenic Fish as a Transient Expression System
   C. Transgenic Fish as a Stably Transformed Line
VII. Concluding Remarks
   References

I. Introduction

Transgenic animals have been generated by the introduction of a cloned gene into fertilized eggs. The exogenously introduced gene can be integrated successfully into the host genome, expressed, and passed on to the next generation. Although the transgenic technique was developed in the early 1980s, the concept of gene transfer can be traced back almost half a century, long before the isolation of a gene. In 1938, Spemann suggested a method for testing nuclear potency by transferring nuclei into enucleated eggs. The first nuclear transplantation experiment was accomplished in the ameba by Commanndon and de Fonbrune (1939) and, a few years later, was done in a complex organism, the frog (*Rana pipiens*), by Briggs and King (1952). Transformation of cultured mammalian cells by the introduction of exogenous DNA was accomplished in the early 1970s (Graham and Van der Eb, 1973). With the advent of recombinant DNA technology, introducing any gene into cultured cells to analyze its function and regulatory elements by transient expression or by the establishment of a stably transformed cell line became feasible. The recent development of transgenic technology is, in a sense, both technically and conceptually based on nuclear transplantation and gene transfer in cultured cells. The first transgenic animal was produced by Gordon *et al.* (1980). By direct injection of a cloned DNA into the pronucleus of mouse eggs, these investigators demonstrated that the foreign injected gene was integrated and passed on to newborn animals. Since then, transgenic animal technology has become an established experimental tool for the study of gene expression and regulation during development. In addition, this technique was widely used for the creation of new strains of animals with desired characteristics, the overproduction of useful proteins, the establishment of animal models for human diseases, and the correction of genetic defects.

Transgenic technology has now been widely used in several animal systems, including mice, *Drosophila, Caenorhabditis elegans*, and sea urchins, for which vast information on the genetics and developmental biology is available. Significant progress has been made in understanding the function of numerous regulatory genes and their regulatory mechanisms in these species. Similarly, transgenic technology has also been applied to fish. The first successful transgenic attempt was reported in goldfish (*Carassius auratus*) by Zhu *et al.* (1985). Since then, the production of transgenic fish has been reported in more than a dozen fish species. At present, there are two major directions in the field of transgenic fish: one focuses on the application of transgenic technology to aquaculture and the other aims at establishing transgenic fish models for studies in developmental biology. In the past few years, transgenic fish have become a rapidly growing research area in both aquaculture and developmental biology. Several reviews on these developments have been written (Maclean *et al.*, 1987; Hew, 1988; Ozato *et al.*, 1989; Chen and Powers, 1990; Chourrout *et al.*, 1990; Guise *et al.*, 1990; Maclean and Penman, 1990; Fletcher and Davies, 1991; Houdebine and Chourrout, 1991; Hew and Gong, 1992). A special issue of *Molecular Marine Biology and Biotechnology* (Vol. 1, No. 4/5, 1992) and a monograph on transgenic fish (Hew and Fletcher, 1992) have been published. In this chapter, we will highlight the current status of this field and suggest some potential applications of transgenic technology in both fish developmental biology and aquaculture.

II. Fish as a Transgenic System

Taxonomically, fish are classified into Class Osteichthyes within the Subphylum Craniata (Vertebrata), the same subphylum in which all mammals, including mice, cows, and humans are found. As an important group of vertebrates, fish have been exploited extensively as model systems for comparative studies in embryology, neurobiology, endocrinology, evolution, environmental biology, and, more recently, molecular and developmental biology. This research has contributed considerably to our knowledge not only of fish but also of biology as a whole (for review, see Powers, 1989). Because fish radiation occurred much earlier evolutionarily than mammalian radiation, the divergence of fish species is much greater than that of mammalian species. Therefore, the structural and morphological differences among various fish species are much greater than those observed among the mammalian species. This divergence of fish species is also reflected in the variety of fish systems used in scientific research. Traditionally, research in fish biology focused primarily on issues important to fisheries and the economy; consequently, fish species used for scientific research are very diverse and tend to have regional preference. For example, carp and related species are popular in Asia, whereas salmon and rainbow trout are widely used in North America and Europe.

The study of transgenic fish is a newly emerging field with an exciting and promising future. In comparison to other well-known transgenic systems, fish, as vertebrates, have obvious advantages over invertebrates such as *C. elegans*, *Drosophila*, and sea urchins. Compared with the mammalian system, the fish offers many attractive features and advantages. One mature female fish can produce several dozen to several thousand eggs, providing a large amount of genetically uniform material. For example, the zebrafish (*Danio rerio*) produces 150–400 eggs whereas the Atlantic salmon (*Salmo salar*) produces 5,000–12,000 and the common carp (*Cyprinus carpio*) produces more than 100,000 eggs (for review, see Fletcher and Davies, 1991). Egg fertilization in most fish is external. Once fertilized, development, at least until hatching, relies completely on the maternal stores in eggs; thus no further manipulation is required for the fertilized fish eggs to hatch. In mammals, however, fertilized eggs require implantation into recipient mothers after microinjection. Thus, the cost for operation and maintenance of a fish embryo facility is low compared with that of a transgenic mouse facility. Most of the embryos in fish are also transparent, as well as easy to observe and manipulate. In several experimental fish, such as zebrafish and Japanese medaka (*Oryzias latipes*), spawning can be induced by simply manipulating the photoperiod and temperature. Spawning can therefore occur daily throughout the year under optimal laboratory conditions. These fish also reach sexual maturity in 3 months. Thus, they are good models for the studies of inheritance, tissue-specific expression, and developmental regulation of the transgenes.

Many fish are economically important species. The improvement of fish strains in aquaculture has long been an important topic, and has been subject to extensive endeavors using classical breeding approaches. With the advent of transgenic technology, gene transfer in economically important fish species provides a new and revolutionary opportunity to expand both our basic knowledge of fish biology and its direct application to the improvement of fish broodstocks in aquaculture. The availability of chromosomal manipulation techniques to produce sterile fish, as well as the production of monosex culture in several species including salmonids, will also be useful in the biological containment and successful farming of these transgenic fish (Devlin and Donaldson, 1992).

III. Production of Transgenic Fish: Methodology

Fish are the largest and most diverse group of vertebrates, with over 20,000 known species. The divergence of fish species is also reflected in their diversities in morphology, spawning behavior, sexual maturation, and developmental programs. Because of these diversities, numerous methods of gene transfer have been developed for different species. Currently, three major approaches have been developed to generate transgenic fish: microinjection, electroporation of eggs, and electroporated sperm as carrier (Table I).

A. Microinjection

Compared with most other transgenic systems, fish eggs are relatively large, ranging from 1 mm (zebrafish, medaka) to 7 mm (salmon) in diameter. However, their size should not imply ease of microinjection. First, fertilized fish eggs consist primarily of yolk, with only a thin layer of cytoplasm beneath the surface. Second, the eggs of many fish species are opaque because of the presence of cortical granules and chorion, which is formed when the egg is released from the body and which is frequently too hard to allow the penetration of a glass needle. On fertilization, the cortical granules disappear and the cytoplasm migrates to the animal pole, where sperm enter through the micropyle to form an obvious blastodisc. Therefore, microinjection into most fish species is performed immediately after fertilization, including direct injection into the blastodisc of channel catfish (*Ictalurus punctatus*; Dunham *et al.*, 1987; Hayat *et al.*, 1991), medaka (Chong and Vielkind, 1989; Winkler *et al.*, 1992), common carp (Zhang *et al.*, 1990),

Table I Summary of Methods for Creating Transgenic Fish

Method	Species	Survival[a]		Foreign DNA persistence[b]	References
		After treatment	At hatching		
Microinjection					
Cytoplasm, direct	Channel catfish	13	NA	20 (2 of 10), 3 wk	Dunham et al. (1987)
	Medaka	NA	70 (90), 2 wk	95, 1–4 wk (based on reporter gene expression)	Chong and Vielkind (1989); Winkler et al. (1991)
	Carp	NA	26–37 (83–100), 4 d	5.5, >90 d	Zhang et al. (1990)
	Channel catfish	NA	3.3–8 (27–65)	1.5– 5.2, fry	Hayat et al. (1991)
	Northern pike	NA	64 (91), 3 d	6, 2 m	Gross et al. (1992)
Cytoplasm via micropyle	Atlantic salmon	NA	80–90	2–6, 8–11 m	Fletcher et al. (1988); Du et al. (1992a)
	Tilapia	NA	66–90 (56–100) 4 d	3–16.6, 90 d	Brem et al. (1988)
Cytoplasm, dechorized					
Manually	Zebrafish	NA	16–43, 10 d	4, 4 m	Stuart et al. (1988)
Enzymatically	Carp	NA	NA	~50, various stages	Zhu et al. (1989)
	Goldfish	NA	50 (<55), 1 m	7, 1 m	Yoon et al. (1990)
Cytoplasm, 2-step	Rainbow trout	NA	77, 30 d	50, 6–12 m	Chourrout et al. (1986); Guyomard et al. (1989)
	Atlantic salmon	NA	80–90	2–6 8–11 m	Fletcher et al. (1988); Du et al. (1992a)
	Tilapia	NA	66–90 (96–100) 4 d	3–15.6, 90 d	Brem et al. (1988)
Cytoplasm, glutathione-treated	Rainbow trout	98.8 (99.3)	66.7 (74.5)	39, 10–12 m	Yoshizaki et al. (1991a)
Nucleus	Medaka	70	<35	50, 7 d	Ozato et al. (1986, 1992)

(continued)

Table I (*Continued*)

Method	Species	Survival[a]		Foreign DNA persistence[b]	References
		After treatment	At hatching		
Electroporation					
	Loach	NA	NA	62.5, 1m	Xie et al. (1989, 1993)
	Medaka	NA	25, 8–20 d	4, ~20 d	Inoue et al. (1990)
	Zebrafish	72	68 (74), 1 wk	65, 1 wk	Buono and Linser (1992); Powers et al. (1992)
	Carp	45	1–24.5 (10–77)	33.8–58.1, 3–5 m	Powers et al. (1992)
	Channel catfish	5	1.5–11.1 (14–42)	50–100, 3–5 m	Powers et al. (1992)
	Zebrafish	>90	NA	70	Zhao et al. (1993)
Sperm as carrier					
With electroporation					
	Carp	NA	NA	2.6, 1–2 wk	Muller et al. (1992)
	Tilapia	NA	NA	3.2, 1 m	Muller et al. (1992)
	African catfish	NA	NA	3.5, 10 d	Muller et al. (1992)
	Chinook salmon	NA	70–90	1.5 (yolk-free fry)	Sin et al. (1993)
	Zebrafish	NA	NA	23–38, 1 wk	Khoo et al. (1992)
Without electroporation					
	Carp	NA	NA	0 (of 94) 1–2 wk	Muller et al. (1992)
	African catfish	NA	NA	0 (of 126) 10 d	Muller et al. (1992)
	Rainbow trout	NA	NA	0, 2 d, 7 wk	Chourrout and Perrot (1992)
Gene bombardment					
	Loach	~70	NA	5, loach larvae (based on reporter gene expression)	Zelenin et al. (1991)
	Zebrafish				
	Rainbow trout				

[a] Percentage of survival; numbers in parentheses indicate the percentage compared with nontreated controls.
[b] Percentage of examined individuals containing foreign DNA and age at the time of examination; d, day; wk, week; m, month; NA, not available.

6. Transgenic Fish 183

and Northern pike (Gross *et al.*, 1992), and injection via the micropyle in Atlantic salmon (Fletcher *et al.*, 1988; Du *et al.*, 1992a) and tilapia (*Oreochromis niloticus*; Brem *et al.*, 1988). In some species such as goldfish, loach (*Misgurnus anguillicaudatus*), and zebrafish, the problem of the hard chorion has been overcome by removing the chorion either manually (Stuart *et al.*, 1988) or by proteinase digestion (Zhu *et al.*, 1985; Yoon *et al.*, 1990; Culp *et al.*, 1991). Several laboratories developed a two-step method by cutting a small hole in the chorion to allow the insertion of a glass needle into the blastodisc in the salmon egg (Chourrout *et al.*, 1986; Guyomard *et al.*, 1989; Rokkones *et al.* 1989). Yoshizaki *et al.* (1991a,b,1992) injected DNA directly into the cytoplasm of fertilized rainbow trout eggs that were treated with reduced glutathione to prevent hardening of the chorion. All these techniques have been used successfully in the generation of transgenic fish.

Researchers have generally assumed that DNA should best be injected into the nucleus or pronucleus of the egg so that it is more readily integrated into the host genome. If the integration occurs before the first cell division, all subsequent cells of an embryo, including the germline, will contain the transgene, thus avoiding the problem of mosaic integration. Nuclear injection has been proven to be more efficient in mice (Brinster *et al.*, 1985). However, nuclear injection is more difficult in fish because the nucleus cannot be seen. Rokkones *et al.* (1989) have reported attempts to view the pronuclei both by centrifugation of the eggs and by vital staining with fluorescent dyes, but both methods were unsuccessful. However, successful nuclear injection was achieved by injection of DNA into the germinal vesicle in oocytes. Following injection, the oocytes are cultured until maturation, fertilized, and allowed to hatch. This approach has been used by Ozato *et al.* (1986) in the medaka. The immature oocytes were obtained surgically from the ovary prior to ovulation. At the time of removal, the cytoplasm is transparent and the large germinal vesicle is clearly visible. However, this method requires sacrificing the adult fish and thus may not be applicable to other fish species. Moreover, the results obtained to date by nuclear injection do not differ significantly from those obtained by direct cytoplasmic injection. For example, 70% of the nucleus-injected oocytes can be fertilized and 50% of these fertilized eggs develop normally. Of these, 50% contain the injected DNA at the 7-day stage (Ozato *et al.*, 1992). In comparison, over 70% of the cytoplasm-injected embryos survived after hatching, of which essentially 100% of the fry contained injected DNA, as evident from the expression of an exogenously introduced reporter gene (Chong and Vielkind, 1989; Winkler *et al.*, 1991).

B. Electroporation

Electroporation utilizes short electrical pulses to temporarily permeate the cell membrane and thus allow the entry of macromolecules such as DNA into the cell. This method has been used routinely for DNA transfection into bacteria and cultured eukaryotic cells. Electroporation offers several advantages over microinjection in the production of transgenic fish: (1) the transference of foreign DNA into a large number of fish eggs in a short period of time is feasible; (2) it provides more uniform, better controlled, and more reproducible experimental conditions for delivering foreign DNA into eggs; and (3) it eliminates the need for the highly skilled personnel required for the microinjection method. As a mass gene transfer method, electroporation is well suited to dealing with a large number of fish eggs. The first successful gene transfer by electroporation was reported in loach by Xie et al. (1989) and later in medaka (Inoue et al., 1990), zebrafish, carp, catfish, and rosy barb (*Barbus conchomus*; Buono and Linser, 1992; Powers et al., 1992; Muller et al., 1993; Xie et al., 1993). The rates for survival and DNA persistence are comparable to those achieved with microinjection. Our preliminary data in medaka indicate that almost 100% of the electroporated embryos express a reporter gene in early stages, but the level of expression is generally lower than that seen in microinjected embryos, presumably because less DNA is delivered by electroporation than by microinjection (D. Liu and C. L. Hew, unpublished observations).

C. Electroporated Sperm as Carrier

The production of transgenic fish using sperm as carrier has been promoted by the controversial report by Lavitrano et al. (1989) that transgenic mice could be generated from sperm pre-incubated with foreign DNA. Since then, several laboratories have tried fish sperm as carrier for the generation of transgenic fish; the results remain controversial. Khoo et al. (1992) reported successful production of transgenic zebrafish with sperm incubated directly with plasmid DNA, but two other laboratories have found no evidence of generation of transgenic rainbow trout, carp, or catfish by the same strategy (Chourout and Perrot, 1992; Muller et al., 1992). However, successful transgenic fish of several species have been reproducibly generated using electroporated sperm as carrier. In these experiments, DNA was first electroporated into sperm, and the sperm were used to fertilize eggs. By this method, Muller et al. (1992) first reported successful transgenic fish in common carp, tilapia, and African catfish (*Clarias gariepinus*). The transgenic rates were 2–4%, generally lower than the rate achieved with electroporated embryos (>30%; Table I). However, because of the high mortality (55–95%) of electroporated embryos, use of electroporated sperm may be a better alternative for the generation of transgenic fish. This new approach causes less damage to embryonic development and is particularly useful for some species with tough chorions, for which the electroporation of eggs or embryos may not be efficient. Recently, successful gene transfer by electroporated sperm has also been performed in the chinook salmon (*Oncorhynchus kisutch*) by Sin et al.

6. Transgenic Fish   185

(1993), as well as in loach (H. J. Tsai, National Taiwan University, personal communication).

D. Gene Bombardment

Gene bombardment is another approach for the introduction of foreign DNA into fish eggs. This method involves the bombardment of cells by high-velocity microprojectiles of small tungsten particles coated with DNA (Klein et al., 1987). One successful application has been reported by Zelemin et al. (1991). Fertilized eggs of loach, rainbow trout, and zebrafish were bombarded, and approximately 70% of these eggs survived this treatment. Both the detection of foreign DNA and the expression of the transgene in early embryos were observed (Table I).

E. Other Potential Methods

All current gene transfer methods have both advantages and limitations. For instance, microinjection is usually the most direct and effective approach to the transfer of DNA, but it requires considerable technical skill and is tedious and time consuming when dealing with a large number of eggs. Electroporation does offer the opportunity to treat a large number of eggs simultaneously; however, the technique is instrument dependent and it may be difficult in eggs with hard chorions. In this regard, electroporated sperm appear to be a good alternative. However, both electroporation methods generally have lower DNA incorporation rates than microinjection; consequently, the frequency of transgenesis is relatively low and would increase the workload for transgenic screening. Therefore, any improvement in gene transfer will be of considerable interest.

One potential method that could be applied to any transgenic animal system is the generation of transgenics through tissue-specific *in vivo* transfection. Researchers have reported that DNA can be delivered specifically into liver through a protein–DNA complex that interacts with receptors on the liver cell membrane (Wu and Wu, 1988). This protein–DNA complex consists of target DNA, a protein ligand interacting specifically with a liver receptor, and cationic molecules such as poly-L-lysine covalently linked to the protein. The complex is formed by the interaction of negatively charged DNA and positively charged protein particles. The successful delivery of DNA into liver cells through the protein receptor has been demonstrated after intravenous injection of the protein–DNA complex (Wu and Wu, 1988). Similarly, this technique may be applied to the transfection of gonadal cells using proteins that interact specifically with receptors present on gonadal cells. A feasible approach may be the transfection of ovaries using vitellogenin as a carrier for DNA delivery. Vitellogenin, a major egg yolk protein, is synthesized in the liver and transported into the oocyte through blood circulation (Tyler et al., 1988). Similarly, researchers have reported that muscle cells can be transfected by direct injection of DNA in mice (Wolff et al., 1990) and in fish (Rahman and Maclean, 1992). Somatic DNA transfection by direct DNA injection is also successful in other tissues such as the brain (Ono et al., 1990) and the heart (Buttrick et al., 1992). It will be interesting to investigate the possibility of generating transgenic fish by the direct injection of DNA into gonadal tissues. Alternatively, DNA may be injected after appropriate treatments based on conventional in vitro transfection methods such as calcium phosphate precipitation, DEAE–dextran treatment, and artificial liposome mediation. These direct gonadal transfection methods, if successful, would be particularly useful in mammalian species, for which in vitro manipulation is complicated, and in species for which in vitro maturation and fertilization techniques are not yet available. More importantly, these techniques may eliminate one breeding cycle needed to establish stably inherited transgenic strains.

Another approach for the introduction of foreign DNA is the use of a laser microbeam. Over the past few years, the laser has been developed to perform selective subcellular microsurgery. Like electroporation, the laser beam can be used to puncture the cell membrane temporarily, allowing the macromolecules such as DNA to enter the cell. The feasibility of using the laser to transfect cultured human cells has been demonstrated by Tao et al. (1987). An obvious advantage of this technique is that the site of DNA entry is controllable. This feature is particularly attractive when introducing foreign DNA into specific types of cells for specific or ectopic gene expression.

F. Other Considerations

Other considerations for the production of transgenic fish include the design of a transgenic DNA construct, the forms of DNA (linearized vs. circular) to be introduced, and the choice of genomic DNA or cDNA. Generally, a transgenic DNA construct should include a gene promoter, a structural gene, and proper transcriptional termination signals consisting of both the polyadenylation signal (AATAAA) and a thymidine-rich region (TTTTTNT) downstream from the polyadenylation signal and present only in the genomic DNA (Du et al., 1992b). No systematic investigation has been done comparing the use of cDNA or genomic DNA in transgenic studies, although some earlier work in transgenic mice indicated a superiority of genomic DNA over cDNA (Brinster et al., 1988). Yoshizaki et al. (1991a,b) have transferred a carp α-globin genomic clone into rainbow trout and reported that 40% of 1-yr-old fish contained the foreign gene. This ratio is considerably higher than that given in most other reports (2–7%) using either growth hormone cDNA clones (e.g., Penman et al., 1990; Du et al., 1992a) or antifreeze protein genomic clones (Fletcher et al., 1988), but the difference may be due to the techniques used in different laboratories. Researchers have demonstrated that the use of linearized DNA results in a higher incorporation frequency in the mouse (Brinster et al., 1988); this may be also true in fish (Chourrout et al., 1986; Penman et al., 1990). Investigators generally recommend that, for the production of stably transmitted transgenic fish, linearized DNA be used; the circular form is a better choice for transient expression (Chong and Vielkind, 1989).

IV. Transgenesis in Fish: Integration, Inheritance, and Expression

A. Persistence and Integration of the Exogenously Introduced DNA

After the introduction of foreign DNA into the embryos, the first thing to be examined is the persistence of the foreign DNA in the host. This goal has been accomplished by DNA hybridization on dot blots or genomic Southern blots, and more recently by polymerase chain reaction (PCR). DNA is commonly prepared from the blood or fin of juvenile fish. The persistence of exogenously introduced DNA is detected usually in only a small percentage of fish developed from DNA-injected eggs; thus, screening for positives represents a major part of the work in transgenic studies, since a large number of fish eggs can be easily available and manipulated. Because of the development of the PCR technique, transgenic fish can now be screened rapidly using less than 1 µl blood after a mild alkaline treatment (Davies and Gauthier, 1992; Du et al., 1992a). Hundreds of samples can be processed easily in a few days. Therefore, the PCR approach has become an important experimental tool for meeting the requirement of handling the large number of samples in transgenic fish research.

The persistence of foreign DNA was demonstrated repeatedly in many species, but the fate of foreign DNA immediately after injection was closely followed only in two model fish, zebrafish (Stuart et al., 1988) and medaka (Chong and Vielkind, 1989). Based on the work of Chong and Vielkind (1989) in the medaka, exogenously introduced DNA underwent an amplification of 10- to 12-fold within 24 hr of development or at the gastrula and neurula stages. The amplification occurred regardless of the form of injected plasmid DNA (linearized or circular supercoiled). The amplification yielded predominantly high molecular weight concatemers of the injected DNA, which was particularly evident with the injection of linearized DNA. The foreign DNA in early embryos was apparently extrachromosomal and gradually declined after the peak accumulation at the neurula stage. By the time the free-swimming stage was reached (4 wk after fertilization), only a fraction of the injected embryos retained the exogenous DNA. When circular plasmid DNA was injected, all possible forms of plasmid DNA were present in early embryos including open circular, multimeric circular, supercoiled, and the high molecular weight forms. By the time of hatching (about 11 days after fertilization), only the high molecular weight form was detected while all other forms disappeared (Chong and Vielkind, 1989). Similar observations on the formation of concatemers, and on the rise and fall of injected DNA, are also reported for zebrafish (Stuart et al., 1988), carp, and loach (Zhu et al., 1989). The fate of electroporated DNA appears to be identical to that of injected DNA (Xie et al., 1989,1993).

Extrachromosomal amplification of injected DNA appears to be a common phenomenon for many transgenic systems, including *Xenopus* (Harland and Laskey, 1980), sea urchins (McMahon et al., 1985), and *C. elegans* (Stinchcomb et al., 1985). However, the mechanism for the amplification remains unknown. The injected DNA is likely to be extrachromosomal and to undergo both DNA replication and degradation. In a rapidly dividing embryo that has high DNA replication activity and little or no degradation activity, the rate of extrachromosomal DNA replication is far in excess of its degradation rate. At later embryonic stages, the rate of cell division slows down, as does the DNA replication activity; thus, the degradation of extrachromosomal DNA becomes apparent. Consistent with this series of events, when DNA is injected into the neural tubes of late-stage embryos of the platyfish, the amount of injected DNA declines rapidly during embryonic development and no amplification of the foreign DNA is observed (Winkler et al., 1992). Based on our model, we predict that the persistence of extrachromosomal DNA should have a preference for the rapidly dividing cells or tissues.

The next question to be addressed about the introduction of foreign DNA is whether it is integrated into the host genome. The reports of transgene integration in earlier studies were usually based on the following evidence: the persistence of the foreign DNA after a long period of time or in the adult fish; the appearance of high-molecular-weight foreign DNA; and the presence of extra restriction fragments of foreign DNA in the genomic Southern analysis. This evidence may not be completely valid, since extrachromosomal high-molecular-weight DNA can be persistent for a long period of time and may be passed on to the next generation (Stinchcomb et al., 1985; Rassoulzadegan et al., 1986). Therefore, in most transgenic fish studies, only the persistence of foreign DNA, rather than its integration into the host chromosome, was demonstrated. Conclusive evidence of transgene incorporation into host genomes comes from inheritance studies and will be discussed later. The percentage of transgene persistence in injected fish varies from one report to another, depending mainly on the developmental stages of the fish or on the time elapsed since introduction of foreign DNA when the assay is performed. Usually, in fish with relatively long life cycles, only a small percentage of the fry developed from DNA-injected eggs retain the foreign gene (for example, 2–6% in Atlantic salmon, Fletcher et al., 1988; Du et al., 1992a; 5.5% in common carp, Zhang et al., 1990; 6% in Northern pike, Gross et al., 1992). Some significantly higher percentages (20–50%) were reported for the rainbow trout (Chourrout *et al.*, 1986; Rokkones *et al.*, 1989; Yoshizaki *et al.*, 1991a).

Although the persistence of foreign DNA does not demonstrate its incorporation into the host genome, it is definitely a prerequisite for DNA integration. The extrachromosomal foreign DNA is conceivably unstable; thus, it is important to increase its integration frequency. Ivics *et al.* (1993) have co-injected the foreign DNA and a purified retroviral integrase into zebrafish embryos to increase the incorporation of transgenic DNA into the genome. The improvement of transgenesis in zebrafish was evident by the presence of a higher level of transgenic DNA and an enhanced expression of the reporter gene. The earliest integration of the transgene into the genome was detected in 14-day-old embryos by linker-mediated PCR. Researchers have reported that the efficient insertion of genes into the germline can be achieved by retroviral vectors in mice (Van der Putten *et al.*, 1985) and the P element transposon in *Drosophila* (Spradling and Rubin, 1982). However, at present, no such retroviral vectors and transposable elements are available for fish. Another potential approach to increasing the transgene integration rate may be the inclusion of a fish repetitive DNA sequence to increase the chance of integration by homologous recombination or transposon mediation. Several fish repetitive DNA sequences have now been identified from carp (Datta *et al.*, 1988), zebrafish (Ekker *et al.*, 1992a; He et al., 1992), medaka (Naruse *et al.*, 1992), and chinook salmon (Du, 1993). However, a preliminary test in zebrafish using an Alu repeat sequence in the transgenic vector showed no evidence of enhanced integration (He *et al.*, 1992). Investigators have reported that in yeast, the rate of homologous recombination is not affected by the copy number of the DNA sequence in the genome (Zheng and Wilson, 1990). Moreover, a transgene integrated by homologous recombination of a repetitive DNA sequence, most of which is located in the transcriptionally inactivated heterochromosomal region, may not be transcriptionally activated.

B. Mosaicism and Inheritance of the Foreign DNA

The inheritance of the transgene is obviously an important subject in transgenic fish. Most of the early transgenic fish reports are limited to the detection of the persistence of the foreign DNA. Inheritance of the transgene has been demonstrated to date in only a few species, including zebrafish (Stuart *et al.*, 1988,1990; Bayer and Campos-Ortega, 1992), rainbow trout (Guyomard *et al.*, 1989; Yoshizaki *et al.*, 1991b), medaka (Inoue *et al.*, 1990), carp (Zhang *et al.*, 1990; Chen *et al.*, 1993), and Atlantic salmon (Shears *et al.*, 1991).

Persistence of the injected DNA does not guarantee the inheritance of the transgene to the next generations since the integration of a transgene, usually a rare event, is likely to happen at late embryogenesis, and independently in different cells. Therefore, tissue mosaicism of the transgene in transgenic fish is common, and only the transgene in the germline is capable of being inherited. Transgene mosaicism is evident by the fact that different tissues contain different copy numbers of the transgene (Guyomard et al., 1989; Winkler et al., 1991; Gross et al., 1992). An unsuccessful attempt to use the drug G-418 to select transgenic fish by transferring the neomycin-resistance gene is probably due to this mosaicism (Yoon et al., 1990). A direct demonstration of transgene mosaicism came from work by Westerfield et al. (1992), who injected fluorescently labeled DNA into zebrafish. The injected DNA immediately formed a spherical bolus that became parceled among a subset of the cellular progeny of the injected cell.

The pattern of transgene inheritance in the F_1 generation also indicates germline mosaicism. When a transgenic fish (P_1) is mated with a nontransgenic control, the rate of F_1 transgenic progeny from different P_1 individuals varies: 36, 54, 12, and 25% in one instance in zebrafish (Stuart et al., 1990) and 17, 15, 33, and 64% in another instance in Atlantic salmon (Shears et al., 1991). However, F_2 transgenic progeny from F_1 transgenic fish crossed with a nontransgenic control showed the typical Mendelian ratio of 50% (Stuart et al., 1990; Shears et al., 1991). Thus, a stably inheriting line of transgenic fish can be established after the F_2 generation. In the zebrafish, such stable lines of transgenic fish showed reproducible patterns of transgene expression (Stuart et al., 1990; Bayer and Campos-Ortega, 1992), indicating the feasibility of the production of transgenic fish lines with stably inherited characteristics.

C. Expression of the Transgene

Although numerous reports have been made on the persistence and inheritance of transgenes, the expression of transgenes has been reported rarely apart from a few studies using transgenic fish as a transient system to express reporter genes in early embryos. These studies detected expression by measuring the activity of reporter enzymes including β-galactosidase (McEvoy et al., 1988; Culp et al., 1991; Winkler et al., 1991; Westerfield et al., 1992), chloramphenicol acetyltransferase (CAT; Stuart et al., 1988,1990; Chong and Vielkind, 1989; Liu et al., 1990; Gong et al., 1991; Winkler et al., 1991; Yoshizaki et al., 1992; Moav et al., 1993), and luciferase (Tamiya et al., 1990; Sato et al., 1992; Muller et al., 1993). Most of these studies were carried out in early embryos when a large amount of extrachromosomal foreign DNA was present. In the medaka, the earliest expression of CAT activity was detected only 12 hr (gastrula stage) after DNA microinjection (Chong and Vielkind, 1989). CAT activity was detected continuously for at least 4 wk when the fish developed into free-swimming fry (Chong and Vielkind, 1989; Winkler et al., 1991). In some stable lines of transgenic zebrafish with the CAT gene, expression of the transgene was found in many tissues of adults including fins, skin, heart, muscle, gills, eyes, and gut (Stuart et al., 1990).

In the case of growth hormone (GH) gene transfer, because of the obvious phenotype, the growth rates and sizes of fish have become the indicators for transgene expression. Frequently, GH transgenic fish are reported to be 20–30% larger than their sibling controls (Zhang et al., 1990; Gross et al., 1992), but no effort has been made to measure the expression of the transgene at either the mRNA level or the protein level. In a few cases, the total GH level in blood was measured in transgenic fish, and was generally higher than that in nontransgenic controls (e.g., Du et al., 1992a), but no direct evidence attributed the higher level of GH to the expression of the transgene or the endogenous GH gene. In some studies using mammalian GH gene constructs, human or bovine GH has been detected by radioimmunoassay using specific antibodies (Rokkones et al., 1989; Gross et al., 1992). In our own studies, the expression of transgenic GH mRNA is too low to be detected by regular Northern blot analysis, but can be detected by a more sensitive assay such as reverse transcription PCR (RT/PCR, Du 1993). Chen et al. (1993) also reported the detection of transgenic GH from various tissues in the common carp. The expression of the GH transgene in loach has also been detected by another sensitive assay, S1 nuclease mapping (Beniumov et al., 1989).

Direct detection of the transgenic protein products was clearly demonstrated in Atlantic salmon transferred with a winter flounder (*Pleuronectes americanus*) antifreeze protein (AFP) gene (Shears et al., 1991). The advantage of this transgenic system is that the salmon does not contain the AFP gene in its genome, so there is no endogenous interference with detection of flounder AFP precursor (proAFP) in salmon sera by antibody. Liver-specific and seasonal expression of AFP mRNA in the transgenic salmon was also detected by standard Northern blot analysis (P. Davies, Queen's University, Kingston, Canada, personal communication). Expression of the chicken δ-crystallin gene was also detected by immunocytochemistry in transgenic medaka, which lacks the δ-crystallin gene (Ozato et al., 1986; Inoue et al., 1989). These reports clearly demonstrate the successful expression of exogenously introduced genes in transgenic fish.

Expression of the transgene might be limited by mosaicism and position effects in the chromosome. Mosaicism, in some cases, can be overcome by subsequent breeding. The position effect is the variability of transgene expression caused by the location of the transgene integrated in the chromosome. One possibility for overcoming the position effect is including a locus activation region (LAR), a DNase I hypersensitive region identified upstream from the mammalian globin gene cluster and responsible for the active expression of the globin gene family. LAR has been demonstrated to enhance transgene expression in transgenic mice independent of the position of transgene integration (Grosveld *et al.*, 1987). Similar LAR DNA can also be included in the study of transgenic fish, but no such LAR region in fish has yet been described.

V. Application of Transgenic Fish in Aquaculture

The rapidly expanding human population demands an increasing amount of food, not only from agriculture but also from natural aquatic sources and aquaculture, including fish, crustaceans, and shellfish. Many commercially important fish species have been employed as research models for decades, and a great deal of knowledge has been accumulated on the cultivation, physiology, and endocrinology of these species. Improvement of fish stocks by transgenic techniques is therefore a logical and feasible development. Any gene transfer technology that increases growth rates, improves feed conversion efficiency, enhances cold and freeze tolerance, improves disease resistance, decreases mortality rates, and reduces the cost of production will be potentially beneficial to the aquaculture industry.

A. Growth Hormone Gene Transfer

The pioneering work of Palmiter *et al.* (1982) in creating the supermouse by transferring GH genes has provided the impetus for the production of other superanimals for livestock improvement. GH gene transfer has been carried out in several domestic animals including rabbits, sheep, and pigs (Hammer *et al.*, 1985). Because of the obvious benefit to aquaculture, the extension of GH gene transfer to fish has been carried out in many laboratories worldwide. The first GH gene transfer was reported by Zhu *et al.* (1985) in goldfish. This technique was later applied to many commercially important species, including rainbow trout (Chourrout *et al.*, 1986; Rokkones *et al.*, 1989; Penman *et al.*, 1990), Atlantic salmon (Rokkones *et al.*, 1989; Du *et al.*, 1992a), carp (Zhu *et al.*, 1989; Zhang *et al.*, 1990), channel catfish (Dunham *et al.*, 1987), tilapia (Brem *et al.*, 1988), and Northern pike (Gross *et al.*, 1992; Guise *et al.*, 1992). In many cases, model fish that have shorter life cycles and can be maintained at lower costs have been used to assess the conditions for GH gene transfer techniques; these fish include loach (Zhu *et al.*, 1986; Beniumov *et al.*, 1989) and medaka (Inoue *et al.*, 1990; Lu *et al.*, 1992). All these studies have demonstrated successful gene transfer into the host fish, but the expected enhancement of growth varied from the modest 20% increase of some reports (Zhang *et al.*, 1990) to the 2- to 6-fold increase of others (Zhu *et al.*, 1986; Du *et al.*, 1992a).

The success of producing commercially useful transgenic fish depends not only on the structural gene used but also on the regulatory sequences for the structural gene. A transgenic DNA construct is generally composed of a promo-

6. Transgenic Fish

ter or regulatory sequence, a structural gene, and proper transcriptional termination signals. Because of the limited availability of fish genes and fish promoters, most of the earlier transgenic fish studies utilized DNA constructs derived from mammals, for example, the mouse metallothionein gene (MT) promoter/human GH gene (Zhu *et al.*, 1985,1986; Guyomard *et al.*, 1989; Rokkones *et al.*, 1989; Penman *et al.*, 1990), the mouse MT promoter/rat GH gene (Guyomard *et al.*, 1989; Penman *et al.*, 1990), the SV40 promoter/human GH gene (Chourrout *et al.*, 1986; Guyomard *et al.*, 1989), and the RSV promoter/bovine GH gene (Gross *et al.*, 1992; Guise *et al.*, 1992; Table I). Although some of these gene constructs are effective in transgenic mice (Palmiter *et al.*, 1982), their effect in fish is largely unknown. No stimulation of growth rate was reported in most of these studies, but Zhu *et al.* (1986,1989) reported a 3- to 5-fold size increase in carp and loach using a mouse MT promoter/human GH gene construct. However, the effectiveness of these gene promoters, such as the mouse MT promoter and the SV40 promoter, in fish cells remains to be evaluated. Strong transcriptional activity from the MT promoter will apparently require heavy metal induction. Several fish GH genes have been isolated (Agellon and Chen, 1986; Hew *et al.*, 1989) and used for gene transfer (Zhang *et al.*, 1990; Du *et al.*, 1992a; Gross *et al.*, 1992). Zhang *et al.* (1990) published the first report on fish GH gene transfer. They used a retroviral promoter (RSV-LTR) linked to rainbow trout GH cDNA (RSV/rtGH) and transferred this gene construct into the common carp. A statistical measurement indicated a 20% increase in size in transgenic fish over the nontransgenic siblings. This value is rather low compared with the average 4- to 5-fold enhancement of growth using another fish GH gene construct (Du *et al.*, 1992a). A potential limitation of the RSV/rtGH construct is its lack of a sequence encoding the signal peptide for secretion of the growth hormone into the circulation, thus the effectiveness of transgenic GH may be restrained.

To create products that are acceptable for human consumption, it is advisable not to use viral promoters and human GH structural genes that might raise consumer concern about gene transfers. Consistent with this notion, we have developed an all-fish gene construct consisting of an ocean pout (*Macrozoarces americanus*) antifreeze protein gene (opAFP) promoter, a chinook salmon GH gene, and the opAFP polyadenylation and transcriptional termination signals (Du *et al.*, 1992a). When this construct was transferred into Atlantic salmon, a dramatic enhancement of growth rate for transgenic fish was observed. In the population of fish developed from microinjected eggs, essentially all large fish (at least 2-fold larger than the average in body weight) contained the exogenous GH gene; the largest individual was 9-fold larger than the average. On average, the transgenic fish were 4- to 5-fold heavier than their sibling controls (Du *et al.*, 1992a). With the same transgenic construct, similar observations have also been made in four other salmonids: coho salmon (*O. kisutch*), chinook salmon (*Onchorhynchus tschawytscha*), rainbow trout (*Onchorhynchus mykiss*), and cutthroat trout (*Onchorhynchus clarki*); the largest transgenic fish was more than 30-

fold larger than the nontransgenic controls (Devlin et al., 1992). This DNA construct was also used in a distantly related species, the loach (*M. anguillicaudatus*). Results in the loach showed that the experimental group injected with the GH construct was 2.5-fold larger than the noninjected control group (H. J. Tsai, National Taiwan University, personal communication). The actual difference between transgenic and nontransgenic fish is probably even greater, since there is no discrimination between the two in the experimental group. Therefore, the success of enhancement is not fish species dependent, but gene construct dependent. The potential of stimulation by GH in fish appears to be much larger than in mice; the supermice generated by GH gene transfer are only about 2-fold larger (Palmiter et al., 1982). This result may not be surprising, in view of the large natural variation in fish size within the same species.

The strong stimulation of growth in GH transgenic fish is only the first step toward the application of transgenesis to aquaculture, and many subsequent efforts should be made. First, it is necessary to establish stable lines of transgenic fish with reproducibly enhanced growth. The feasibility of having a stable transgenic line with a reproducible pattern of transgene expression was demonstrated in zebrafish (Stuart et al., 1990). Second, the physiology of GH transgenic fish needs to be studied. In some GH transgenic livestock, although the growth rate is stimulated, many animals develop numerous diseases including gastric ulcers, arthritis, cardiomegaly, dermatitis, renal disease, and infertility (Pursel et al., 1989). Devlin et al. (1993) noticed that some of the largest transgenic fish displayed a phenotypic syndrome consisting of head, fin, jaw, and opercular abnormalities arising from what superficially appeared to be excess cartilage and bone growth; thus, detrimental stimulation of growth could occur in GH transgenic fish. Obviously, the level of GH transgene expression will need to be fine tuned. Third, it is necessary to select a stable line of transgenic fish, with an adjustment between growth enhancement and fish health. This selection is possible in view of the many phenotypic variants that are presumably caused by the position effect of the transgene in the host genome. Other considerations, such as ecological impact and safety requirements, will be discussed later.

Our success in GH gene transfer appears to be largely dependent on the use of the proper gene regulatory sequence, the opAFP promoter. This promoter offers many attractive features for transgenic studies. First, it is expressed predominantly in the liver (Gong et al., 1992), a tissue that has large synthetic and secretory capacities. The expression of the transgene in the liver is one of the most common approaches in gene transfer studies. Second, the AFP gene is present only in a small number of fish species. Its expression will not interfere with homologous endogenous genes. Third, several positive and negative *cis*-acting elements are found in the opAFP gene promoter (Gong and Hew, 1993), providing an opportunity to modify this vector to meet the different requirements of transgene expression. An all-fish gene cassette for gene transfer in aquaculture has been described by Du et al. (1992b). This all-fish gene cassette is based on the opAFP/GH gene construct, in which the GH gene has been replaced by a short linker sequence to facilitate the insertion of other potentially interesting genes. This cassette should be applicable for any gene to be controlled by a liver gene promoter. Another all-fish transgenic vector was previously reported using the carp actin gene promoter and a salmon GH polyadenylation signal (Liu *et al.*, 1990), but this construct lacked a transcriptional termination signal; thus, its effectiveness as a transgenic vector is less convincing. Moreover, the exogenously introduced actin gene promoter may interfere with endogenous actin gene expression. Such homologous interference of exogenous promoter and endogenous gene expression has been observed previously in transgenic mice (McGrane *et al.*, 1990).

B. Antifreeze Protein Gene Transfer

Another useful system for generating commercially useful transgenic fish is AFP gene transfer. AFP genes are present in only a limited number of species of fish living in subzero freezing environments (Davies and Hew, 1990). Salmonids and most other commercially important fish do not have AFP genes and freeze to death if they come into contact with icy seawater at temperatures below −0.7°C. Therefore, the production of stable lines of freeze-tolerant salmon and other fish species will greatly facilitate the development of aquaculture in many regions where the only limiting factor is the freezing temperature. The feasibility of protecting salmonids against freezing in ice-laden seawater has been demonstrated by infusing seawater-acclimated rainbow trout with AFP purified from the winter flounder (Fletcher *et al.*, 1986). A complete genomic AFP gene from the winter flounder has been incorporated successfully into Atlantic salmon (Fletcher *et al.*, 1988). Research to date indicates low levels of expression of the AFP transgene in about 3% of the salmon developed from microinjected eggs. Inheritance of the AFP gene by offspring (F_1) from crosses between P_1 transgenic and wild-type salmon revealed that transgenic founders (P_1) were germline mosaic. Low levels of AFP could be detected in the blood of all these transgenic offspring (F_1). Approximately 50% of the progeny produced by crosses between transgenic F_1 and wild-types contained the AFP gene, indicating the possibility of the establishment of stable germline-transformed Atlantic salmon (Shears *et al.*, 1991). The current problem with AFP gene transfer is the low level of production of AFP in transgenic salmon. The winter flounder has 40–50 copies of AFP genes in the genome (Scott *et al.*, 1985) to produce 2–5 mg/ml (functional concentrations) AFP in the blood. However, only 10–50 μg/ml proAFP is produced in transgenic salmon. Current research focuses on improving AFP production in transgenic fish. The possible approaches include increasing the copy number of the transgene by crossing different stable lines of transgenic fish, performing another round of microinjection using transgenic fish as recipients, using strong promoters/enhancers in the transgenic construct, or using engineered AFP genes with stronger AFP activity (Hew et al., 1992).

C. Other Potential Genes for Gene Transfer

Other potential genes used in fish aquaculture include those to enhance fish growth, such as the growth hormone-releasing hormone (GHRH), the growth hormone receptor, and the insulin-like growth factor I (IGF-I). The GHRH gene (Parker et al., 1993) and the IGF-I gene (Cao et al., 1989; Shamblott and Chen, 1992) were recently isolated from fish. The stimulation of growth of transgenic mice by transfer of the GHRH or IGF-I gene has been demonstrated (Hammer et al., 1985; Mathews et al., 1988). Overexpression of other peptide hormone genes in fish may also be useful. For example, prolactin is involved in osmoregulation and the transition of fish from seawater to fresh water (Pickford and Phillips, 1959). It might be possible to transform some marine species into freshwater species by prolactin gene transfer. Changing the pattern of expression of gonadotropin genes might be useful for shortening/lengthening reproductive cycles. Other candidate genes include the gene for complex carbohydrate hydrolases to generate a fish that can use a low-cost carbohydrate diet; the gene(s) for fatty acid unsaturase to increase the level of unsaturated fatty acids; and the recently discovered Nramp gene to improve host resistance to a variety of pathogens (Vidal et al., 1993).

D. Ecological Concerns

Several concerns arise regarding the use of transgenic fish for aquaculture. Unlike most transgenic animals used for animal husbandry or experimental research, transgenic fish pose the danger of accidental release from fish farms. Although unsubstantiated, some individuals have suggested that "superior" transgenic fish might invade and dominate wild-types, and thus destroy the delicate ecological balance of aquatic environments (Hallerman and Kapuscinski, 1992). However, several options are available that will minimize any potential environmental hazards. The first concerns physical containment: the selection of a site, its monitoring, and the security of the facility. The second concerns the biological containment of transgenic fish. Transgenic fish can be made sterile by chromosomal manipulation and transgenesis (Devlin and Donaldson, 1992). Except for the broodstock, the fish used in aquaculture would be sterile, thus minimizing the ecological impact due to accidental escape. Cell lineage ablation and antisense techniques might also be applicable for making the transgenic fish sterile. For example, fusing gonadotropin gene promoters to toxin genes might be useful in destroying the gonadotrophs in the pituitary. The lack of circulating gonadotropins would render the animal reproductively incompetent. Similarly, antisense genes against gonadotropin-releasing hormone (GnRH) mRNA would inhibit GnRH action on gonadal development. The hypogonadal broodstock could be rescued by GnRH injection (Alestrom et al., 1992).

VI. Application of Transgenic Fish in Developmental Biology

A. Fish as a Model in Developmental Biology

Fish have long been a favorite model for the study of embryogenesis. Detailed embryologies of many species, such as killifish (*Fundulus heteroclitus*; Oppenheimer, 1937), platyfish (*Platypoecilus maculatus*; Tavolga and Rugh, 1947), medaka (*O. latipes*; Matsui, 1949), chum salmon (*Oncorhynchus keta*; Mahon and Hoar, 1956), zebrafish (*B. rerio*), and blue gourami (*Trichogaster trichoperus*; Hisaoka and Firlit, 1962) have been documented. Extensive nuclear transplantation experiments have been carried out on fish to address one of the fundamental questions in developmental biology: the relationship between and importance of the nucleus and cytoplasm (for review, see Yan, 1989). In the 1970s, Tung and Niu (1977a,b) performed a series of experiments to inject mRNAs or purified genomic DNA from one fish species to another; they found that the recipient fish developed some morphological characteristics of the donor species and concluded that these features were inherited through the injected mRNAs. Although the conclusion was controversial, the technique and concept are similar, if not identical, to those for transgenic fish.

Fish developmental biology lags greatly behind several well-developed model systems such as the mouse, *Drosophila*, and *C. elegans*. This lag is partly because research in fish molecular biology has been diluted across many different species, with an emphasis on economically important species such as salmon, rainbow trout, and carp. These fish have a long life cycle and are expensive to maintain, and thus are not suitable for the analysis of developmental events. Tremendous efforts have been made to promote the zebrafish as a new model system in developmental biology in vertebrates (Barinaga, 1990). The zebrafish offers the possibility of combining excellent embryology and genetics. The first detailed zebrafish embryology was described by Hisaoka and Battle (1958). Complete cell lineages were established (Kimmel, 1989; Kimmel et al., 1990; Strehlow and Gilbert, 1993). Elegant cell transplantation was performed to determine the fate of a single cell (Ho and Kane, 1990; Ho and Kimmel, 1993). Many genetic mutants were accumulated, and new mutants can be induced by exposure to γ-ray radiation (Chakrabarty et al., 1983; Walker and Streisinger, 1983), ultraviolet light (Grunwald and Streisinger, 1992a), or several mutagens (Grunwald and Streisinger, 1992b). The molecular mechanism for some of the mutants is beginning to be understood (Moven et al., 1990). An increasing number of zebrafish genes, particularly those important in early development (e.g., the homeobox genes), have been described and the pattern of their expression has been well documented

Table II Highlights of Transgenic Fish[a]

Gene constructs (promoter/gene)	Fish species	Expression	Inheritance	Biological effects	Comments	References
GH gene transfer						
mouse MT/hGH	Goldfish, loach, carp	Yes	Yes	3 to 4-fold size increase	First transgenic fish report	Zhu et al. (1985, 1989)
SV40/hGH, SV40-MMTV/rat GH, mMT/hGH, mMT/rat GH, H-2k antigen/hGH	Rainbow trout	ND	Yes	ND		Chourrout et al. (1986); Guyomard et al. (1989)
mMT/hGH	Channel catfish, tilapia	ND	ND	ND		Dunham et al. (1987); Brem et al. (1988)
mMT/hGH	Atlantic salmon, rainbow trout	hGH detected in early embryos	ND	ND		Rokkones et al. (1989)
RSV-LTR/trout GH	Carp	Yes, in various tissues	Yes	~20% size increase	First fish GH gene transfer	Zhang et al. (1990); Chen et al. (1993)
opAFP/salmon GH	Atlantic salmon	Overall increase in circulating GH	ND	4 to 6-fold size increase	All-fish gene transfer	Du et al. (1992a)
carp actin/salmon GH, RSV/bovine GH	Northern pike	Bovine GH	ND	~20% size increase		Gross et al. (1992)
AFP gene transfer						
wf AFP gene	Atlantic salmon	Yes	Yes	ND	ProAFP synthesized; low-level expression (10–50 μg/ml)	Fletcher et al. (1988); Shears et al. (1991)
Reporter gene transfer						
SV40/CAT	Zebrafish	Yes	Yes	ND	Stable transgenic line expressed	Stuart et al. (1988, 1990)

Construct	Species	Expression	Phenotype	Purpose	Reference
SV40-RSV/CAT	Medaka	Yes	ND	Transient expression system	Chong and Vielkind (1989)
carp actin/CAT	Zebrafish, goldfish	Yes	ND	Promoter analysis	Liu et al. (1990); Moav et al. (1993)
various fish AFP/CAT	Medaka	Yes	ND	Temporal expression and promoter analysis	Gong et al. (1991)
SV40-RSV/CAT, SV40/LaCZ, RSV/LacZ	Medaka	Yes	ND	Temporal and spatial expression	Winkler et al. (1991)
mHox.1/CAT, hHox 3.3/CAT	Zebrafish	Yes	ND	Spatial expression	Westerfield et al. (1992)
RSV/LacZ, m heat shock/LacZ	Zebrafish	Yes	Yes	Enhancer trap	Bayer and Campos-Ortega (1992)
RSV/luciferase	Medaka	Yes	ND	Expression detected in fixed embryos with spatial image	Tamiya et al. (1990)
mMT/luciferase	Medaka	Yes	ND	Expression detected in embryonic extract	Sato et al. (1992)
Other gene transfers					
chicken δ-crystalin gene	Medaka	Yes	ND	Temporal and spatial expression	Ozato et al. (1986); Inoue et al. (1989)
RSV/neomycin	Goldfish	Yes	Not resistart to neomycin ND		Yoon (1990)
carp globin gene	Rainbow trout	ND	Yes		Yoshizaki et al. (1991a,b)
mouse tyrosinase gene	Medaka	Yes	Yes. Skin based on pigmentation phenotype	Gene function study	Matsumoto et al. (1993)

[a] Abbreviations: ND, not determined; AFP, antifreeze protein; CAT, chloramphenicol acetyltransferase; GH, growth hormone; Lac Z, β-galactosidase; LTR, long terminal repeats; MT, metallothionein; RSV, Rous sarcoma virus, SV40, Simian virus 40; h, human; m, mouse; op, ocean pout; wf, winter flounder.

(e.g., Hatta et al., 1990; Njolstad et al., 1990; Ekker et al., 1992b; Fjose et al., 1992; Krauss et al., 1992). A stable line of transgenic fish with reproducible patterns of transgene expression was established (Stuart et al., 1988,1990; Culp et al., 1991). Furthermore, hundreds of parthenogenetic progeny can be generated from a single mother by using pressure treatments or temperature shocks to inhibit either the second meiotic or the first mitotic division of eggs fertilized with inactivated sperm (Streisinger et al., 1981). This procedure effectively saves one generation in the production of homozygous mutant individuals. An equally good model for developmental studies is the Japanese medaka, which shares many of the advantages of the zebrafish. The medaka has been explored as a biological model since the beginning of the century in Japan, and many mutants and genetic linkages are available (Yamamoto, 1975; Egami et al., 1990).

Fish, as popular developmental models at the molecular level, have several drawbacks. Notably, there is no genetic linkage map and the number of cloned genes is relatively small. Therefore, much work is needed to expand the repertoire of genetic information in fish. A fish genome project was launched for the fish *Fugu rubripes*, which has a small genome (Brenner et al., 1993). We have recently developed a procedure called gene tagging to increase rapidly the availability of fish gene probes. Our approach is to partially sequence random cDNA clones from cDNA libraries, as done by Adams et al. (1991) in the human genome project. The partial sequences obtained act as gene tags for each cDNA clone used to search for sequence homology in DNA databases. Over one-third of the random clones share significant homology with known sequences in DNA databases, and most of the identified homologous clones have not been reported in fish (Gong et al., 1994). Thus, gene tagging is a highly efficient approach to rapidly increasing the availability of fish gene resource. Our short-term goal is to establish a tagged zebrafish gene library with thousands of tagged cDNA clones, which should be a worthy investment at this stage of fish developmental biology.

B. Transgenic Fish as a Transient Expression System

Most transgenic fish research to date focuses on the methodology and its application in aquaculture; relatively few reports have been made on the establishment of a transgenic fish model for the analysis of developmental mechanisms. Two freshwater fish, the medaka and zebrafish, have received intensive attention as transient expression systems for the study of the early events of transgene expression. Such systems are useful for the functional analysis of the regulatory elements of a promoter and their roles in regulation of the temporal and spatial pattern of gene expression in early embryos. The feasibility of these studies was explored initially by Stuart et al. (1988,1990) in zebrafish and by Chong and Vielkind (1989) in medaka, using nonfish promoters. In both fish, researchers have demonstrated the ability to express the exogenously introduced genes.

Using transgenic medaka as a transient expression system, we have analyzed the promoter regions of several AFP genes from several cold water fish (Gong et al., 1991). A number of positive and negative regulatory regions were defined that generally agreed with our transfection data in cultured fish cells (Gong and Hew, 1993). The significant increase of the AFP promoter activity correlated temporally with the development of the embryonic liver, and was consistent with the fact that AFP genes are expressed predominantly in the liver (Gong et al., 1992). Similar studies have also been carried out for an MT promoter isolated from a platyfish (*Xiphophorus maculatus*) in both medaka and platyfish embryos (Winkler et al., 1991,1992), and for a carp actin gene promoter in zebrafish and goldfish (Moav et al., 1993).

The spatial pattern of gene expression has also been analyzed in transgenic fish. Inoue et al. (1989) introduced a chicken δ-crystallin gene into medaka embryos and observed that the δ-crystallin gene was expressed transiently in the central fiber of the lens at an early stage after injection. However, at later stages, expression of the δ-crystallin gene in many other tissues was noticed (Ozato et al., 1986; Inoue et al., 1989). In another experiment, Winkler et al. (1991) tested two LacZ constructs with either the SV40 early promoter or the RSV promoter and observed differential patterns of expression in the medaka embryos. With the SV40 promoter, expression occurred mainly in the developing neural organs of the fish, consistent with the findings in transgenic mice (Brinster et al., 1984). The RSV promoter led to a preferential expression in epidermal derivatives and epithelial tissues derived from the mesoderm, a finding that was also in agreement with results in transgenic mice (Overbeek et al., 1986). Using transgenic zebrafish as a transient expression system, Westerfield et al. (1992) transferred several LacZ constructs to examine the activation of two mammalian Hox genes, mouse Hox-1.1 and human Hox-3.3. These researchers found that these Hox promoters are activated in specific regions and tissues, mainly in the mesodermal or neuroectodermal tissues. Furthermore, the specificity of the Hox genes in zebrafish required the same regulatory elements that specified the spatial expression of these genes in mice. Thus, zebrafish can faithfully express these two mammalian genes, and the pattern of their expression may reflect that of the corresponding Hox genes in the zebrafish.

Analysis of exogenously introduced genes in developing embryos has become a powerful tool for studying the temporal and spatial regulation of gene expression. In sea urchin embryos, exogenously introduced genes are usually expressed correctly in terms of the timing and tissue specificity (for review, see Davidson, 1989). In *Xenopus,* however, the injected foreign genes, including the adult globin gene, are all transcribed at the midblastula stage when embryonic RNA transcription becomes active (Etkin and Balcells, 1985, and references therein), although Bendig and Williams (1984) noted that a tadpole globin gene was activated only at relatively late stages of tadpole development. At present, there are not enough data, particularly for the comparison of the developmental patterns of expression of endogenous and exogenous fish genes, to evaluate transgenic fish as a model for gene expression. In view of the large amount of gene amplification that occurs immediately after the introduction of the foreign gene, some important transacting transcription factors are likely to be limited and saturated, resulting in the incorrect expression of some genes. In some cases, the faithful expression of a transgene might be overshadowed by a high level of ectopic or ubiquitous expression (Ozato et al., 1986; Inoue et al., 1989). The spatial pattern of expression may also be misled by transgene mosaicism, although it may be minimized by the analysis of a sufficient number of samples. All these problems may be partially circumvented by the analysis of transgene expression in later stages of embryogenesis when excess foreign DNA does not exist. A more reliable assay should be performed in stable lines of transgenic fish. Nevertheless, the ability of early fish embryos to express different promoter DNA constructs differentially is a good justification for using transgenic fish embryos as transient gene expression systems (Gong et al., 1991; Winkler et al., 1991; Westerfield et al., 1992; Moav et al., 1993).

C. Transgenic Fish as a Stably Transformed Line

1. Overexpression

To date, most transgenic fish are designed to overexpress the transgene, for example, the GH or the AFP gene, to acquire a new characteristic. The successful growth stimulation of GH transgenic fish indicates the feasibility of using transgenic fish to study gene function. Matsumoto et al. (1992) transferred a mouse gene encoding a tyrosinase into the medaka to study its role in pigmentation of the skin. The orange-colored variants of the host medaka, which is genetically devoid of skin melanophores, gained the ability to exhibit melanogenesis in skin after gene transfer, indicating a role of the tyrosinase gene in skin pigmentation.

At present, at least two groups of fish genes can be studied by ectopic expression or overexpression. One group is made up of the genes important in development, such as the homeobox genes. An increasing number of genes encoding transcription factors, including dozens of homeobox genes and some zinc finger genes, has been isolated. The early patterns of expression of some of these genes, such as Hox genes (Njolstad et al., 1990), paired-box homeobox genes (Krauss et al., 1992), engrailed genes (Hatta et al., 1990; Ekker et al., 1992b; Fjose et al., 1992), and others, have been well documented. These genes usually have restricted patterns of expression in certain cell lineages and at certain times. Overexpression or ectopic expression of these genes in transgenic fish will provide insights into their function and the developmental programs they control. Another group of genes, the oncogenes, may transform normal cells into tumor cells by overexpression in certain tissues. Increasing numbers of oncogenes such as *ras* and *myc* have been cloned based on sequence similarity with the homologous genes from other animals, but little is known about their functions in fish (for review, see Beneden, 1993). Analysis of the functions of these genes in transgenic fish will provide important insight into basic mechanisms on carcinogenesis, as well as a means to detect environmental toxins.

2. Cell Lineage Ablation

The function of a gene can also be analyzed by the loss-of-function study. One of the techniques for this study is cell lineage ablation. The principle of cell lineage ablation is to transfer and express a toxin gene in a specific cell type; the toxin molecules subsequently destroy the host cells without affecting neighboring cells. Such a technique is useful for studying cell lineage, its fate, and its function. The first cell lineage ablation system was developed by Palmiter *et al.* (1987) using the elastase I promoter fused to the diphtheria toxin A chain (DTA) gene. DTA inhibits protein synthesis once it is inside the cell, but without a B chain it has no ability to cross the cell membrane to affect the neighboring cells. Transgenic mice with the elastin–toxin construct lack normal pancreas. This approach was also used to destroy somatotrophs in the pituitary of transgenic mice; such mice become dwarfs (Behringer *et al.*, 1988). This technique is similarly very attractive for studying transgenic fish. However, one potential limitation of this technique is the relatively high nonspecific promoter activity during early fish embryogenesis (Z. Gong and C. L. Hew, unpublished data); thus the embryos harboring the toxin gene may not develop normally. This problem may be overcome by using inducible promoters such as heat shock promoters. Some temperature-sensitive cell lineage ablation systems have been developed. Bellen *et al.* (1992) have isolated a temperature-sensitive DTA gene that is functional at 18°C but not at 30°C. However, this finding is not directly applicable to fish since most fish species live below 18°C. In contrast, Moffat *et al.* (1992) developed a cold-sensitive ricin A chain system for cell ablation. The mutant ricin A chain is functional at 29°C but not at 18°C. Another conditional cell lineage system uses the herpes simplex virus thymidine kinase (HSV-1-tk), capable of phosphorylating specific nucleoside analogs to the nucleoside monophosphate. The incorporation of such nucleosides into DNA leads to inhibition of DNA synthesis and cell death; the ablation is induced by treating transgenic animals expressing HSV-1-tk with certain nucleoside analogs (Heyman *et al.*, 1989). However, the administration of a specific nucleoside analog requires implantation of a minipump, which might limit its application in fish. The obvious advantage of conditional cell lineage ablation is the possibility of maintaining a transgenic line. Examples of cell lineage ablation in fish include the study of the unresolved functions of many pituitary hormones. The genes encoding most of these pituitary hormones have been isolated. For some of these genes such as prolactin, growth hormone, somatolactin, gonadotropin α, and gonadotropin IIβ, the promoter regions have been characterized (Xiong et al., 1993). Cell-type-specific gonadotropin gene promoters linked to a toxin gene will destroy the gonadotropin-producing cell lineage, resulting in sterile fish useful in aquaculture.

3. Antisense RNA

Disruption of a gene function can also be achieved by the expression of an antisense RNA. Presumably, the antisense RNA forms hybrids *in vivo* with the complementary sense mRNA to prevent translation of the latter. In transgenic mice, antisense RNA was demonstrated to function as a gene suppresser. For example, by expressing an antisense RNA for the myelin basic protein (MBP) gene in transgenic mice, a shiverer phenotype was observed that was similar to a mutant with an MBP gene defect (Katsuki et al., 1988). Researchers also reported that the GH antisense transgenic mice developed into dwarfs (Matsumoto et al., 1993). The technique of expressing an antisense RNA transgene can be easily applied in fish. Unlike cell lineage ablation, antisense RNA does not completely abolish gene expression but partly suppresses it. Antisense RNA transgenes, therefore, provide a complementary and often advantageous strategy for suppressing the expression of a gene, if the appropriate tissue-specific promoter is used.

4. Gene Targeting

A more sophisticated transgenic technique, gene targeting, was developed in mice (for review, see Joyner, 1991). By homologous recombination or replacement, a normal gene can be deleted or replaced by a mutated one, or vice versa. The consequence of such manipulation can be analyzed accordingly. In the mouse, gene targeting requires transfection of target DNA into the cultured stem cells derived from the inner cell mass of the blastocyst. The transfected stem cells are then screened for correct incorporation of foreign DNA, and transferred into the embryos. The feasibility of applying this delicate technique to fish has been exploited by Collodi et al. (1992). These researchers have successfully prepared blastula-derived cell cultures from the zebrafish, introduced plasmid DNA into the cells, and transferred them into blastula-stage embryos. The transferred cultured cells survived, and may contribute to the developing organism. Similarly, rainbow trout chimeras have been produced by the injection of blastomeres into recipient blastulae (Nilsson and Cloud, 1992). However, the availability of large numbers of fish eggs and the relative ease of generation of transgenic fish may prove the use of *in vitro* stem cells to perform gene targeting in fish to be unnecessary, since thousands of transgenic zebrafish or medaka can be produced easily using the techniques of electroporation and PCR screening.

5. Gene Disruption

Gene disruption is an important phenomenon in transgenic research. Because the site of foreign DNA integration is likely to be random, except in gene targeting through homologous recombination, integration may occur in nongenic regions or in the regions of protein coding and gene regulatory sequences. In higher eukaryotes, most of the genomic sequences are nongenic or functionally less important; therefore, the integration of a foreign DNA in most cases has no obvious effect on the host. However, in rare cases, the integration of a foreign DNA has interrupted a functionally important gene, causing a mutant phenotype. This phenomenon has been used as a technique for creating genetic mutants (Joyner, 1991; Rossant and Hopkins, 1992). The mutants can be analyzed and the mutant gene isolated by means of the transgene marker. Fish obviously provide a good system for gene disruption studies because of the relatively simple transgenic technique, but the large size of the genome makes mutagenesis screening difficult. The recently characterized pufferfish is probably the best model for mutagenesis by gene disruption, since it has the smallest genome in fish and has fewer nongenic sequences (Brenner et al., 1993). However, the knowledge of embryology and genetics of this species is limited, and the transgenic technique has not been established. Another potential application of gene disruption is the identification of endogenous genomic regulatory elements (O'Kane and Gehring, 1987). A reporter gene without a promoter (promoter trap) or with a truncated weak promoter (enhancer trap) can be introduced into the embryo; the expression of the reporter gene requires promoter or enhancer from the neighbor close to the position at which it has integrated into the host genome. Such a technique is useful in isolating tissue-specific promoters and enhancers. Bayer and Campos-Ortega (1992) have successfully isolated an enhancer trap line from transgenic zebrafish after the injection of a LacZ gene construct with a truncated mouse heat shock promoter. The enhancer trap line showed a spatially and temporally restricted expression of LacZ in some neuronal cells.

VII. Concluding Remarks

The techniques for generating transgenic fish have been reasonably established for numerous species, including some species suitable for developmental studies, such as the zebrafish and medaka, as well as many commercially important species. Transgenic fish models are useful for testing the promoter activity and for analyzing the pattern of gene expression. Compared with many well-established transgenic systems such as transgenic mice, the transgenic fish system is still in its infancy in developmental biology. Many advanced transgenic techniques developed in mice, such as gene targeting and cell lineage ablation, are likely to be applicable to fish, but a big gap between mice and fish in terms of the availability of gene resources must be filled before a thorough analysis of the molecular mechanism of fish development becomes possible. The zebrafish is likely to play a central role in the development of model fish in this respect (Rossant and Hopkins, 1992). In contrast, tremendous efforts and progress have already been made in the development of commercially useful fish stocks by the transgenic technique. Several laboratories have successfully produced transgenic fish with enhanced growth rates. Transgenic fish are closer to commercial application than any other transgenic livestock, and are likely to be the first marketable transgenic animal for human consumption.

Acknowledgments

We thank Linda Gardiner for her assistance in typing and proofreading the manuscript. This investigation is supported by the Natural Sciences and Engineering Research Council of Canada, and the Department of Fisheries and Oceans, Government of Canada.

References

Adams, M. D., Kelley, J. M., Gocayne, J. D., Dubnick, M., Polymeropoulos, M. H., Xiao, H., Merril, C. R., Wu, A., Olde, B., Moreno, R. F., Kerlavage, A. R., McCombie, W. R., and Venter, J. C. (1991). Complementary DNA sequencing: Expressed sequence tags and human genome project. *Science* **252,** 1651–1656.

Agellon, L. B., and Chen, T. T. (1986). Rainbow trout growth hormone: Molecular cloning and expression in *E. coli. DNA* **5,** 463–471.

Alestrom, P., Klungland, H., Kisen, G., and Andersen, O. (1992). Fish gonadotropin-releasing hormone gene and molecular approaches for control of sexual maturation: Development of a transgenic fish model. *Mol. Marine Biol. Biotechnol.* **1,** 376–379.

Askins, J., and Townes, T. M. (1987). Transfer of the metallothionein-human growth hormone fusion gene into channel catfish. *Am. Fish. Soc.* **116,** 87–91.

Barinaga, M. (1990). Zebrafish: Swimming into the development mainstream. *Science* **250,** 34–35.

Bayer, T. A., and Campos-Ortega, J. A. (1992). A transgene containing lacZ is expressed in primary sensory neurons in zebrafish. *Development* **115,** 421–426.

Behringer, R. R., Matthews, L. S., Palmiter, R. D., and Brinster, R. L. (1988). Dwarf mice produced by genetic ablation of growth hormone-expressing cells. *Genes Dev.* **2 (4),** 453–461.

Bellen, H. J., D'Evelyn, D., Harvey, M., and Elledge, S. J. (1992). Isolation of temperature-sensitive diphtheria toxins in yeast and their effects on *Drosophila* cells. *Development* **114,** 787–96.

Bendig, M. M., and Williams, J. G. (1984). Differential expression of the *Xenopus laevis* tadpole and adult β-globin genes when injected into fertilized *Xenopus laevis* eggs. *Mol. Cell Biol.* **4,** 567–570.

Beneden, R. J. V. (1993). Oncogenes. *In* "Biochemistry and Molecular Biology of Fishes" (P. W. Hochachka and T. P. Mommson, eds.), Vol. 2, pp. 113–136. Elsevier, Amsterdam.

6. Transgenic Fish
207

Beniumov, A. O., Enikolopov, G. N., Barmintsev, V. A., Zelenina, I. A., and Sleptsova, L. A. (1989). Integration and expression of the human somatotropic hormone gene in Teleostei. *Genetika* **25**, 24–35.

Brem, G., Brenig, B., Horstgen-Schwark, G., and Winnacker, E.-L. (1988). Gene transfer in tilapia (*Oreochromis niloticus*). *Aquaculture* **68**, 209–219.

Brenner, S., Elgar, G., Sanford, R., Macrae, A., Venkatesh, B., and Aparicio, S. (1993). Characterization of the pufferfish (*Fugu*) genome as a compact model vertebrate genome. *Nature* **366**, 265–270.

Briggs, R., and King, T. J. (1952). Transplantation of living nuclei from blastula cells into enucleated frog's eggs. *Proc. Natl. Acad. Sci. USA* **38**, 455–463.

Brinster, R. L., Chen, H. Y., Messing, A., van Dyke, T., Levine, A. J., and Palmiter, R. D. (1984). Transgenic mice harboring SV40 T antigen genes develop characteristic brain tumors. *Cell* **37**, 367–379.

Brinster, R. L., Chen, H. Y., Trumbauer, M. E., Yagle, M. K., and Palmiter, R. D. (1985). Factors affecting the efficiency of introducing foreign DNA into mice by microinjecting eggs. *Proc. Natl. Acad. Sci. USA* 4438–4442.

Brinster, R. L., Allen, J. A., Behringer, R. R., Gelinas, R. E., and Palmiter, R. D. (1988). Introns increase transcriptional efficiency in transgenic mice. *Proc. Natl. Acad. Sci. USA* **85**, 836–840.

Brinster, R. L., Sandgren, E. P., Behringer, R. R., and Palmiter, R. D. (1989). No simple solution for making transgenic mice. *Cell* **59**, 239–241.

Buono, R. J., and Linser, P. J. (1992). Transient expression of RSVCAT in transgenic zebrafish made by electroporation. *Mol. Marine Biol. Biotechnol.* **1**, 271–275.

Buttrick, P. M., Kass, A., Kitsis, R. N., Kaplan, M. L., and Leinwand, L. A. (1992). Behavior of genes directly injected into the rat heart *in vivo*. *Circ. Res.* **70**, 193–98.

Cao, Q. P., Duguay, S. J., Plisetskaya, E., Steiner, D. F., and Chan, S. J. (1989). Nucleotide sequence and growth hormone-regulated expression of salmon insulin-like growth factor I mRNA. *Mol. Endocrinol.* **3**, 2005–2010.

Chakrabarti, S., Streisinger, G., Singer, F., and Walker, C. (1983). Frequency of gamma-ray induced specific site and recessive lethal mutations in mature germ cells of the zebrafish (*Brachydanio rerio*). *Genetics* **103**, 109–123.

Chen, T. T., and Powers, D. A. (1990). Transgenic fish. *Trends Biotechnol.* **8**, 209–216.

Chen, T. T., Kight, K., Lin, C. M., Powers, D. A., Hayat, M., Chatakondi, N., Ramboux, A. C., Duncan, P. L., and Dunham, R. A. (1993). Expression and inheritance of RSVLTR-rtGH1 complementary DNA in the transgenic common carp, *Cyprinus carpio*. *Mol. Marine Biol. Biotechnol.* **2**, 88–95.

Chong, S. S. C., and Vielkind, J. R. (1989). Expression and fate of CAT reporter gene microinjected into fertilized medaka (*Oryzias latipes*) eggs in the form of plasmid DNA, recombinant phage particles and its DNA. *Theoret. Appl. Genet.* **78**, 369–380.

Chourrout, D., and Perrot, E. (1992). No transgenic rainbow trout produced with sperm incubated with linear DNA. *Mol. Marine Biol. Biotechnol.* **1**, 257–265.

Chourrout, D., Guyomard, R., and Houdebine, L. M. (1986). High efficiency gene transfer in rainbow trout (*Salmo gairdneri* rich by microinjection into egg cytoplasm. *Aquaculture* **51**, 143–150.

Chourrout, D., Guyomard, R., and Houdebine, L. M. (1990). Techniques for the development of transgenic fish: A review. *In* "Transgenic Models in Medicine and Agriculture" (R. B. Church, ed.), pp. 89–99. Wiley-Liss, Toronto.

Collodi, P., Kaemi, Y., Sharps, A., Weber, D., and Barnes, D. (1992). Fish embryo cell for derivation of stem cells and transgenic chimeras. *Mol. Marine Biol. Biotechnol.* **1**, 257–265.

Commandon, J., and De Fonbrune, F. (1939). Greffe nucleaire total, simple ou multiple, chez une amibe. *C.R. Soc. Biol. Paris* **130**, 744–748.

Culp, P., Nusslein-Volhard, C., and Hopkins, N. (1991). High frequency germ-line transmission of plasmid DNA sequences injected into fertilized zebrafish eggs. *Proc. Natl. Acad. Sci. USA* **88,** 7953–7957.

Datta, U., Dutta, P., and Mandall, R. K. (1988). Cloning and characterization of a highly repetitive fish nucleotide sequence. *Gene* **62,** 331–336.

Davidson, E. H. (1989). Lineage-specific gene expression and the regulative capacities of the sea urchin embryo: A proposed mechanism. *Development* **105,** 421–445.

Davies, P. L., and Gauthier, S. Y. (1992). The application of PCR to aquaculture. *In* "Transgenic Fish" (C. L. Hew and G. L. Fletcher, eds.), pp. 61–71. World Scientific, Singapore.

Davies, P. L., and Hew, C. L. (1990). Biochemistry of fish antifreeze proteins. *FASEB J.* **4,** 2460–2468.

Devlin, R. H., and Donaldson, E. M. (1992). Containment of genetically altered fish with emphasis on salmonids. In "Transgenic Fish" (C. L. Hew and G. L. Fletcher, eds.), pp. 229–265. World Scientific, Singapore.

Devlin, R. H., Yesaki, T. Y., Donaldson, E. M., Du, S. J., and Hew, C. L. (1994). Production of germline transgenic Pacific salmonids with dramatically increased growth performance. *Can. J. Fish Aquat. Sci. (in press).*

Du, S. J. (1993). The isolation and characterization of chinook salmon GH genes and the creation of fast-growing Atlantic salmon by GH gene transfer. Ph.D. Thesis. Department of Biochemistry, University of Toronto, Toronto, Canada.

Du, S. J., Gong, Z., Fletcher, G. L., Shears, M. A., King, M. J., Idler, D. R., and Hew, C. L. (1992a). Growth enhancement in transgenic Atlantic salmon by the use of an "all fish" chimeric growth hormone gene construct. *Bio/Technology* **10,** 176–181.

Du, S. J., Gong, Z., Hew, C. L., Tan, C. H., and Fletcher, G. L. (1992b). Development of an all-fish gene cassette for gene transfer in aquaculture. *Mol. Marine Biol. Biotechnol.* **1,** 290–300.

Dunham, R. A., Eash, J., Askins, J., and Townes, T. M. (1987). Transfer of the metallothionein-human growth hormone fusion gene into channel catfish. *Trans. Am. Fish Soc.* **116,** 87–91.

Egami, N., Yamagami, K., and Shima, A. (1990). "Biology of *Medaka.*" Tokyo University Press, Tokyo.

Ekker, M., Fritz, A., and Westerfield, M. (1992a). Identification of two families of satellite-like repetitive DNA sequences from the zebrafish (*Brachydanio rerio*). *Genomics* **13,** 1169–1173.

Ekker, M., Wegner, J., Akimenko, A. A., and Westerfield, M. (1992b). Coordinate embryonic expression of three zebrafish engrailed genes. *Development* **116,** 1001–1010.

Etkin, L. D., and Balcells, S. (1985). Transformed *Xenopus* embryos as a transient expression system to analyze gene expression at the midblastula transition. *Dev. Biol.* **108,** 173–178.

Fjose, A., Njolstad, P. R., Nornes, S., Molven, A., and Krauss, S. (1992). Structure and early embryonic expression of the zebrafish engrailed-2 gene. *Mech. Dev.* **39,** 51–62.

Fletcher, G. L., and Davies, P. L. (1991). Transgenic fish for aquaculture. *Genetic Eng.* **13,** 331–370.

Fletcher, G. L., Kao, M. H., and Fourney, R. M. (1986). Antifreeze peptides confer freezing resistance to fish. *Can. J. Zool.* **64,** 1897–1901.

Fletcher, G. L., Shears, M. A., King, M. J., Davies, P. L., and Hew, C. L. (1988). Evidence for antifreeze protein gene transfer in Atlantic salmon (*Salmo salar*). *Can. J. Fish. Aquat. Sci.* **45,** 352–357.

Fletcher, G. L., Davies, P. L., and Hew, C. L. (1992). Genetic engineering of freeze-resistant Atlantic salmon. *In* "Transgenic Fish" (C. L. Hew and G. L. Fletcher, eds.), pp. 190–208. World Scientific, Singapore.

Franks, R. R., Hough-Evans, B. R., Britten, R. J., and Davidson, E. H. (1988). Direct introduction of cloned DNA into the sea urchin zygote nucleus, and fate of injected DNA. *Development* **102,** 287–299.

Gong, Z., and Hew, C. L. (1993). Promoter analysis of fish antifreeze protein genes. *In* "Biochemistry and Molecular Biology of Fishes" (P. W. Hochachka and T. P. Mommson, eds.), Vol. 2, pp. 307–324. Elsevier, Amsterdam.

Gong, Z., Vielkind, J. R., and Hew, C. L. (1991). Functional analysis and temporal expression of promoter regions from fish antifreeze protein genes in transgenic Japanese medaka embryos. *Mol. Marine Biol. Biotechnol.* **1,** 64–72.

Gong, Z., Fletcher, G. L., and Hew, C. L. (1992). Tissue distribution of fish antifreeze protein mRNAs. *Can. J. Zool.* **70,** 810–814.

Gong, Z., Hu, Z., Gong, Z. Q., Kitching, R., and Hew, C. L. (1994). Bulk isolation and identification of fish genes by cDNA clone tagging. *Mol. Marine Biol. Biotech.* (in press).

Gordon, J. W., Scangos, G. A., Plotkin, D. J., Barbosa, J. A., and Ruddle, F. H. (1980). Genetic transformation of mouse embryos by microinjection of purified DNA. *Proc. Natl. Acad. Sci. USA* **77,** 7380–7384.

Graham, F., and van der Eb, A. (1973). A new technique for the assay of infectivity of human adenovirus 5 DNA. *Virology* **52,** 456–457.

Gross, M. T., Schneider, J. F., Moav, N., Moav, B., Alvarez, C., Myster, S. H., Liu, Z., Hallerman, E. M., Hackett, P. B., Guise, K. S., Faras, A. J., and Kapuscinski, A. R. (1992). Molecular analysis and growth evaluation of northern pike (*Esox lucius*) microinjected with growth hormone genes. *Aquaculture* **103,** 253–273.

Grosveld, F., van Assendelft, G. B., Greaves, D. R., and Kollias, G. (1987). Position-independent, high level expression of the human β-globin gene in transgenic mice. *Cell* **49,** 369–378.

Grunwald, D. J., and Streisinger, G. (1992a). Induction of mutations in the zebrafish with ultraviolet light. *Genet. Res.* **59,** 93–101.

Grunwald, D. J., and Streisinger, G. (1992b). Induction of recessive lethal and specific locus mutations in the zebrafish with ethyl nitrosourea. *Genet. Res.* **59,** 103–116.

Guise, K. S., Kapuscinski, A. A. R., Hackett, P. B., and Faras, A. J. Jr. (1990). Gene transfer in fish. *In* "Transgenic Animals" (N. First and F. P. Haseltine, eds.), pp. 295–306. Butterworth-Heinemann, Boston.

Guyomard, R., Chourrout, D., Leroux, C., Houdebine, L. M., and Pourrain, F. (1989). Integration and germ line transmission of foreign genes microinjected into fertilized trout eggs. *Biochimie* **71,** 857–863.

Hallerman, E. M., and Kapuscinski, A. R. (1992). Ecological and regulatory uncertaintles associated with transgenic fish. *In* "Transgenic Fish" (C. L. Hew and G. L. Fletcher, eds.), pp. 209–228. World Scientific, Singapore.

Hammer, R. E., Pursel, V. G., Rexroad, C. E. Jr., Wall, R. J., Bolt, D. J., Ebert, K. M., Palmiter, R. D., and Brinster, R. L. (1985). Production of transgenic rabbits, sheep and pigs by microinjection. *Nature* **315,** 680–683.

Harland, R., and Laskey, R. (1980). Regulated replication of DNA microinjected into eggs of *Xenopus laevis. Cell* **21,** 761–771.

Hatta, K., Schilling, T. F., BreMiller, R. A., and Kimmel, C. B. (1990). Specification of jaw muscle identity in zebrafish. Correlation with engrailed-homeoprotein expression. *Science* **250,** 802–805.

Hayat, M., Joyce, C. P., Townes, T. M., Chen, T. T., Powers, D. A., and Dunham, R. A. (1991). Survival and integration rate of channel catfish and common carp embryos microinjected with DNA at various developmental stages. *Aquaculture* **99,** 249–255.

He, L., Zhu, Z., Faras, A. J., Guise, K. S., Hackett, P. B., and Kapuscinski. (1992). Characterization of AluI repeats of zebrafish (*Brachydanio rerio*). *Mol. Marine Biol. Biotechnol.* **1,** 125–135.

Hew, C. L. (1988). Transgenic fish: Present status and future directions. *Fish Physiol. Biochem.* **7,** 409–413.

Hew, C. L., and Fletcher, G. L. (1992). "Transgenic Fish." World Scientific, Singapore.
Hew, C. L., and Gong, Z. (1992). Transgenic fish: A new technology for fish biology and aquaculture. *Biol. Int.* **24**, 2–10.
Hew, C. L., Trinh, K. Y., Du, S. J., and Song, S. (1989). Molecular cloning and expression of salmon pituitary hormones. *Fish Physiol. Biochem.* **7**, 375–380.
Hew, C. L., Davies, P. L., Shears, M., and Fletcher, G. L. (1992). Antifreeze protein gene transfer in Atlantic salmon. *Mol. Marine Biol. Biotechnol.* **1**, 309–317.
Heyman, R. A., Borreli, E., Lesley, J., Anderson, D., Richman, D. D., Baird, S. M., Hyman, R., and Evans, R. M. (1989). Thymidine kinase obliteration: Creation of transgenic mice with controlled immune deficiency. *Proc. Natl. Acad. Sci. USA* **86**, 2698–2702.
Hisaoka, K. K., and Battle (1958). The normal developmental stages of the zebrafish, *Brachydanio rerio*. *J. Morph.* **102**, 311–328.
Hisaoka, K. K., and Firlit, C. F. (1962). The embryology of the blue gourami, *Trichogaster trichopterus* (Pallas). *J. Morphol.* **111**, 239–253.
Ho, R. K., and Kane, D. A. (1990). Cell-autonomous action of zebrafish spt-1 mutation in specific mesodermal precursors. *Nature* **348**, 728–730.
Ho, R. K., and Kimmel, C. B. (1993). Commitment of cell fate in the early zebrafish embryo. *Science* **261**, 109–111.
Houdebine, L. M., and Chourrout, D. (1991). Transgenesis in fish. *Experientia* **47**, 891–897.
Inoue, K., Ozata, K., Kondoh, H., Iwamatsu, T., Wakamatsu, Y., Fijita, T., and Okada, T. S. (1989). Stage-dependent expression of the chicken δ-crystallin gene in transgenic fish embryos. *Cell Diff. Dev.* **27**, 57–68.
Inoue, K., Yamashita, S., Hata, J., Kabeno, S., Asada, S., Nagahisa, E., and Fujita, T. (1990). Electroporation as a new technique for producing transgenic fish. *Cell Diff. Dev.* **29**, 123–128.
Ivics, Z., Izsvak, Z., and Hackett, P. B. (1993). Enhanced incorporation of transgenic DNA into zebrafish chromosomes by a retroviral integration protein. *Mol. Marine Biol. Biotechnol.* **2**, 162–173.
Joyner, A. L. (1991). Gene targeting and gene trap screens using embryonic stem cells: New approaches to mammalian development. *BioEssays* **13**, 649–656.
Katsuki, M., Sato, M., Kimura, M., Yokoyama, M., Kobayashi, K., and Nomura, T. (1988). Conversion of normal behavior to shiverer by myelin basic protein antisense cDNA in transgenic mice. *Science* **241**, 593–595.
Khoo, H.-W., Ang, L.-H., Lim, H.-B., and Wong, K.-Y. (1992). Sperm cells as vectors for introducing foreign DNA into zebrafish. *Aquaculture* **107**, 1–19.
Kimmel, C. B. (1989). Genetics and early development of zebrafish. *Trends Genet.* **5**, 283–288.
Kimmel, C. B., Warga, R. M., and Schilling, T. D. (1990). Origin and organization of the zebrafish fate map. *Development* **108**, 581–594.
Klein, T. M., Wolf, E. D., Wu, R., and Sanford, J. C. (1987). High-velocity microprojectiles for delivering nucleic acids into living cells. *Nature* **327**, 70–73.
Krauss, S., Maddeen, M., Holder, N., and Wilson, S. W. (1992). Zebrafish pax[b] is involved in the formation of the midbrain-hindbrain boundary. *Nature* **360**, 87–89.
Lavitrano, M. L., Camaioni, A., Fazio, V. M., Dolci, S., Farace, M. G., and Spadafora, C. (1989). Sperm cells as vectors for introducing foreign DNA into eggs: Genetic transformation of mice. *Cell* **57**, 717–723.
Liu, Z., Moav, B., Faras, A. J., Guise, K. S., Kapuscinski, A. R., and Hackett, P. B. (1990). Development of expression vectors for transgenic fish. *Biotechnology* **8**, 1268–1272.
Lu, J.-K., Chrisman, C. L., Andrisani, O. M., Dixon, J. E., and Chen, T. T. (1992). Intergration, expression, and germ-line transmission of foreign growth hormone genes in medaka, *oryzias latipes*. *Mol. Marine Biol. Biotech.* **1**, 366–375.
Maclean, N., and Penman, D. (1990). The application of gene manipulation to aquaculture. *Aquaculture* **85**, 1–20.

6. Transgenic Fish

Maclean, N., Penman, D., and Zhu, Z. (1987). Introduction of novel genes into fish. *Bio/Technology* **5**, 257–261.

Mahon, H. F., and Hoar, W. S. (1956). The early development of the chum salmon, *Oncorhynchus keta*. *J. Morph.* **98**, 1–47.

Mathews, L. S., Hammer, R. E., Behringer, R. R., D'Ercole, A. J., Bell, G. I., Brinster, R. L., and Palmiter, R. D. (1988). Growth enhancement of transgenic mice expressing insulin-like growth factor. I. *Endocrinology* **123**, 2827.

Matsui, K. (1949). Illustration of the normal course of development in the fish, *Oryzias latipes*. *Japanese J. Exp. Morph.* **5**, 33–42.

Matsumoto, J., Akiyama, T., Hirose, E., Nakamura, M., Yamamoto, H., and Takeuchi, T. (1992). Expression and transmission of wild-type pigmentation in the skin of transgenic orange-colored variants of medaka (*oryzias latipes*) bearing the gene for mouse tyrosinase. *Pigment Cell Res.* **5**, 322–327.

Matsumoto, K., Kakidani, H., Takahashi, A., Nakagata, N., Anzai, M., Matsuzaki, Y., Takahashi, Y., Miyata, K., Utsumi, K., and Iritani, A. (1993). Growth retardation in rats whose growth hormone gene expression was suppressed by antisense RNA transgene. *Mol. Reprod. Dev.* **36**, 53–58.

McEvoy, T., Stack, M., Keane, B., Barry, T., Sreenan, J., and Gannon, F. (1988). The expression of a foreign gene in salmon embryos. *Aquaculture* **68**, 27–37.

McGrane, M. M., Yun, J. S., Moorman, A. F. M., Lamers, W. H., Hendrick, G. K., Arafah, B. M., Park, E. A., Wagner, T. E., and Hanson, R. W. (1990). Metabolic effects of developmental tissue-, and cell-specific expression of a chimeric phosphoenolpyruvate carboxykinase (GTP)/bovine growth hormone gene in transgenic mice. *J. Biol. Chem.* **265**, 22371–22379.

McMahon, A. P., Flytzanis, C. N., Hough-Evans, B. R., Katula, K. S., Britten, R. J., and Davidson, E. H. (1985). Introduction of cloned DNA into sea urchin egg cytoplasm: Replication and persistence during embryogenesis. *Dev. Biol.* **108**, 420–430.

Moav, B., Liu, Z., Caldovic, L. D., Gross, M. L., Faras, A. J., and Hackett, P. B. (1993). Regulation of expression of transgenes in developing fish. *Transgenic Res.* **2**, 153–161.

Moffat, K. G., Gould, J. H., Smith, H. K., and O'Kane, C. J. (1992). Inducible cell ablation in *Drosophila* by cold-sensitive ricin A chain. *Development* **114**, 681–687.

Moven, A., Wright, C. V. E., BreMiller, R., de Robertis, E. M., and Kimmel, C. B. (1990). Expression of a homeobox gene product in normal and mutant zebrafish embryos: Evolution of the tetrapod body plan. *Development* **109**, 271–277.

Muller, F., Ivics, Z., Erdelyi, F., Papp, T., Varadi, L., Horvath, L., Maclean, N., and Orban, L. (1992). Introducing foreign genes into fish eggs with electroporated sperm as a carrier. *Mol. Marine Biol. Biotechnol.* **1**, 276–281.

Muller, F., Lele, Z., Varadi, L., Menczel, L., and Orban, L. (1993). Efficient transient expression system based on square pulse electroporation and in vitro luciferase assay of fertilized fish eggs. *FEBS Lett.* **324**, 27–32.

Naruse, K., Mitani, H., and Shima, A. (1992). A highly repetitive interspersed sequence isolated from genomic DNA of the medaka, *Oryzias latipes*, is conserved in three other related species within the genus *Oryzias*. *J. of Exp. Zool.* **262**, 81–86.

Nilsson, E. E., and Cloud, J. G. (1992). Rainbow trout chimeras produced by injection of blastomeres into recipient blastulae. *Proc. Natl. Acad. Sci. USA* **89**, 9425–9428.

Njolstad, P. R., Molven, A., Apold, J., and Fjose, A. (1990). The zebrafish homeobox gene hox-2.2: Transcription unit, potential regulatory regions and in situ localization of transcripts. *EMBO J.* **9**, 515–524.

O'Kane, C. J., and Gehring, W. J. (1987). Disruption of regulatory elements in *Drosophila*. *Proc. Natl. Acad. Sci. USA* **84**, 9123–9127.

Ono, T., Fujino, Y., Tsuchiya, T., and Tsuda, M. (1990). Plasmid DNAs directly injected into mouse brain with lipofectin can be incorporated and expressed by brain cells. *Neurosci. Lett.* **117,** 259–263.

Oppenheimer, J. M. (1937). The normal stages of *Fundulus heteroclitus*. *Anat. Rec.* **68,** 1–15.

Overbeek, P. A., Lai, S. P., Van Quill, K. R., and Westphal, H. (1986). Tissue-specific expression in transgenic mice of a fused gene containing RSV terminal sequences. *Science* **231,** 1574–1577.

Ozato, K., Kondoh, H., Inohara, H., Iwamatsu, T., Wakamatsu, Y., and Okada, T. S. (1986). Production of transgenic fish: Introduction and expression of chicken delta-crystallin gene in medaka embryos. *Cell Diff.* **19,** 237–244.

Ozato, K., Inoue, K., and Wakamatsu, Y. (1989). Transgenic fish: Biological and technical problems. *Zool. Sci.* **27,** 57–68.

Ozato, K., Inoue, K., and Wakamatsu, Y. (1992). Medaka as a model of transgenic fish. *Mol. Marine Biol. Biotechnol.* **1,** 346–354.

Palmiter, R. D., Brinster, R. L., Hammer, R. E., Trumbauer, M. E., Rosenfeld, M. G., Birnberg, N. C., and Evans, R. M. (1982). Dramatic growth of mice that develop from eggs microinjected with metallothionein-growth hormone fusion genes. *Nature* **300,** 611–615.

Palmiter, R. D., Behringer, R. R., Quaife, C. J., Maxwell, F., Maxwell, I. H., and Brinster, R. L. (1987). Cell lineage ablation in transgenic mice by cell-specific expression of a toxin gene. *Cell* **50 (3),** 435–443.

Palmiter, R. D., Sandgren, E. P., Avarbock, M. R., Allen, D. D., and Brinster, R. L. (1991). Heterologous introns can enhance expression of transgenes in mice. *Proc. Natl. Acad. Sci. USA* **88,** 478–482.

Parker, D. B., Coe, I. R., Dixon, G. H., and Sherwood, N. M. (1993). Two salmon neuropeptides encoded by one brain cDNA are structurally related to members of the glucagon superfamily. *Eur. J. Biochem.* **215,** 439–448.

Penman, D. J., Beeching, A. J., Iyengar, A., and Maclean, N. (1988). Introduction of metallothionein-somatotropin fusion gene into rainbow trout: Analysis of adult transgenics. *Aquaculture Assn. Can. Bull.* **88,** 137–139.

Penman, D. J., Beeching, A. J., Penn, S., and Maclean, N. (1990). Factors affecting survival and integration following microinjection of novel DNA into rainbow trout eggs. *Aquaculture* **85,** 35–50.

Pickford, G. E., and Phillips, J. G. (1959). Prolactin, a factor in promoting survival of hypophysectomized killifish in fresh water. *Science* **130,** 454–455.

Powers, D. A. (1989). Fish as model systems. *Science* **246,** 352–358.

Powers, D. A., Hereford, L., Cole, T., Creech, K., Chen, T. T., Lin, C. M., Kight, K., and Dunham, R. (1992). Electroporation: A method for transferring genes into the gametes of zebrafish (*Brachydanio rerio*), channel catfish (*Ictalurus punctatus*), and common carp (*Cyprinus carpio*). *Mol. Marine Biol. Biotechnol.* **1,** 301–308.

Pursel, V. G., Pinkert, C. A., Miller, K. F., Bolt, D. J., Campbell, R. G., Palmiter, R. D., Brinster, R. L., and Hammer, R. E. (1989). Genetic engineering of livestock. *Science* **244,** 1281–1288.

Rahman, A., and Maclean, N. (1992). Fish transgenic expression by direct injection into fish muscle. *Mol. Marine Biol. Biotechnol.* **1,** 286–289.

Rassoulzadegan, M., Leopold, P., Vailly, J., and Cuzin, F. (1986). Germ line transmission of autonomous genetic elements in transgenic mouse strains. *Cell* **46,** 513–519.

Rokkones, E., Alestrom, P., Skjervold, H., and Gautvik, K. M. (1989). Microinjection and expression of a mouse metallothionein human growth hormone fusion gene in fertilized salmonid eggs. *J. Comp. Physiol. Biol.* **158,** 751–785.

Rossant, J., and Hopkins, N. (1992). Of fin and fur: Mutational analysis of vertebrate embryonic development. *Genes Dev.* **6,** 1–13.

Sato, A., Aoki, K., Komura, J., Mashahito, P., Matsukuma, S., and Ishikawa, T. (1992). Firefly luciferase gene transmission and expression in transgenic medaka (*Oryzias latipes*). *Mol. Marine Biol. Biotechnol.* **1,** 318–325.

Scott, G. K., Hew, C. L., and Davies, P. L. (1985). Antifreeze protein genes are tandemly linked and clustered in the genome of the winter flounder. *Proc. Natl. Acad. Sci. USA* **82,** 2613–2617.

Shamblott, M. J., and Chen, T. T. (1992). Identification of a second insulin-like growth factor in a fish species. *Proc. Natl. Acad. Sci. USA* **89,** 8913–8917.

Shears, M. A., Fletcher, G. L., Hew, C. L., Gauthier, S., and Davies, P. L. (1991). Transfer, expression, and stable inheritance of antifreeze protein genes in Atlantic salmon (*Salmo salar*). *Mol. Marine Biol. Biotechnol.* **1,** 58–63.

Sin, F. Y. T., Bartley, A. L., Walker, S. P., Sin, I. L., Symonds, J. E., Hawke, L., and Hopkins, C. L. (1993). Gene transfer in chinook salmon (*Oncorhynchus tshawytscha*) by electroporating sperm in the presence of pRSV-lacZ DNA. *Aquaculture* **117,** 57–69.

Spemann, H. (1938). "Embryonic Development and Induction." Yale University Press, New Haven, Connecticut. Reprinted by Hafner Press (Macmillan, Inc.), New York (1962).

Spradling, A. C., and Rubin, G. M. (1982). Transposition of cloned P elements into *Drosophila* germ line chromosomes. *Science* **218,** 341–347.

Stinchcomb, D. T., Shaw, J. E., Carr, S. H., and Hirsh, D. (1985). Extrachromosomal DNA transformation of *Caenorhabditis elegans*. *Mol. Cell Biol.* **5,** 3484–3496.

Strehlow, D., and Gilbert, W. (1993). A fate map for the first cleavages of the zebrafish. *Nature* **361,** 451–453.

Streisinger, G., Walker, C., Dower, N., Knauber, D., and Singer, F. (1981). Production of clones of homozygous diploid zebra fish (*Brachydanio rerio*). *Nature* **291,** 293–296.

Stuart, G. W., McMurray, J. V., and Westerfield, M. (1988). Replication, integration and stable germ-line transmission of foreign sequences injected into early zebrafish embryos. *Development* **103,** 403–412.

Stuart, G. W., Vielkind, J. R., McMurray, J. V., and Westerfield, M. (1990). Stable lines of transgenic zebrafish exhibit reproducible patterns of transgene expression. *Development* **109,** 577–584.

Tamiya, E., Sugiyama, T., Masaki, K., Hirose, A., Okoshi, T., and Karube, I. (1990). Spatial imaging of luciferase gene expression in transgenic fish. *Nucleic Acids Res.* **18,** 1072.

Tao, W., Wilkinson, J., Stanbridge, E. J., and Berns, M. W. (1987). Direct gene transfer into human cultured cells facilitated by laser micropuncture of the cell membrane. *Proc. Natl. Acad. Sci. USA* **84,** 4180–4184.

Tavolga, W. N., and Rugh, R. (1947). Development of the platyfish, *Platypoecilus maculatus*. *Zoologica* **32,** 1–15.

Tung, T. C., and Niu, M. C. (1977a). Organ formation caused by nucleic acid from different class. Urodele DNA mediated balancer formation in goldfish. *Sci. Sin.* **20,** 56–58.

Tung, T. C., and Niu, M. C. (1977b). The effect of carp EGG-mRNA on the transformation of goldfish tail. *Sci. Sin.* **20,** 59–63.

Tyler, C. R., Sumpter, J. P., and Bromage, N. R. (1988). In vivo ovarian uptake and processing of vitellogenin in the rainbow trout, *Salmo gairdneri*. *J. Exp. Zool.* **246,** 171–179.

Van Der Putten, H., Botteri, F. M., Miller, A. D., Rosenfeld, M. G., Fan, H., Evans, R. M., and Verma, I. M. (1985). Efficient insertion of genes into the mouse germ line via retroviral vectors. *Proc. Natl. Acad. Sci. USA* **82,** 6148–6152.

Vidal, S. M., Malo, D., Vogan, K., Skamene, E., and Gros, P. (1993). Natural resistance to infection with intracellular parasites: Isolation of a candidate for *Bcg*. *Cell* **7,** 469–485.

Walker, C., and Streisinger, G. (1983). Induction of mutations by gamma rays in pregonial germ cells of zebrafish embryos. *Genetics* **103,** 125–136.

Westerfield, M., Wegner, J., Jegalian, B. G., DeRobertis, E. M., and Puschel, A. W. (1992). Specific activation of mammalian Hox promoters in mosaic transgenic zebrafish. *Genes Dev.* **6,** 591–598.

Winkler, C., Vielkind, J. R., and Schartl, M. (1991). Transient expression of foreign DNA during embryonic and larval development of the medaka fish (*Oryzias latipes*). *Mol. Gen. Genet.* **226**, 129–140.

Winkler, C., Hong, Y., Wittbrodt, J., and Schartl, M. (1992). Analysis of heterologous and homologous promoters and enhancers in vitro and in vivo by gene transfer into Japanese medaka (*Oryzias latipes*) and *Xiphophorus*. *Mol. Marine Biol. Biotechnol.* **1**, 326–337.

Wolff, J., Malone, R. W., Williams, P., Chong, W., Acsadi, G., Jani, A., and Felgner, P. L. (1990). Direct gene transfer into mouse muscle in vivo. *Science* **247**, 1465–1468.

Wu, G. Y., and Wu, C. H. (1988). Receptor-mediated gene delivery and expression in vivo. *J. Biol. Chem.* **263**, 14621–14624.

Xie, Y., Liu, D., Zou, J., Li, G., and Zhu, Z. (1989). Gene transfer in the fertilized eggs of loach via electroporation. *Acta Hydrobiol. Sin.* **13**, 387–389.

Xie, Y., Liu, D., Zou, J., Li, G., and Zhu, Z. (1993). Gene transfer via electroporation in fish. *Aquaculture* **111**, 207–213.

Xiong, F., Chin, R., Gong, Z., Suzuki, K., Kitching, R., Majumdar-Sonnylal, S., Elsholtz, H. P., and Hew, C. L. (1993). Control of salmon pituitary hormone gene expression. *Fish Physiol. Biochem.* **11**, 63–70.

Yamamoto, T. (1975). "Medaka (Killifish): Biology and Strains." Keigaku, Tokyo.

Yan, S. (1989). The nucleo-cytoplasmic interactions as revealed by nuclear transplantation in fish. In "Primers in Developmental Biology" (G. M. Malacinski, ed.), Vol. 5, pp. 61–81. McGraw-Hill, New York.

Yoon, S. J., Hallerman, E. M., Gross, M. L., Liu, Z., Schneider, J. F., Faras, A. J., Hackett, P. B., Kapuscinski, A. R., and Guise, K. S. (1990). Transfer of the gene for neomycin resistance into goldfish, *Carassius auratus*. *Aquaculture* **85**, 21–33.

Yoshizaki, G., Oshiro, T., and Takashima, F. (1991a). Introduction of carp α-globin gene into rainbow trout. *Nippon Suisan Gakkaishi* **57**, 819–824.

Yoshizaki, G., Oshiro, T., Takashima, F., Hirono, I., and Aoki, T. (1991b). Germ line transmission of carp α-globin gene introduced in rainbow trout. *Nippon Suisan Gakkaishi* **57**, 2203–2209.

Yoshizaki, G., Kobayashi, S., Oshiro, T., and Takashima, F. (1992). Introduction and expression of CAT gene in rainbow trout. *Nippon Suisan Gakkaishi* **58**, 1659–1665.

Zelenin, A. V., Alimov, A. A., Barmintzev, V. A., Beniumov, A. O., Zelenina, I. A., Krasnov, A. M., and Kolesnikov, V. A. (1991). The delivery of foreign genes into fertilized fish eggs using high-velocity microprojectiles. *FEBS Lett.* **287**, 118–120.

Zhang, P. J., Hayat, M., Joyce, C., Gonzalez-Villasenor, L. I., Lin, C. M., Dunham, R. A., Chen, T. T., and Powers, D. A. (1990). Gene transfer, expression and inheritance of pRSV-rainbow trout-GH cDNA in the common carp, *Cyprinus carpio* (Linnaeus). *Mol. Reprod. Dev.* **25**, 3–13.

Zhao, X., Zhang, P. J., and Wong, T. K. (1993). Application of baekonization: A new approach to produce transgenic fish. *Mol. Marine Biol. Biotechnol.* **2**, 63–69.

Zheng, H., and Wilson, J. (1990). Gene targeting in normal and amplified cell lines. *Nature* **344**, 170–173.

Zhu, Z., Li, G., He, L., and Chen, S. (1985). Novel gene transfer into the fertilized eggs of goldfish (*Carassius auratus* L. 1758). *Z. Angew. Ichthyol* **1**, 31–34.

Zhu, Z., Xu, K., Li, G., Xie, Y., and He, L. (1986). Biological effects of human growth hormone gene microinjected into the fertilized eggs of loach *Misgurnus anguillicaudatus*. *Kexue Tongbao Acad. Sin.* **31**, 988–990.

Zhu, Z., Xu, K., Xie, Y., Li, G., and He, L. (1989). A model of transgenic fish. *Sci. Sin. (B)* **2**, 147–155.

7
Axis Formation during Amphibian Oogenesis: Reevaluating the Role of the Cytoskeleton

David L. Gard
Department of Biology
University of Utah
Salt Lake City, Utah 84112

I. Introduction
II. Stage VI *Xenopus* Oocytes Are Structurally and Functionally Polarized along the A–V Axis
  A. *Xenopus* Oocytes Contain Three Major Cytoskeletal Networks
  B. Intermediate Filament Organization in Stage VI Oocytes
  C. F-Actin Is a Component of Both Nuclear and Cytoplasmic Cytoskeletons in *Xenopus* Oocytes
  D. Stage VI *Xenopus* Oocytes Contain an Extensive Network of Cytoplasmic Microtubules
  E. Cytoplasmic Microtubules Play an Important Role in Maintaining the Polarity of Stage VI Oocytes
III. Cytoskeletal Organization and Axis Specification during Early Oogenesis
  A. Postmitotic *Xenopus* Oocytes Are Polarized along a Nuclear/Cytoplasmic Axis
  B. Does Centrosome Position in Early Oocytes Define the Orientation of the A–V Axis?
  C. Little Evidence Remains of the Initial Oocyte Axis during the Early Diplotene
  D. Is the "Initial" Axis of Early Oocytes Maintained in Latent Form by a Polarized Distribution of F-Actin?
  E. The Mitochondrial Mass Defines a Cytoplasmic Axis in Stage I *Xenopus* Oocytes
  F. Stage I Oocytes Contain a Complex Cytoskeleton That Serves as a Framework for Cytoplasmic Organization
IV. Cytoplasmic and Cytoskeletal Polarity Are Established during Stage IV of Oogenesis
  γ-Tubulin Is Asymmetrically Distributed in the Cortex of *Xenopus* Oocytes
V. Ectopic Spindle Assembly during Maturation of *Xenopus* Oocytes: Evidence for Functional Polarization of the Oocyte Cortex
VI. Concluding Remarks
References

I. Introduction

Amphibians have long been popular models for studying the specification of developmental axes and pattern formation during vertebrate embryogenesis. In large part, this popularity stems from the large size of amphibian eggs and embryos, and the ease with which eggs can be fertilized and embryos manipulated *in vitro*. However, specification of the body plan of amphibian embryos actually begins with the formation of the characteristic animal–vegetal (A–V) axis of amphibian oocytes and eggs during oogenesis (Fig. 1). Although formed during oogenesis, the A–V axis plays a major role in later embryonic development. The anterior–posterior axis of the developing embryo can be roughly superimposed on the A–V axis. Moreover, fate mapping of early embryos has shown that, during normal development, cells derived from the animal hemisphere form predominantly ectodermal structures whereas those from the vegetal hemisphere form endodermal structures (Keller, 1975; Gimlich and Gerhart, 1980; Heasman et al., 1984b).

Despite its status as the first developmental axis to be established (the A–V

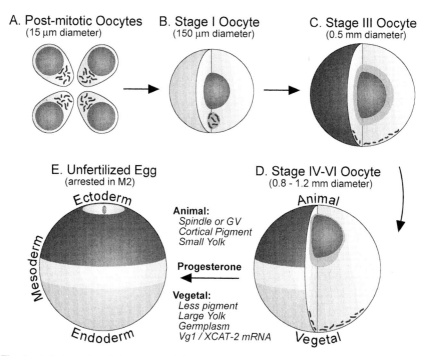

Fig. 1 Axis formation during oogenesis in *Xenopus laevis*. (A) Nests of 16 oocytes (only four oocytes are shown) are derived from a single oogonium by four sequential mitotic divisions. These oocytes are polarized along an "initial" nuclear–cytoplasmic axis, here indicated by the clustering of cytoplasmic organelles toward the proximal end of each cell. (B) The position of the "mitochondrial mass" in midstage I may relate to the A–V axis formed later in oogenesis (see text). (C) By stage III, the components of the mitochondrial mass have dispersed to the future vegetal cortex (see text). (D) The definitive A–V axis is established during stage IV of oogenesis. Markers of A–V polarity include differential pigmentation of the animal and vegetal cortex, the position of the GV or spindle in the animal hemisphere, a gradient of yolk, and the localized distribution of maternal mRNAs (see text). (E) The second meiotic spindle and white "maturation spot" mark the animal pole of unfertilized *Xenopus* eggs. [Compiled from Al-Mukhtar and Webb, 1971; Billet and Adams, 1976; Coggins, 1973; Czolowska, 1969; Dumont, 1972; Gerhart, 1980; Hausen and Riebesell, 1990; Mosquera et al., 1993; Rebagliati et al., 1985; Keller, 1975; and others.]

7. Axis Formation during *Xenopus* Oogenesis

axis has been called the "primary" axis; Gerhart et al., 1986) and its important role in later development, surprisingly little is known about the mechanisms underlying formation of the A–V axis of amphibian oocytes and eggs. Several observations suggest that A–V polarization is not significantly influenced by external factors. First, the orientation of the A–V axis does not appear to be dependent on the orientation of the oocyte with respect to gravity. In addition, the polarity of the A–V axis is not dependent on the orientation of oocytes with respect to the follicle stalk, which supplies nutrients to the growing oocyte (for reviews, see Gerhart, 1980; Gerhart et al., 1983, and references therein). These observations suggest that the formation of the A–V axis is dependent on factors that are entirely intrinsic to the oocyte. However, the nature of these factors and the mechanism by which polarity is established remain largely unknown.

In recent years, it has become clear that the cytoskeleton plays an important role in establishing the structural and functional polarity of eukaryotic cells, leading to speculation that the cytoskeleton plays similar roles during the formation of important developmental axes. Indeed, evidence indicates that the microtubule cytoskeleton of fertilized eggs is a key player in specification of the dorsal–ventral axis of early amphibian embryos (Scharf and Gerhart, 1983; Vincent et al., 1986; Elinson and Rowning, 1988). In contrast, less is known about the role of the cytoskeleton during A–V axis formation in amphibian oocytes, and only recently has a global view of the cytoskeleton in amphibian oocytes begun to emerge.

In the following discussion, I summarize our current view of the cytoskeleton and its role in the formation of the A–V axis of amphibian oocytes. For the sake of discussion, formation of the A–V axis will be regarded as three processes: (1) specification, in which A–V polarity, although perhaps not yet evident at a structural level, is first determined; (2) establishment, in which the specified axis is used to distribute developmental information and organelles asymmetrically between the animal and vegetal hemispheres; and, finally, (3) maintenance, in which the asymmetries generated during establishment of the axis are reinforced and maintained throughout later oogenesis. Although these processes are conceptually separable, it is entirely possible (and quite probable) that they overlap both temporally and mechanistically. Furthermore, although elements of the cytoskeleton play important roles in axis formation, they also respond to the processes by which the axis is established, making assignment of causal relationships difficult.

To organize and simplify this potentially confusing topic, I have subdivided the discussion into four sections including: (1) a brief overview of the polarized architecture of fully grown, stage VI *Xenopus* oocytes (oocyte stages are according to Dumont, 1972), including discussion of the composition and structure of the oocyte cytoskeleton and its role in maintaining the A–V axis; (2) a review of cytoskeletal organization during early oogenesis, addressing some of the existing hypotheses regarding the specification and establishment of the A–V axis and its relationship to cytoskeletal organization; (3) a discussion of a mechanism by which oocytes might regulate microtubule organization during axis formation; and (4) a brief discussion of the functional relationship between spindle location and the underlying A–V polarization of the oocyte during oocyte maturation. Although I focus on recent studies of cytoskeletal organization during oogenesis in the African frog, *Xenopus laevis*, examples from other amphibian species will be included when they provide additional insight into the role of the cytoskeleton during oogenesis.

II. Stage VI *Xenopus* Oocytes Are Structurally and Functionally Polarized along the A–V Axis

The A–V axis of fully grown, stage VI *Xenopus* oocytes is readily apparent due to the differential distribution of pigmentation in the oocyte cortex: the animal cortex is heavily pigmented whereas the vegetal cortex is unpigmented or only lightly pigmented (Dumont, 1972). However, the polarization of *Xenopus* oocytes is not limited to pigment distribution; many structural, morphological, and functional features of the oocyte are polarized along the A–V axis (see Gerhart, 1980; Gerhart *et al.*, 1986; Hausen and Riebesell, 1990). For example, most of the soluble cytoplasmic components and organelles, including the large oocyte nucleus (also called the germinal vesicle, GV), are found in the animal hemisphere. By comparison, the vegetal hemisphere is relatively deficient in cytoplasm and cytoplasmic organelles, but contains two-thirds of the yolk protein, stored as large, membrane-bound yolk platelets (Dumont, 1972; Gerhart *et al.*, 1986). A–V differences in the size and distribution of the yolk platelets arise not from differential endocytosis, but from differential transport of yolk platelets within the oocyte (Danilchik and Gerhart, 1987). The differential translocation of yolk from the animal to the vegetal hemispheres thus provides evidence for functional polarity of the oocyte cytoplasm (Gerhart *et al.*, 1986; Danilchik and Gerhart, 1987), perhaps as a result of the polarized organization of the oocyte cytoskeleton (see subsequent discussion).

The animal and vegetal hemispheres of *Xenopus* oocytes also differ in their content of developmental information. For example, islands of germinal cytoplasm, or "germplasm," a putative cytoplasmic determinant for the germline of anurans (see Nieuwkoop and Sutasurya, 1979, Smith *et al.*, 1983), are specifically localized in the vegetal cortex of stage VI *Xenopus* oocytes. In addition, molecular analysis has revealed that the distribution of maternal mRNA differs both qualitatively and quantitatively along the A–V axis (Capco and Jefferies, 1982; Carpenter and Klein, 1982; King and Barklis, 1985).

Recent studies have identified several maternal mRNA species that are spatially restricted to the animal or vegetal hemisphere of oocytes and eggs (Rebagliati *et al.*, 1985; Weeks and Melton, 1987; Gururajan *et al.*, 1991; Mosquera *et al.*, 1993). Vg1 mRNA, for example, is concentrated in the vegetal cortex of stage VI *Xenopus* oocytes (Melton, 1987; Pondel and King, 1988; Yisraeli *et al.*, 1990). The protein product of the Vg1 mRNA, a member of the TGF-β family of growth factors (Weeks and Melton, 1987), has been shown to play an important role in mesoderm induction during early embryogenesis (Thomsen and Melton, 1993). XCAT-2, another maternal mRNA, is associated with islands of germplasm in the vegetal cortex of stage VI oocytes (Mosquera *et al.*, 1993). Although the function of XCAT-2 protein has yet to be established, the predicted protein sequence bears significant similarity to the product of the *nanos* gene of *Drosophila melanogaster* (Mosquera *et al.*, 1993). Interestingly, the *nanos* mRNA is spatially localized in *Drosophila* oocytes, and encodes a protein required for proper establishment of the anterior–posterior axis during *Drosophila* development (Irish *et al.*, 1989; Lehmann and Nüsskin-Volhard, 1991; Wang and Lehmann, 1991).

A. *Xenopus* Oocytes Contain Three Major Cytoskeletal Networks

Vg1 and XCAT-2, along with other spatially localized maternal mRNAs, undoubtedly contribute to the distinct functions or fates of cells derived from the animal and vegetal hemispheres during later development (Keller, 1975; Gimlich and Gerhart, 1980; Heasman *et al.*, 1984b; Snape *et al.*, 1987; Wylie *et al.*, 1987). However, they are unlikely to play an important determinative role in the specification and establishment of the underlying A–V axis.

Several studies suggest that the polarized architecture of stage VI *Xenopus* oocytes, including features such as nuclear position and mRNA localization, is dependent on the oocyte cytoskeleton. Fully grown (stage VI) *Xenopus* oocytes contain examples of all three major cytoskeletal networks found in somatic cells, including intermediate filaments (IF; Franz *et al.*, 1983; Godsave *et al.*, 1984a,b; Klymkowsky *et al.*, 1987; Tang *et al.*, 1988; Dent and Klymkowsky, 1989; Dent *et al.*, 1989; Herman *et al.*, 1989; Torpey *et al.*, 1990,1992), actin microfilaments (Franke *et al.*, 1976; Roeder and Gard, 1994); and microtubules (MTs; Palecek *et al.*, 1985; Wylie *et al.*, 1985; Jessus *et al.*, 1987; Huchon *et al.*, 1988; Yisraeli *et al.*, 1990; Gard, 1991a). Until recently, however, little was known of the organization of the cytoskeleton during amphibian oogenesis.

In large part, our lack of knowledge resulted from one of the very features that made amphibian eggs attractive to embryologists and cell biologists: the large size of amphibian oocytes and eggs (more than 1 mm in diameter) severely hampered visualization of the oocyte cytoskeleton by both light and electron microscopy. Over the last 5 years, however, the refinement of whole-mount immunocytochemical techniques for large cells (Dent and Klymkowsky, 1989; Dent *et al.*, 1989; Gard, 1991a), used in conjunction with conventional or confocal microscopy (Yisraeli *et al.*, 1990; Gard, 1991a,1992), has greatly facilitated the study of the cytoskeleton in amphibian oocytes and eggs. Results from these studies have provided new insight into the organization of intermediate filaments, actin microfilaments, and microtubules during oogenesis (Fig. 2). However, the images obtained have also raised several questions regarding the roles these cytoskeletal components play during specification and formation of the A–V axis, and the mechanisms by which cytoskeletal organization is itself determined.

B. Intermediate Filament Organization in Stage VI Oocytes

Xenopus oocytes have been reported to contain two biochemically and spatially distinct populations of intermediate filaments, composed of cytokeratins (Franz *et al.*, 1983; Godsave *et al.*, 1984b; Klymkowsky *et al.*, 1987; Dent and Klymkowsky, 1989; Dent *et al.*, 1989; Herman *et al.*, 1989) and vimentin (Godsave *et al.*, 1984a; Tang *et al.*, 1988; Torpey *et al.*, 1990,1992). Using immunofluorescence microscopy and antibodies against mammalian vimentin, Godsave *et al.* observed vimentin in yolk-free "tracts" or "radii" that extend from the GV to the animal cortex. Subsequent studies using antisera against mammalian or *Xenopus* vimentin revealed concentrations of vimentin associated with islands of germplasm in the vegetal cortex (Tang *et al.*, 1988; Torpey *et al.*, 1990). Unfortunately, the trichloro acetic acid (TCA) fixation used in these studies did not adequately preserve cytoplasmic structure, and individual vimentin-containing filaments could not be identified. In addition, several laboratories have failed to identify vimentin or vimentin-containing filaments in stage VI oocytes by both immunocytochemical and Western blotting techniques (Franz *et al.*, 1983; Dent and Klymkowsky, 1989; Dent *et al.*, 1989,1992; Herman *et al.*, 1989). The presence of vimentin in *Xenopus* oocytes thus remains controversial.

In contrast to vimentin, the presence of cytokeratins (CK) in *Xenopus* oocytes has been well established (Franz *et al.*, 1983; Godsave *et al.*, 1984b; Klymkowsky *et al.*, 1987; Dent and Klymkowsky, 1989; Torpey *et al.*, 1992). Stage VI oocytes contain at least three cytokeratin isoforms, two type I cytokeratins and a single type II cytokeratin, that are most similar to the cytokeratins of simple epithelia (Franz *et al.*, 1983; Fouquet *et al.*, 1988; Torpey *et al.*, 1992). A network of radially oriented CK filaments is found in the yolk-free radii extending from near the GV toward the animal cortex of stage VI oocytes (Franz *et al.*, 1983). Although CK filaments in the vegetal cytoplasm appear to be less numerous (Franz *et al.*, 1983), Klymkowsky and colleagues (1987) have described an extensive system of CK filaments in the cortex of stage VI *Xenopus* oocytes. This network of CK filaments exhibits striking polarization along the A–V axis: CK filaments form a complex, anastomosing meshwork in the vegetal cortex that is normally absent, or very much diminished, in the animal cortex.

7. Axis Formation during *Xenopus* Oogenesis 221

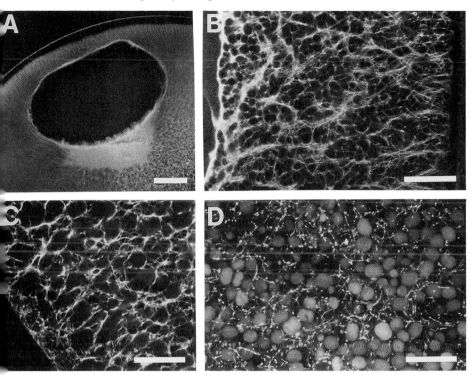

Fig. 2 Stage VI *Xenopus* oocytes contain a polarized cytoskeleton. (A) Confocal immunofluorescence microscopy using anti-α-Tb reveals a dense network of cytoplasmic MTs in stage VI oocytes (Gard, 1991a). (B) At higher magnification, loosely organized bundles can be seen extending from the GV to the animal cortex. (C) MTs appear more disordered in the vegetal hemisphere. (D) A complex three-dimensional meshwork of actin cables extends throughout the vegetal cytoplasm (Roeder and Gard, 1994). Bar: (A) 100 μm; (B–D) 25 μm.

Dent *et al.* (1992) have suggested that the A–V polarization of the cortical CK filament network is due to a specific inhibitory factor in the animal cortex. CK filaments in both the animal and the vegetal cortex are disassembled during oocyte maturation, as a result of maturation-promoting factor (MPF)-dependent phosphorylation of their protein components (Klymkowsky *et al.*, 1991).

The function of intermediate filaments (containing cytokeratins or vimentin) during oogenesis is unclear. Unlike actin filaments and microtubules (see subsequent discussion), intermediate filaments are not readily disrupted *in vivo*, making analysis of their specific functions during oogenesis more difficult. Based on the association of maternal mRNAs with a cytokeratin-rich cytoskeletal fraction from stage VI oocytes, Pondel and King (1988) speculated that CK filaments play a role in RNA localization. This hypothesis was supported further by a temporal correlation between the release of Vg1 mRNA from the vegetal cortex and breakdown of the cortical CK network during oocyte maturation (Klymkowsky et al., 1987). However, subsequent studies revealed that breakdown of the CK network is neither necessary nor sufficient for release of Vg1 from the vegetal cortex (Klymkowsky et al., 1991). The extensive store of cytokeratins found in oocytes thus may play other, as yet undefined, roles in maintaining the structural organization of *Xenopus* oocytes. Alternatively, as suggested by Franz et al. (1983) and Klymkowsky et al. (1991), cytokeratins stored in the oocyte might provide a subunit pool used during later differentiation of embryonic epithelium.

C. F-Actin Is a Component of both Nuclear and Cytoplasmic Cytoskeletons in *Xenopus* Oocytes

Stage VI *Xenopus* oocytes contain an abundance of maternally supplied actin, with cytoplasmic concentrations exceeding with 4 mg/ml (Clark and Merriam, 1978; Merriam and Clark, 1978). Using immunofluorescence and electron microscopy, Franke et al. (1976) identified two classes of actin-containing structures in the cortex of amphibian oocytes. First, bundles of actin filaments formed the cores of microvilli that cover the surface of amphibian oocytes. In addition, poorly organized whorls and bundles of actin filaments were apparent in the oocyte cortex. More recently, confocal microscopy of stage VI *Xenopus* oocytes labeled with rhodamine–phalloidin revealed an extensive network of actin cables extending throughout the cytoplasm of stage VI *Xenopus* oocytes (Roeder and Gard, 1994). Actin cables were observed surrounding the GV, in the yolk-free radii extending between the GV and the animal cortex, and forming a dense three-dimensional network in the vegetal cytoplasm.

Several lines of evidence suggest that filamentous actin (F-actin) plays an important role in the maintenance of oocyte polarity. For example, treatment with cytochalasins (which disrupt cortical and cytoplasmic actin cables) results in displacement and rotation of the GV from its normal position in the animal hemisphere (Coleman et al., 1981; Roeder and Gard, 1994). Interestingly, long-term treatment with cytochalasins also disrupts microtubule organization in stage VI *Xenopus* oocytes, suggesting a functional link between the actin and microtubule cytoskeletons (see subsequent discussion). Treatment with cytochalasins also releases Vg1 mRNA from the vegetal cortex, indicating that cortical F-actin plays an important role in anchoring maternal mRNA (Yisraeli et al., 1990).

Actin is also a major protein component of the GV (nucleus) of amphibian oocytes, constituting approximately 6% of the total nuclear protein in stage VI *Xenopus* oocytes (Clark and Merriam, 1977; Clark and Rosenbaum, 1979). Both

7. Axis Formation during *Xenopus* Oogenesis 223

biochemical and cytological analyses indicate that a substantial fraction of the actin present in the GV of stage VI oocytes is in the form of a polymer (Clark and Merriam, 1977; Roeder and Gard, 1994). Although the function of nuclear actin has not been established conclusively, the association of actin with isolated lampbrush chromosomes (Karsenti *et al.*, 1978) suggests that actin plays a role in the structure of these transcriptionally active chromosomes. The collapse of lampbrush chromosome loops, inhibition of transcription, and disruption of chromosome condensation after microinjection of actin antibodies into the GV of amphibian oocytes (Rungger *et al.*, 1979; Scheer *et al.*, 1984) further support the hypothesis that actin plays a significant role in chromosome structure in amphibian oocytes. However, care must be taken in interpreting the results of antibody injection because of the bivalent nature of IgG antibodies. Indeed, although cytochalasins disrupt nuclear actin and result in distortion of GV morphology in stage VI oocytes (Roeder and Gard, 1994), treatment of maturing oocytes with cytochalasins has little effect on the condensation of meiotic chromosomes (Ryabova *et al.*, 1986; D. L. Gard, A. D. Roeder, and B.-J. Cha, manuscript in preparation). The role of nuclear actin thus remains uncertain.

Little is known about the mechanisms by which actin assembly, distribution, and function are regulated in stage VI oocytes. Oocyte homologs of several actin-associated proteins found initially in somatic cells have been identified, including the actin-severing protein gelsolin (Ankenbauer *et al.*, 1988) and the adhesion plaque proteins talin and vinculin (Evans *et al.*, 1990). Interestingly, the distributions of talin and vinculin, components of adherens junctions that often colocalize in somatic cells, do not overlap in stage VI *Xenopus* oocytes: talin is found in the cortex whereas vinculin is found in the cytosol (Evans *et al.*, 1990). Further investigation will be required to identify the proteins and mechanisms responsible for regulating the assembly, distribution, and function of F-actin in stage VI oocytes.

D. Stage VI Xenopus Oocytes Contain an Extensive Network of Cytoplasmic Microtubules

The apparent roles of microtubules (MTs) and microtubule organizing centers (MTOCs) in establishing the polarity of many somatic cells have led to considerable speculation regarding the role that these elements play in the polarization of amphibian oocytes. As recently as 5 years ago, however, the presence of MTs in *Xenopus* oocytes was a topic of some controversy. Several lines of evidence suggested that oocytes contained few, if any, MTs. First, initial attempts to visualize oocyte MTs by electron microscopy and immunocytochemistry were largely unsuccessful (Heidemann and Gallas, 1980; Heidemann *et al.*, 1985; Palecek *et al.*, 1985). Furthermore, stage VI *Xenopus* oocytes lack conventional

MTOCs such as centrosomes, which are contributed to the zygote by the sperm (Heidemann and Kirschner, 1975; Gerhart, 1980). Finally, cytoplasmic extracts prepared from *Xenopus* oocytes inhibited MT assembly *in vitro* (Jessus *et al.*, 1984; Gard and Kirschner, 1987a), suggesting that oocytes contain an inhibitor of MT assembly. On the other hand, *Xenopus* oocytes contain a substantial pool of MT subunits, including multiple isoforms of both α- and β-tubulin (Thibier *et al.*, 1992). The α- and β-tubulins (Tbs) constitute nearly 3% of the total soluble protein of stage VI oocytes (Pestell, 1975; Gard and Kirschner, 1987a; Jessus *et al.*, 1987) and a single oocyte contains sufficient tubulin to assemble 1.5–2 *kilometers* of polymer. Analysis of the monomer–polymer pools of stage VI oocytes indicated that nearly 20% of the tubulin in oocytes actually is present in the form of polymer (Jessus *et al.*, 1987; Gard, 1991a), *corresponding to more than 300 meters of MTs per oocyte.*

The presence of individual MTs in *Xenopus* oocytes was first demonstrated by immunofluorescence using cortical preparations from stage VI oocytes (Huchon *et al.*, 1988). Subsequently, improved techniques of whole-mount immunocytochemistry and confocal microscopy have made visualization of the three-dimensional organization of oocyte MTs possible, revealing an extensive array of MTs in the cytoplasm of stage VI oocytes (Gard, 1991a). In the animal hemisphere, loosely organized MT bundles are apparent in the yolk-free radii that extend from the GV to the oocyte cortex, linked together by a disordered array of individual MTs. Enumeration of MTs in tangential optical sections of the subcortical cytoplasm suggested that as many as 0.5–1 million MTs extend from the cytoplasm to the oocyte cortex, consistent with estimates of the polymer mass obtained by biochemical techniques. A meshwork of MTs also surrounds the GV, and numerous MTs are observed in a perinuclear cap of yolk-free cytoplasm at the basal (vegetal) surface of the GV. In the vegetal hemisphere, the MT array appears more disordered, as individual MTs wend their way between the large, densely packed yolk platelets. Many MTs in the vegetal cytoplasm also extend to the cortex, forming a tightly woven network in the cortical cytoplasm.

A substantial fraction of the MTs in stage VI oocytes contain α-Tb that has been modified by covalent addition of an acetate group (Gard, 1991). In many cell types, acetylation of α-Tb on the ε-amino group of Lys-40 has been correlated with nondynamic, or stable, MTs (Piperno *et al.*, 1987; Schulze *et al.*, 1987; Webster and Borisy, 1989). In the animal hemisphere of stage VI oocytes, acetylated MTs are common in radially oriented bundles that extend from the GV to the animal cortex, and in the perinuclear cap. Although acetylated MTs are less common in deeper regions of the vegetal hemisphere, a substantial fraction of MTs in the vegetal cortex is acetylated. The large population and polarized distribution of acetylated, presumably stable microtubules present in stage VI oocytes suggests that MT stabilization may play an important role in the establishment and maintenance of oocyte polarity (see subsequent discussion).

E. Cytoplasmic Microtubules Play an Important Role in Maintaining the Polarity of Stage VI Oocytes

Disassembly of the cytoplasmic MT array of stage VI *Xenopus* oocytes, by treatment with either cold or nocodazole, results in a dramatic displacement of the GV from its normal position in the animal hemisphere (Gard, 1991a,1993). In the most extreme cases, inversion of cold or nocodazole-treated oocytes shifts the GV to the vegetal pole, a technique that has been used to examine the capacity of different cytoplasmic regions to support spindle assembly (Gard, 1993). The displacement of the GV after MT disassembly is much more dramatic in rate and extent than that observed after the disruption of actin with cytochalasins (Roeder and Gard, 1994; B.-J. Cha and D. L. Gard, unpublished observations), suggesting that MTs play a more direct role in nuclear positioning. Disassembly of the MT array of stage VI oocytes does not affect the distribution of yolk, pigment, or maternal RNA (Yisraeli *et al.*, 1990; Gard, 1993), suggesting that these aspects of the oocyte architecture are maintained by interactions with other cytoskeletal elements such as filaments composed of actin or cytokeratins.

A further indicator of the importance of MTs in maintaining the polarized organization of the oocyte cytoplasm is the evolution of mechanisms to preserve or reestablish the cytoplasmic MT array should it become disrupted by environmental conditions. Amphibians have developed at least two mechanisms for preserving the information residing in the cytoplasmic MT array of polarized oocytes. In *Xenopus,* stage VI oocytes can reassemble their polarized MT array during recovery from cold-induced MT disassembly (B.-J. Cha and D. L. Gard, unpublished observations). Recent results suggest that the information required to reestablish the polarized MT array of *Xenopus* oocytes resides in a polarized distribution of γ-Tb in the oocyte cortex (Gard, 1994; see subsequent discussion).

Fully grown oocytes from the North American frog *Rana pipiens* also contain an extensive polarized array of cytoplasmic MTs (Wang *et al.*, 1993; D. L. Gard, unpublished observations). However, unlike the cold-sensitive MTs of *Xenopus* oocytes (Gard, 1991a,1993), MTs in fully grown *Rana* oocytes remain intact after incubation at 0–2°C for up to 2 wk. MTs in *Rana* oocytes are sensitive to nocodazole and other anti-microtubule drugs, and drug-induced disassembly of the MT network in Rana oocytes results in displacement of the GV from its normal location in the animal hemisphere (Lessman, 1987; Wang *et al.*, 1993; D. L. Gard, unpublished observations). Although the biochemical basis for cold-stable MTs in *Rana* oocytes has yet to be determined (for a discussion of cold-stable MTs in Antarctic fishes, see Williams *et al.*, 1985; Dietrich *et al.*, 1987) the evolution and developmental regulation of cold-stable MTs during oogenesis in *Rana pipiens*, in which the oocyte polarity established during summer must be maintained through a period of winter hibernation, underscore the importance of MTs in maintaining the A–V axis.

III. Cytoskeletal Organization and Axis Specification during Early Oogenesis

The evidence just presented supports the conclusion that F-actin and MTs (and perhaps cytokeratins) provide a structural framework required for localizing developmentally important molecules and organelles, and thus play important roles in maintaining oocyte polarity once it has been established. However, the data presented do not address the significant question, "Does the oocyte cytoskeleton play a determinitive role in specification of the A–V axis?" One hypothesis suggests that the A–V axis of amphibian oocytes and eggs is determined by a preexisting polarization of early oocytes along a cytoplasmic axis defined by the oocyte centrosome or other components of the cytoskeleton (Coggins, 1973; Gerhart et al., 1983,1986; Heasman et al., 1984a). Although this model is attractive, recent results place several constraints on the role of the cytoskeleton in axis specification.

A. Postmitotic Xenopus Oocytes Are Polarized along a Nuclear/Cytoplasmic Axis

Although formation of the definitive A–V axis is not evident until much later in oogenesis, postmitotic *Xenopus* oocytes are polarized along an "initial" axis established during the last oogonial division. Four sequential oogonial divisions produce a "nest" of 16 oocytes that remain joined by cytoplasmic bridges resulting from incomplete cytokinesis (Al-Mukhtar and Webb, 1971; Coggins, 1973). Most of the cytoplasmic organelles—including numerous mitochondria, elements of the endoplasmic reticulum, and the golgi apparatus—are clustered in a cap of cytoplasm at the narrow end of each roughly pear-shaped oocyte, whereas the large rounded oocyte nucleus (later to be called the GV) nearly fills the broad end (Al-Mukhtar and Webb, 1971; Coggins, 1973; D. L. Gard, D. G. Affleck, and B. M. Error, manuscript in preparation). Even the nuclear organization of early oocytes is polarized along the initial oocyte axis. During the zygotene–pachytene stages of prophase, oocyte chromosomes are arranged in a characteristic "bouquet," with the synaptonemal complexes clustered toward the proximal side of the nucleus nearest the cytoplasmic cap and centrosome (Fig. 3; Al-Mukhtar and Webb, 1971; Coggins, 1973).

The distribution of cytoplasmic MTs in early postmitotic oocytes is also po-

---

Fig. 3 Postmitotic oocytes are polarized along a nuclear–cytoplasmic axis. (A) Clusters of mitochondria in the perinuclear cytoplasm (arrows), visualized here by confocal microscopy and rhodamine 123, represent one indicator of the "initial" axis (denoted by the direction of the arrows) of early postmitotic oocytes. (B) A nest of early postmitotic oocytes stained with propidium iodide. Nuclear polarity is evident in the "bouquet" organization of synaptonemal complexes during the early stages

7. Axis Formation during *Xenopus* Oogenesis

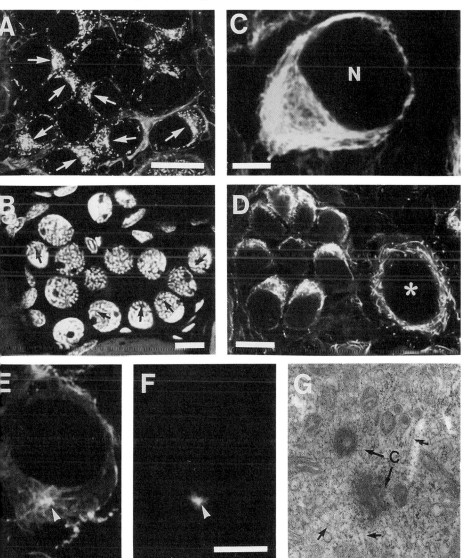

of meiotic prophase. (C) The polarized distribution of MTs in postmitotic oocytes is evident in this small oocyte (17 × 21 mm in diameter) stained with TUB1A2 antiserum against tyrosinated α-Tb. (D) A subpopulation of acetylated MTs is apparent in the cytoplasmic caps of these small postmitotic oocytes, whereas an early stage I oocyte (asterisk) contains a perinuclear ring of acetylated MTs. (E–F) The maternal centrosome (arrows in E and F) of this postmitotic oocyte was identified both by functional assays (MT nucleation during recovery from cold-induced MT disassembly) and by immunofluorescence using anti-γ-Tb (arrow in F). (G) Electron microscopy reveals a pair of centrioles (C) in this small postmitotic oocyte. Several MTs (arrows) are apparent in the pericentriolar region. Bar: (A, B, D) 10 μm, (C) 5 μm.

larized along this initial cytoplasmic/nuclear axis. Electron microscopy of early postmitotic oocytes reveals many MTs and a typical eukaryotic centrosome in the organelle-filled cytoplasmic cap (Al-Mukhtar and Webb, 1971; Coggins, 1973; D. L. Gard, D. G. Affleck, and B. M. Error, manuscript in preparation; for a discussion of centrosome structure, see McIntosh, 1983; Brinkley, 1985). Confocal immunofluorescence microscopy using antibodies against α-Tb also dramatically illustrates the polarization of the MT cytoskeleton in early oocytes, revealing a dense network of MTs in the cytoplasmic cap (D. L. Gard, D. G. Affleck, and B. M. Error, manuscript in preparation) and a sparse network of MTs extending from the cytoplasmic cap to surround the oocyte nucleus. Despite evidence that early oocytes contain centrosomes (Al-Mukhtar and Webb, 1971; Coggins, 1973; D. L. Gard, D. G. Affleck, and B. M. Error, manuscript in preparation), immunofluorescence microscopy with antisera against either α-Tb (which stains all MTs) or acetylated α-Tb (which stains stable MTs) reveals no evidence of radial organization or discrete MTOCs (see below). However, early postmitotic oocytes do contain a subpopulation of acetylated MTs that are largely restricted to the cytoplasmic cap. The presence of acetylated MTs in these early oocytes contrasts with their absence in mitotic and interphase oogonia, and is the first indication of what will become a large population of stable MTs during later oogenesis.

B. Does Centrosome Position in Early Oocytes Define the Orientation of the A–V Axis?

In somatic cells, a cytoplasmic axis defined by the relative positions of the centrosome (or other MTOC) and the cell nucleus often can be correlated directly with a morphological or functional polarization of the cell (see McIntosh, 1983). Because stage VI *Xenopus* oocytes lack identifiable centrosomes (see Gerhart, 1980, and references therein; Gard, 1991a) it is impossible to correlate the structural and functional polarity of the A–V axis with centrosome position during later oogenesis. However, *Xenopus* oogonia and early postmitotic oocytes do contain a typical centrosome (see preceding discussion). As in somatic cells, the location of the centrosome can be used to define, and may be causally related to, the "initial" cytoplasmic polarity of postmitotic *Xenopus* oocytes.

Does the "initial axis" defined by the oocyte centrosome, or even an axis defined by centrosome position in the primordial germ cell or oogonium, influence or specify the A–V polarity established later in oogenesis? Although this question cannot yet be answered with certainty, the influence of the oocyte centrosome position on A–V axis formation is constrained by its transient nature: examination of MT nucleation and the distribution of centrosomal antigens indicates that the oocyte centrosome is inactivated very early in the process of oocyte differentiation (D. L. Gard, D. G. Affleck, and B. M. Error, manuscript in

7. Axis Formation during *Xenopus* Oogenesis   229

preparation). Functional centrosomes (centers of MT nucleation after release from cold-induced disassembly) can be identified in the majority (>70%) of oocytes with mean diameters less than 25 μm. However, this fraction falls to less than 40% for oocytes with diameters between 25 and 50 μm, less than 25% for oocytes between 50 and 75 μm, and less than 5% for oocytes with diameters greater than 75 μm.

The mechanisms of centrosome inactivation and the fate of the maternal centrioles remain largely unknown. Centrosome inactivation temporally corresponds to dispersal of γ-Tb into multiple foci throughout the oocyte cytoplasm during early diplotene (D. L. Gard, D. G. Affleck, and B. M. Error, manuscript in preparation). These dispersed foci of γ-Tb are unable to nucleate MTs during recovery from cold-induced MT disassembly, suggesting that γ-Tb alone is not sufficient to nucleate MTs. Several lines of evidence suggest that cell-cycle-dependent phosphorylation of centrosomal components regulates centrosome function in other systems (Vandre and Borisy, 1989; Verde *et al.*, 1991). Conceivably, inactivation of the maternal centrosome during early *Xenopus* oogenesis might result from modification by phosphorylation or dephosphorylation on entry into the prolonged prophase arrest of meiosis.

C. Little Evidence Remains of the Initial Oocyte Axis during Early Diplotene

After completion of the leptotene through pachytene stages of meiotic recombination, *Xenopus* oocytes arrest in a prolonged diplotene stage characterized by extensive growth (oocyte volume increases more than 25,000-fold) and biosynthesis. In *Xenopus*, the diplotene stage of meiotic prophase has been divided by Dumont (1972) into a series of six morphologically and functionally defined stages, beginning with previtellogenic stage I oocytes (ranging from less than 50 to 300 μm in diameter) and culminating with fully grown, maturation-competent, stage VI oocytes (1.2–1.3 mm in diameter). In contrast to the early stages of prophase (which are most easily found in subadult animals), diplotene oocytes ranging from midstage I through stage VI of oogenesis are well represented in the ovaries of mature adult females (Dumont, 1972), making these the most commonly examined stages of oocyte differentiation.

Stage I encompasses the entire previtellogenic period of diplotene, including growing oocytes with diameters ranging from less than 50 μm to approximately 300 μm. During early stage I (oocyte diameters between 35 and 50 μm), little evidence remains of the "initial" axis that characterizes oocytes in the leptotene-through-pachytene stages of meiotic prophase (Fig. 4). By early diplotene, oocytes have adopted a more rounded morphology, the GV is more centrally located, and mitochondria, which were concentrated in the perinuclear cap during

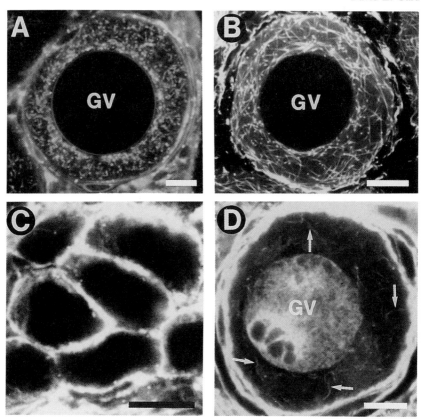

Fig. 4 Little evidence of the cytoplasmic axis remains in early stage I oocytes. (A) Confocal microscopy of early stage I oocytes stained with rhodamine 123 reveal that mitochondria are dispersed throughout the cytoplasm (Heasman *et al.*, 1984; D. L. Gard, D. G. Affleck, and B. M. Error, manuscript in preparation). (B) MTs also are distributed throughout the cytoplasm during early stage I, with no evidence of a cytoplasmic MTOC. (C,D) Confocal microscopy of oocytes stained with rhodamine–phalloidin reveals that F-actin is restricted to a cortical shell in small postmitotic oocytes (C). However in larger oocytes (D), F-actin is present in the GV, and actin cables are sparsely distributed throughout the cytoplasm (Roeder and Gard, 1994). Bar: 10 μm.

earlier stages of oogenesis, are dispersed in the perinuclear cytoplasm (Heasman *et al.*, 1984a; D. L. Gard, D. G. Affleck, and B. M. Error, manuscript in preparation). Despite their lack of a definable MTOC, oocytes of this size contain a substantial array of MTs distributed throughout their cytoplasm. However, aside from concentrations of MTs (including a subpopulation of acetylated MTs) in the perinuclear and cortical cytoplasm, there is little evidence for spatial organization or polarization of the MT distribution that is seen earlier in oogenesis.

D. Is the "Initial" Axis of Early Oocytes Maintained in Latent Form by a Polarized Distribution of F-Actin?

Inactivation of the oocyte centrosome during early oogenesis suggests that centrosome position cannot contribute directly to specification of the A–V axis later in oogenesis. Conceivably, the initial oocyte axis defined by the position of the maternal centrosome (or other spatial determinant) could be maintained in "latent" form by other cytoskeletal systems such as intermediate filaments or F-actin, only later to be elaborated to form the definitive A–V axis. Although little information is available regarding the presence and organization of intermediate filaments (composed of cytokeratins or vimentin) prior to midstage I of oogenesis, recent results suggest that polarization of F-actin during early oogenesis does not contribute to later establishment of the A–V axis (Roeder and Gard, 1994).

Confocal microscopy of oocytes stained with rhodamine-conjugated phalloidin revealed that F-actin was restricted to a thin shell (about 1 μm thick) in the cortex of the smallest oocytes examined (those less than 35 μm in diameter). This cortical actin network exhibited little obvious polarity. In fact, despite indications that oocytes of this size are connected by cytoplasmic bridges (Coggins, 1973), evidence of cytoskeletal specializations containing either MTs or F-actin associated with intercellular contacts was not observed. This result contrasts with the structure of the "ring canals" observed linking nurse cells and oocytes during *Drosophila* oogenesis; these canals contain prominent rings of actin filaments (Warn *et al.*, 1985; Theurkauf *et al.*, 1992).

Larger oocytes (with diameters >35 μm) contain two additional actin networks: a diffuse network of F-actin in the GV and an extensive system of cytoplasmic actin cables. Nuclear actin is first apparent in oocytes with diameters of 35–40 μm, roughly coincident with entry into the diplotene stage of meiotic prophase. The appearance of F-actin in the nucleus during early diplotene supports the hypothesis that actin plays an important role in nuclear and chromosomal organization during this extended period of oogenesis, and implies that actin is not required for recombination events during the zygotene–pachytene stages of prophase. Cytoplasmic actin cables are first observed sparsely scattered through the cytoplasm of oocytes with diameters of 35–40 μm, and later forming a dense interconnected meshwork. No evidence of specific organizing centers is apparent, and neither the network of cytoplasmic actin cables nor the nuclear F-actin matrix exhibits any evidence of polarization during early oogenesis. Therefore, polarization of the actin cytoskeleton is unlikely to contribute to the specification of an early progenitor of the A–V axis. However, this result does not preclude the possibility that the actin cytoskeleton plays an important, as yet undefined, role in axis formation.

E. The Mitochondrial Mass Defines a Cytoplasmic Axis in Stage I *Xenopus* Oocytes

One of the most striking features of midstage I *Xenopus* oocytes (75–150 μm in diameter) is a tightly compacted aggregate containing many thousands of mitochondria. Variously called the "Balbiani body," the "yolk nucleus," the "mitochondrial cloud," or the "mitochondrial mass" (the term used hereafter), this unique structure is readily visible in a good stereomicroscope and can be specifically stained with dyes such as rhodamine 123 (Fig. 5; Czolowska, 1969; Dumont, 1972; Coggins, 1973; Billet and Adams, 1976; Heasman *et al.*, 1984a). In addition to mitochondria, the mitochondrial mass of stage I oocytes contains components of the germplasm, other membranous organelles, and elements of the cytoskeleton (Billet and Adams, 1976; Heasman *et al.*, 1984a; see subsequent discussion).

The importance of the mitochondrial mass in specification of the A–V axis and the relationship between the mitochondrial mass of stage I oocytes and the initial axis of early oogenesis have been topics of considerable speculation. During stage II, components of the mitochondrial mass (including mitochondria, germplasm, and XCAT-2 mRNA) disperse to the adjacent region of the oocyte cortex, which will later be situated in the vegetal hemisphere (Heasman *et al.*, 1984a; Hausen and Riebesell, 1990; King *et al.*, 1993). The position of the mitochondrial mass in the cytoplasm of stage I *Xenopus* oocytes thus represents one of the earliest known markers of the definitive A–V axis.

The distinctive aggregations of mitochondria observed in the cytoplasmic cap of early postmitotic oocytes and again later during stage I have led to suggestions that the cytoplasmic axis defined by the mitochondrial mass of stage I oocytes is directly descended from the initial axis of early postmitotic oocytes (Coggins, 1973; Gerhart *et al.*, 1983,1986; Heasman *et al.*, 1984a). This ultimately leads to the hypothesis that the initial axis of early postmitotic oocytes, as defined by the position of the centrosome and mitochondria in the cytoplasmic cap, is a direct progenitor of the A–V axis. Although this hypothesis has yet to be addressed directly, several results are inconsistent with its most literal interpretation. First, inactivation of the maternal centrosomes occurs very early in oogenesis, and is largely completed prior to formation of the definitive mitochondrial mass (D. L. Gard, D. G. Affleck, and B. M. Error, manuscript in preparation; see preceding discussion). Although elements of the oocyte cytoskeleton are concentrated around the mitochondrial mass of stage I oocytes (see subsequent text), no evidence exists that the mitochondrial mass functions as a specific organizing center for any cytoskeletal element. Furthermore, traditional and confocal fluorescence microscopy of oocytes stained with rhodamine 123 indicates that mitochondria are dispersed throughout the perinuclear cytoplasm of *Xenopus* oocytes during early diplotene prior to appearance of the true mitochondrial mass in midstage I (Heasman *et al.*, 1984a; D. L. Gard, D. G. Affleck, and B. M. Error, manuscript in preparation). Also, some stage I oocytes contain two or more mitochondrial masses, which is inconsistent with their functioning as a singular determinant of oocyte polarity. Finally, aggregation of mitochondria into a compact mass is not a common feature of oogenesis in all amphibian species. Thus, although the mitochondrial mass serves as an early indicator of the A–V polarity of *Xenopus* oocytes, its relationship to the initial oocyte axis and its importance to the general mechanisms of axis specification during amphibian oogenesis require further evaluation.

F. Stage I Oocytes Contain a Complex Cytoskeleton That Serves as a Framework for Cytoplasmic Organization

The overall distribution of cytoskeletal elements in stage I *Xenopus* oocytes provides one clue to their possible function: CK filaments, MTs, and actin cables course throughout the cytoplasm, linking the mitochondrial mass and GV to the oocyte cortex (Godsave *et al.*, 1984, Wylie *et al.*, 1985; Gard, 1991a; Roeder and Gard, 1994; D. L. Gard, D. G. Affleck, and B. M. Error, manuscript in preparation). Moreover, loosely organized bundles of acetylated, presumably stable, MTs link the GV, mitochondrial mass, and cortex of stage I oocytes (Gard, 1991a; D. L. Gard, D. G. Affleck, and B. M. Error, manuscript in preparation). Immunocytochemistry and electron microscopy also reveal significant concentrations of cytoskeletal components, including CK filaments, MTs (including acetylated MTs), and actin cables within and surrounding the mitochondrial mass of stage I *Xenopus* oocytes (Godsave, 1984b; Gard, 1991a; Roeder and Gard, 1994; D. L. Gard, D. G. Affleck, and B. M. Error, manuscript in preparation). These distributions again support the notion that the cytoskeleton (composed of IFs, MTs, and F-actin) forms a structural framework that is responsible for maintaining cytoplasmic organization during oogenesis. Indeed, disassembly of the MT cytoskeleton in stage I oocytes treated with des-acetylcolchicine disrupts cytoplasmic organization, allowing the mitochondrial mass and other organelles to float free in the cytoplasm (Wylie *et al.*, 1985). Interestingly, disassembly of neither F-actin nor MTs has any apparent effect on cohesion of the mitochondrial mass itself, suggesting that other cytoskeletal elements (cytokeratins?) are responsible for maintaining this characteristic structure (Wylie *et al.*, 1985; Roeder and Gard, 1994).

It is tempting to speculate that, during their dispersal to the oocyte cortex, the components of the mitochondrial mass (mitochondria, germplasm, and XCAT-2 mRNA, for example) are all actively transported along MTs or other elements of the cytoskeleton. However, evidence supporting this hypothesis is, at this time, limited to inferences from observations that MTs link the mitochondrial mass to the cortex and the GV (Gard, 1991a). The molecular mechanisms by which components of the mitochondrial mass are targeted and transported to the cortex, and the signals regulating the timing of aggregation and dispersal of the mito-

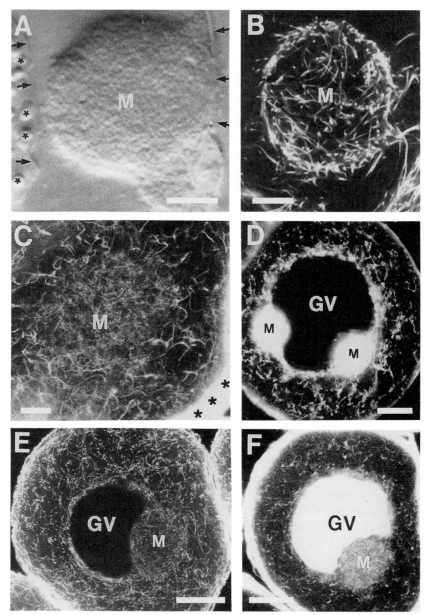

Fig. 5 Microtubules and actin cables are concentrated in the mitochondrial mass of midstage I oocytes. (A) The mitochondrial mass (M) of midstage I *Xenopus* oocytes is readily apparent by Nomarski DIC microscopy. Arrows denote the boundary of the cell (*right*) and the GV (*left*). Many peripheral nucleoli are apparent in the GV (asterisks). (B) Numerous MTs, including a subpopulation containing acetylated α-Tb (as shown here), are apparent within and surrounding the mitochondrial

chondrial mass during oogenesis in *Xenopus*, remain subjects for future investigation.

Aside from the observed concentrations of cytoskeletal elements in and around the mitochondrial mass, there is little evidence that the cytoskeleton of stage I *Xenopus* oocytes is polarized: MTs, actin cables, and CK filaments are present throughout the oocyte cytoplasm. This organization is consistent with the distributions of MTs and mitochondria in previtellogenic oocytes from two other genera of amphibians. Examination of previtellogenic oocytes from both *Rana pipiens* and *Ambystoma* sp. after staining with rhodamine 123 reveals that oocytes from these species lack the mitochondrial mass characteristic of stage I *Xenopus* oocytes (D. L. Gard, unpublished observations). Confocal immunofluorescence microscopy reveals that MTs are distributed throughout the cytoplasm of previtellogenic oocytes in both *Rana* and *Ambystoma*, with no evidence of cytoplasmic polarity (Wang *et al.*, 1993; D. L. Gard, unpublished observations).

IV. Cytoplasmic and Cytoskeletal Polarity Are Established during Stage IV of Oogenesis

Although elements of the mitochondrial mass mark the future vegetal cortex of *Xenopus* oocytes as early as stage II of oogenesis, other aspects of the architecture of oocytes largely remain unpolarized throughout stages II–III. The GV of stage III oocytes is normally found near the oocyte center, and yolk and pigment are equally distributed in the animal and vegetal hemispheres (Dumont, 1972). In addition, little evidence exists for polarization of the oocyte cytoskeleton. During stage III, *Xenopus* oocytes contain a dense network of cytoplasmic MTs and MT bundles, which appear to radiate from the GV toward all regions of the oocyte cortex (Gard, 1991a).

The A–V polarization of *Xenopus* oocytes first becomes outwardly apparent during stage IV of oogenesis, as pigment granules are differentially segregated to the animal cortex (Dumont, 1972). At the same time, the internal polarity of oocytes is established: the GV moves from the center of the oocyte to a position in the animal hemisphere; the A–V yolk gradient is established by the differential transport of yolk platelets into the animal and vegetal hemispheres; and Vg1

---

mass (M) (Gard, 1991a; D. L. Gard, D. G. Affleck, and B. M. Error, manuscript in preparation). (C) Cables of F-actin (stained here with rhodamine-conjugated phalloidin) are also apparent in the mitochondrial mass, and link the mass to the oocyte cortex (asterisks) (Roeder and Gard, 1994). (D) Multiple mitochondrial masses (M) are evident in a small fraction of stage I oocytes (shown here after staining with rhodamine 123). (E,F) Stage I oocytes contain an extensive cytoskeleton composed of MTs (E; Gard, 1991a); actin cables (F; Roeder and Gard, 1994), and cytokeratins (not shown; Godsave *et al.*, 1984b). Bar: (A–C) 10 μm; (D–F) 25 μm.

mRNA (and other maternal RNAs) are transported to the vegetal cortex (Danilchik and Gerhart, 1987; Dumont, 1972; Yisraeli et al., 1990). Polarization of the cytoplasmic MT array and the cortical network of CK filaments also develops during stage IV of oogenesis. By stage VI, loosely organized bundles of MTs extend from the GV to the cortex of the animal hemisphere while a more disordered MT array extends to the cortex of the vegetal hemisphere (Gard, 1991a); the complex cortical CK network has been established as well (Klymkowsky, 1987).

The polarization of *Xenopus* oocytes during stages IV–VI of oogenesis can mechanistically be considered as two processes: (1) the translocation of organelles (such as the GV, yolk, and pigment) and developmentally important molecules (such as the Vg1 mRNA) to their final positions along the A–V axis and (2) the underlying polarization of the oocyte cytoskeleton. Several lines of evidence suggest that the overall polarization of the oocyte is dependent on the cytoskeleton. The dense arrays of cytoplasmic MTs present throughout oogenesis, and their dramatic reorganization from stage I to stage VI (summarized in Fig. 6), suggest that MTs play an important structural role during establishment of the A–V axis (Gard, 1991a). Direct and indirect evidence support a role for MTs and actin filaments in positioning the GV in the animal hemisphere of amphibian oocytes (Lessman, 1987; Gard, 1991a,1993; Roeder and Gard, 1994). In addition, Yisraeli et al. (1990) demonstrated that translocation and anchoring of Vg1 mRNA to the vegetal cortex occurs in a two-step process that requires both MTs and actin filaments: translocation to the cortex during stages III–IV requires MTs whereas F-actin is required to anchor the Vg1 mRNA once it has reached the cortex.

The precise mechanisms responsible for establishing the polarized distribution of these elements remain unclear. Since mRNA localization and GV position are dependent on MTs, it is tempting to speculate that microtubule-dependent motors (such as members of the dynein or kinesin families; Skoufias and Scholey, 1993) power the directed translocation of RNA and organelles during A–V polarization. In the case of Vg1 mRNA, molecular analysis reveals that the translocation signal lies in the 3' untranslated region (Mowry and Melton, 1992); a protein that specifically binds to this region of the RNA has been identified (Schwartz et al., 1992). However, the manner by which targeting signals are interpreted, and the mechanisms by which RNAs such as Vg1 are specifically targeted to the vegetal cortex along MTs at a time when MTs extend to all regions of the oocyte cortex, remain unknown.

The mechanism by which the oocyte cytoskeleton becomes polarized during axis formation also remains largely unknown. In the case of MTs, we might ask, "How, in the absence of centrosomes, do *Xenopus* oocytes regulate MT organization and reorganization during formation of the A–V axis?" As in somatic cells, both microtubule-associated proteins (MAPs) and MTOCs are likely to play important roles in determining MT organization. Several MAPs have been identified in *Xenopus* oocytes and eggs (Gard and Kirschner, 1987b; Fellous et al.,

7. Axis Formation during *Xenopus* Oogenesis 237

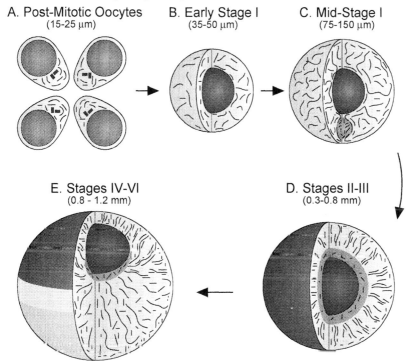

Fig. 6 Summary of MT reorganization during oogenesis. (A) MTs are concentrated in the cytoplasmic cap (see text) of small postmitotic oocytes. Oocytes of this size contain a functional centrosome, including centrioles and γ-Tb. (B) By early stage I, MTs are distributed throughout the cytoplasm, with little evidence of order. MTOCs cannot be identified by optical sectioning or in functional assays, and γ-Tb has dispersed to multiple cytoplasmic foci. (C) In midstage I, oocytes contain an extensive MT network, with concentrations of mitochondria in the cortical and perinuclear cytoplasm and in the mitochondrial mass. No MTOC is evident. (D) By stage III, MTs extend from the GV to the cortex, forming a loosely organized radial array. (E) During stages IV–VI, the MT array becomes polarized along the A–V axis. Loosely organized bundles of MTs extend from the GV to the animal cortex. MTs in the vegetal hemisphere appear more disordered.

1991; Shiina *et al.*, 1992). Interestingly, these proteins are all cell-cycle-dependent phosphoproteins, suggesting that phosphorylation of MAPs might regulate MT assembly dynamics during oogenesis and early embryogenesis.

γ-Tubulin Is Asymmetrically Distributed in the Cortex of *Xenopus* Oocytes

Although they lack discrete centrosomes, *Xenopus* oocytes and eggs contain sufficient cytoplasmic pools of centrosomal components, including the pericentriolar protein γ-tubulin (γ-Tb), to assemble several thousand centrosomes in the absence of most protein synthesis (Gard et al., 1990; Stearns et al., Verde et al., 1991). This maternally supplied pool of centrosome components undoubtedly plays an important role during the rapid cell cycles of early cleavage. However, centrosome components stored in the oocyte might also play an important role in regulating MT organization during oogenesis.

As a first step in determining the role played by this pool of centrosome components during oogenesis, we examined the distribution of γ-Tb in stage VI *Xenopus* oocytes (Fig. 7; Gard, 1994). Confocal immunofluorescence microscopy revealed that γ-Tb was concentrated in the cytoplasm surrounding the GV, consistent with earlier results that suggested that the GV serves as an MTOC during later oogenesis (Gard, 1991a). However, prominent concentrations of γ-Tb were also apparent in the oocyte cortex, raising a number of interesting questions regarding the mechanisms of MT nucleation and the polarity of MTs in *Xenopus* oocytes.

Individual MTs are both structurally and functionally polarized, a property that is evident in their dynamics of assembly (see Kirschner and Mitchison, 1986, for a review of the effects of MT polarity on assembly dynamics), their interactions with MT-dependent motor proteins (Skoufias and Scholey, 1993), and their association with cytoplasmic MTOCs (Euteneuer and McIntosh, 1981a,b; Mitchison and Kirschner, 1984). γ-Tb, a component of centrosomes and spindle poles in a variety of eukaryotic cells (Oakley and Oakley, 1989; Oakley et al., 1990; Stearns et al., 1991; Joshi et al., 1992), has been postulated to function as a nucleation or capping factor specific for the minus end of MTs (Stearns et al., 1991; Joshi et al., 1992). The observation of γ-Tb in the cortex of stage VI *Xenopus* oocytes then suggests that a substantial fraction of the oocyte MTs have their minus ends anchored in the cortex. This result contrasts with the MT orientation found in most somatic cells, in which MTs are anchored to the centrosome by their minus ends and have their plus ends extending into the cytoplasm (Euteneuer and McIntosh, 1981a,b; McIntosh, 1983; Mitchison and Kirschner, 1984). However, apical MTOCs and γ-Tb have recently been identified in the cortex of polarized epithelia in *Drosophila* and in vertebrates (Troutt and Burnside, 1988; Mogensen et al., 1989; Tucker et al., 1992), where they are thought to play an important role in establishing the functional polarity of these cells (Bre et al., 1990). By analogy, the orientation of MTs may be a significant factor in establishing the functional polarity of the A–V axis during oogenesis. Further analysis will be required to identify the polarity of MTs in *Xenopus* oocytes conclusively.

γ-Tb may also play a significant role in establishing the spatially polarized distribution of MTs in *Xenopus* oocytes. Immunofluorescence microscopy reveals that the distribution of γ-Tb in the cortex of stage VI *Xenopus* oocytes is polarized along the A–V axis. Moreover, polarization of the cortical distribution of γ-Tb temporally coincides with the reorganization of the MT array during

7. Axis Formation during *Xenopus* Oogenesis

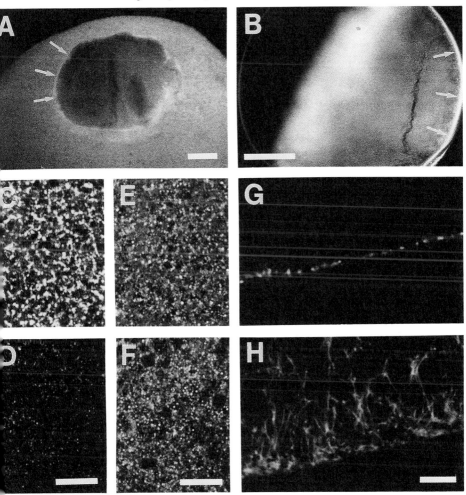

Fig. 7 γ-Tb is asymmetrically distributed in the cortex of stage VI *Xenopus* oocytes. Confocal immunofluorescence microscopy using antibodies against *Xenopus* γ-Tb reveal that this centrosomal protein is concentrated in the perinuclear (A) and cortical (B) cytoplasm of stage VI *Xenopus* oocytes (Gard, 1994). (C) Grazing optical sections of vegetal cortex from stage VI oocytes reveal brightly stained foci and aggregates of γ-Tb. (D) Sections of animal cortex from the same oocyte reveal a more even distribution of γ-Tb foci. (E,F) Optical sections of opposing regions of cortex of a stage III oocyte stained with γ-Tb antibodies reveal similar staining patterns, indicating that the polarized distribution of γ-Tb is established coincident with oocyte polarization during stage IV. (G,H) Cross sections of the vegetal cortex reveal that γ-Tb (G) is restricted to a narrow region of the oocyte cortex whereas MTs (H) extend into the subcortical cytoplasm. Bar: (A) 100 μm; (B) 250 μm; (C–H) 10 μm.

stages III–IV of oogenesis, consistent with the hypothesis that the distribution of γ-Tb in the oocyte cortex is responsible for reorganization of the MT cytoskeleton during oogenesis. Cortical γ-Tb in the oocyte cortex could influence MT organization in several ways. Differential nucleation of MTs from the animal and vegetal cortex could lead to an increased density of MTs in the animal hemisphere. Alternatively, the selective stabilization of MTs nucleated elsewhere in the oocyte (from the GV?) by γ-Tb in the oocyte cortex could result in polarization of the MT array (see Kirschner and Mitchison, 1986, for a discussion of MT stabilization and its potential role in morphogenesis).

Although these models have yet to be tested directly, circumstantial evidence suggests that MT stabilization plays an important role in establishing the polarized organization of MTs in *Xenopus* oocytes. First, populations of MTs containing acetylated α-Tb, a marker for stable or nondynamic MTs, are apparent throughout postmitotic oogenesis (Gard, 1991a; D. L. Gard, D. G. Affleck, and B. M. Error, manuscript in preparation). During later oogenesis, acetylated MTs are more numerous in the animal hemisphere, where they are found in loosely organized bundles extending from the GV to the animal cortex. Furthermore, the time course of MT reassembly and polarization during recovery from cold-induced disassembly is more consistent with models of MT stabilization: although nucleation and reassembly of MTs is apparent within minutes of return to room temperature (Gard, 1991a), more than 12 hr are required to reestablish the polarized MT array (B.-J. Cha and D. L. Gard, unpublished observations). Finally, reestablishment of the polarized distribution of MTs is accompanied by reacetylation of α-Tb (a marker for MT stability), and appears to be dependent on interactions with the oocyte cortex (B.-J. Cha and D. L. Gard, unpublished observations). These observations suggest that stabilization of oocyte MTs, through interactions with the oocyte cortex, plays an important role in polarization of the MT array during A–V axis formation.

γ-Tb in the oocyte cortex may also integrate components of the oocyte cytoskeleton. The distribution of γ-Tb in the cortex of stage VI oocytes is disrupted by treatment with cytochalasin B (Gard, 1994); long-term treatment with cytochalasin appears to release MTs from the cortex and the GV of stage VI oocytes (B.-J. Cha and D. L. Gard, unpublished observations). Interestingly, γ-Tb is also released from the oocyte cortex during progesterone-induced maturation (D. L. Gard, unpublished observations), coincident with a dramatic remodeling of the MT cytoskeleton. Together, these observations suggest that γ-Tb serves as a bridge between the MT cytoskeleton and networks of F-actin in the perinuclear and cortical cytoplasm (Fig. 8).

Although polarization of γ-Tb in the oocyte cortex during stages III–IV of oogenesis may direct the reorganization and polarization of cytoplasmic MTs (thereby contributing to the structural and functional polarization of the growing oocyte), we are left with an unanswered question: "How does the distribution of γ-Tb in the cortex become polarized?" Presumably, the polarized distribution of

7. Axis Formation during *Xenopus* Oogenesis 241

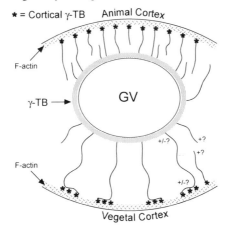

Fig. 8 γ-Tb may link cytoplasmic MTs to a cortical cytoplasm composed of F-actin. Treatment with cytochalasin B disrupts both the organization of γ-Tb in the vegetal cortex of stage VI oocytes and the organization of cytoplasmic MTs (Gard, 1994). Together, these observations suggest that γ-Tb may play a role in anchoring cytoplasmic MTs to the cortical cytoskeleton. The presence of γ-Tb in the oocyte cortex also raises a number of questions regarding the polarity of MTs in *Xenopus* oocytes (see text).

cortical γ-Tb reflects an underlying structural and functional polarity of the oocyte cortex.

V. Ectopic Spindle Assembly during Maturation of *Xenopus* Oocytes: Evidence for Functional Polarization of the Oocyte Cortex

The resumption of meiosis during oocyte maturation is accompanied by a dramatic reorganization of the oocyte cytoskeleton. The cortical network of CK filaments is disassembled as a result of MPF-dependent phosphorylation of its component proteins (Klymkowsky *et al.*, 1987,1991). Vg1 mRNA and γ-Tb are released from the vegetal cortex, presumably because of changes in the organization of cortical actin (Dictus *et al.*, 1984; Yisraeli *et al.*, 1990; Gard, 1994; D. L. Gard, unpublished observations). Finally, the oocyte MT array is extensively reorganized (Gard, 1991b), culminating in the assembly of the meiotic spindles at the animal pole of the unfertilized egg (Jessus *et al.*, 1986; Huchon *et al.*, 1988; Gard, 1992).

Despite significant reorganization of its cytoskeleton during maturation, the maturing oocyte maintains a distinct A–V polarity (for reviews, see Gerhart, 1980; Gerhart *et al.*, 1986; Hausen and Riebesell, 1990). This polarity is perhaps most evident in the formation of the distinctive white "maturation spot" in the animal cortex (resulting from breakdown of the GV) and assembly of the meiotic spindles in close proximity to the animal cortex (Jessus et al., 1986; Huchon et al., 1988; Gard, 1992). In fact, the final location of the second meiotic spindle often is used as a cytological marker to define the animal pole of the unfertilized *Xenopus* egg (Gerhart, 1980; Hausen and Riebesell, 1990).

Assembly of the meiotic spindles in *Xenopus* oocytes begins with the nucleation of a transient MT array (TMA) by a novel, discoidal MTOC that itself is assembled near the basal (vegetal) surface of the GV (Brachet et al., 1970; Huchon et al., 1981; Gard, 1992). This MTOC–TMA complex rapidly transports the condensed meiotic chromosomes (at an estimated rate of 15–25 µm per min) to the animal pole, where it serves as the immediate precursor of the first meiotic spindle (Gard, 1992). The observation of similar structures in oocytes from other species of anurans and uredeles (Beetschen and Gautier, 1989, and references therein) suggests that the MTOC–TMA complex is a common precursor of spindle assembly during the maturation of amphibian oocytes. Assembly of the first (M1) and second (M2) meiotic spindles then follows a pathway consisting of four characteristic stages: (1) aggregation; (2) formation of a bipolar spindle axis, transverse to the A–V axis; (3) spindle elongation during prometaphase; and (4) rotation of the spindle into alignment with the A–V axis (Gard, 1992). This novel aggregation–elongation pathway of spindle assembly may, in part, result from the lack of discrete centrosomes or MTOCs at the poles of the meiotic spindles (Huchon et al., 1981). After completion of M1, which results in cytokinesis and formation of the first polar body, the maturing oocyte arrests in metaphase of M2 with an axially oriented spindle (Fig. 9).

The final location of the meiotic spindles, then, results from the directed translocation of the MTOC–TMA complex, carrying its burden of chromosomes, to the animal cortex. However, the machinery underlying this process remains unknown. Neither gravity nor buoyancy of the MTOC–TMA complex in the yolk-filled cytoplasm plays an important role, since inversion of the oocytes with respect to gravity does not result in misdirection of the MTOC–TMA complex or displacement of spindle assembly from the animal pole (Gard, 1993). In fact, oocyte maturation in the ovary occurs in random orientations with respect to gravity. Ryabova et al. (1986) reported that migration of the nascent spindle during the maturation of *Xenopus* oocytes was blocked by cytochalasins, and suggested that F-actin is a required component of the translocation machinery. However, Beetschen and Gautier (1989) reported that translocation of similar complexes in *Ambystoma* oocytes required MTs but not F-actin. Since it is unlikely that migration of the MTOC–TMA complexes in these two amphibian species involves different mechanisms, we have reevaluated the effect of cytochalasin on MTOC–TMA migration and spindle assembly during maturation of *Xenopus* oocytes (D. L. Gard, B.-J. Cha, and A. D. Roeder, manuscript in preparation). Our results, consistent with those obtained with *Ambystoma*, indicate that F-actin is not strictly required for either translocation or spindle assem-

7. Axis Formation during *Xenopus* Oogenesis

A. Meiosis 1 in Animal Hemisphere.
(time after WSF)

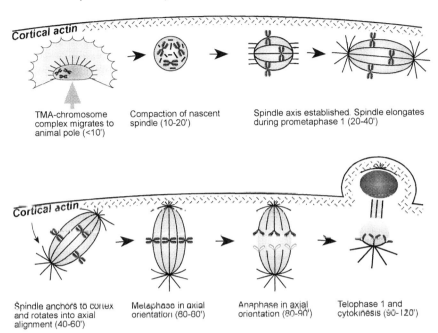

B. Meiosis 1 in CB-Treated or Vegetal Hemisphere.

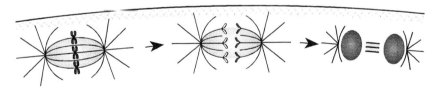

Fig. 9 Spindle rotation is dependent on F-actin in the cortex of the animal hemisphere. (A) Assembly of the first meiotic spindle during maturation of *Xenopus* oocytes follows a complex pathway of aggregation, elongation, and rotation (Gard, 1992). (B) Spindles assembled in the vegetal hemisphere, or in the animal hemisphere of cytochalasin-treated oocytes, assemble normally but cannot anchor to the cortex, and thus do not undergo rotation. These spindles complete anaphase in a transverse orientation, and twin M2 spindles are assembled during M2.

bly per se, but suggest that F-actin is required for maintaining the structural integrity of the MTOC–TMA complex during its migration to the animal pole.

The directed translocation of the MTOC–TMA complex to the animal pole during oocyte maturation suggests that a specific mechanism exists for targeting spindle assembly to the animal pole. We tested this hypothesis by examining MTOC–TMA migration and spindle location after experimental displacement of the GV from its normal position in the animal hemisphere of stage VI *Xenopus* oocytes (Gard, 1993). Results from these experiments indicated that spindle assembly is not targeted to a specific location in the oocyte cortex. Instead, the MTOC–TMA complex migrates to the region of cortex in closest proximity to the GV prior to GV breakdown. The location of the meiotic spindles is thus a direct consequence of the position of the GV in the animal hemisphere. Furthermore, the observed assembly of ectopic spindles in equatorial or vegetal regions of oocytes after displacement of the GV demonstrates that spindle assembly is not restricted to the animal pole, but can occur (although with lowered efficiency) nearly anywhere in the oocyte cytoplasm (Fig. 10).

Assembly of both M1 and M2 spindles during maturation of *Xenopus* oocytes initially occurs in a transverse orientation (parallel to the surface, and perpendicular to the A–V axis), followed by rotation into alignment with the A–V axis (Gard, 1992). Rotation of the meiotic spindles in *Xenopus* oocytes is inhibited by cytochalasin (D. L. Gard, B.-J. Cha, and A. D. Roeder, manuscript in preparation), suggesting that astral MTs are interacting, either directly or indirectly, with F-actin in the oocyte cortex. However, in contrast to spindle assembly, which can occur over a wide area of the oocyte cortex, spindle rotation is dependent on the association of spindles with the animal cortex (Gard, 1993). Whereas ectopic M1 spindles assembled in the animal hemisphere undergo normal rotation and progress through anaphase and cytokinesis to form the first polar body, M1 spindles assembled in more vegetal locations (more than 120° from the animal pole) fail to rotate and do not complete cytokinesis. These observations suggest that the cortex of the maturing oocyte is functionally polarized along the A–V axis. Although the molecular basis for this cortical polarity has yet to be established, the dependence of spindle rotation on cortical actin suggests that determinants specifying the functional identity of the animal and vegetal cortex are components of, or are otherwise bound to, the actin cytoskeleton.

VI. Concluding Remarks

The last 5 years have seen significant advances in our understanding of cytoskeletal organization during amphibian oogenesis and its role in formation of the characteristic animal–vegetal axis of amphibian oocytes. The studies cited here provide substantial support for the hypothesis that cytoskeletal networks composed of F-actin and MTs (and perhaps CK filaments) play significant roles in

7. Axis Formation during *Xenopus* Oogenesis

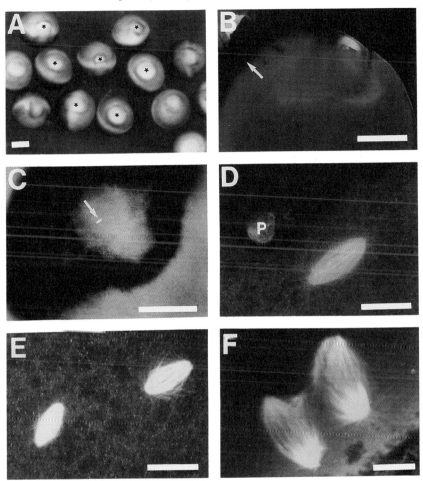

Fig. 10 Ectopic spindle assembly provides evidence for functional polarization of the oocyte cortex. (A) Maturation spots form in the lateral and vegetal regions of oocytes after experimental displacement of the GV (see Gard, 1993). (B) An MTOC–TMA complex near the equatorial cortex of a maturing oocyte after displacement of the GV (arrow marks the approximate position of the animal pole). (C,D) An ectopic spindle in the animal hemisphere, approx. 65° from the animal pole. Note the normal-appearing polar body (P) in D. (E) Twin M2 spindles about 170° from the animal pole, resulting from failure of cytokinesis in the vegetal hemisphere. (F) Twin M2 spindles resulting from failure of cytokinesis in a cytochalasin B-treated oocyte (25 µg/ml during maturation). Bar: (A) 500 µm; (B,C) 250 µm; (D,E) 25 µm; (F) 10 µm.

establishing and maintaining the polarized architecture of amphibian oocytes and eggs, including such features as nuclear position, the localization of maternal mRNA, and the location and orientation of the meiotic spindles.

Although establishment and maintenance of A–V polarity are clearly dependent on MTs and F-actin, the role of the cytoskeleton in the determinitive events of axis specification, and the relationship of the A–V axis to the initial cytoplasmic polarity of postmitotic oocytes, still remains uncertain. However, recent observations place constraints on involvement of the cytoskeleton in axis specification, suggesting that the spatial polarization contributed by the oocyte centrosome, MT array, and F-actin network of postmitotic oocytes cannot contribute directly to determination of the A–V axis later in oogenesis. Finally, the studies cited provide evidence that the cytoskeleton itself undergoes a remarkable reorganization and polarization during formation of the A–V axis, and suggest that polarization of the oocyte cortex contributes to the structural and functional polarization of the oocyte.

Much remains to be learned about the molecular mechanisms by which assembly and organization of the cytoskeleton are regulated during oocyte differentiation. Many of the proteins involved, including MAPs, actin-binding proteins, and intermediate-filament-associated proteins, will undoubtedly have functions similar to those of their counterparts in somatic cells. However, the unique requirements imposed on amphibian oocytes may have resulted in novel solutions to the problems of cytoskeletal regulation. A more thorough understanding of the assembly and regulation of the cytoskeleton during oogenesis will provide an essential foundation for understanding the mechanisms by which the structural and functional axis of amphibian oocytes is first specified during early oogenesis, and how this axis is subsequently translated into the developmental polarization of amphibian oocytes, eggs, and embryos.

Acknowledgments

I thank all the members of my lab (past and present) who have contributed to the work and ideas presented in this commentary; Drs. G. Piperno, M. Kirschner, M. Klymkowsky, and T. Stearns for providing antibodies; and Dr. E. King for his assistance with the confocal microscope facility and many thought-provoking discussions. Work described in this manuscript was supported by grants from the National Institutes of General Medical Sciences, the National Science Foundation, and the University of Utah Research Committee.

References

Al-Mukhtar, K. A. K., and Webb, A. (1971). An ultrastructural study of primordial germ cells, oogonia, and early oocytes in *Xenopus laevis. J. Embryol. Exp. Morphol.* **26,** 195–217.

Ankenbauer, T., Kleinschmidt, J., Vanderkerckhove, J., and Franke, W. (1988). Proteins regulating actin assembly in oogenesis and early embryogenesis of *Xenopus laevis:* Gelsolin is the major cytoplasmic actin-binding protein. *J. Cell Biol.* **107,** 1489–1498.

Beetschen, J.-C., and Gautier, J. (1989). Oogenesis. *In* "Developmental Biology of the Axolotl"

7. Axis Formation during *Xenopus* Oogenesis 247

(J. B. Armstrong and G. M. Malacinski, eds.), pp. 25–35. Oxford University Press, New York.
Bement, W. M., and Capco, D. G. (1990). Transformation of the amphibian oocyte into the egg: Structural and biochemical events. *J. Electron Microsc.* **16**, 202–234.
Billet, F. S., and Adams, E. (1976). The structure of the mitochondrial cloud of *Xenopus laevis* oocytes. *J. Embryol. Exp. Morphol.* **33**, 697–710.
Brachet, J., Hanocq, F., and Van Gansen, P. (1970). A cytochemical and ultrastructural analysis of in vitro maturation in amphibian oocytes. *Dev. Biol.* **21**, 157–195.
Bre, M.-H., Pepperkok, R., Hill, A. M., Levilliers, N., Ansorge, W., Stelzer, E. H. K., and Karsenti, E. (1990). Regulation of microtubule dynamics and nucleation during polarization in MDCK II Cells. *J. Cell Biol.* **111**, 3013–3021.
Brinkley, B. R. (1985). Microtubule organizing centers. *Annu. Rev. Cell Biol.* **1**, 145–172.
Campanella, C., Chaponnier, C., Quaglia, L., Gualtieri, R. M., and Gabbiani, G. (1990). Different cytoskeletal organization in maturation stages of *Discoglossus pictus* (Anura) oocytes: Thickness and stability of actin microfilaments and tropomyosin localization. *Mol. Reprod. Dev.* **25**, 130–139.
Capco, D. G., and Jeffery, W. R. (1982). Transient localizations of messenger RNA in *Xenopus laevis*. *Dev. Biol.* **89**, 1–12.
Carpenter, C. D., and Klein, W. (1982). A gradient of poly (A)+ RNA sequences in *Xenopus laevis* eggs and embryos. *Dev. Biol.* **89**, 1–12.
Clark, T. G., and Merriam, R. W. (1977). Diffusible and bound actin in nuclei of *Xenopus laevis* oocytes. *Cell* **12**, 883–891.
Clark, T. G., and Merriam, R. W. (1978). Actin in *Xenopus* oocytes. I. Polymerization and gelation in vitro. *J. Cell Biol.* **77**, 427–438.
Clark, T. G. and Rosenbaum, J. L. (1979). An actin filament matrix in hand-isolated nuclei of *X. laevis* oocytes. *Cell* **18**, 1101–1108.
Coggins, L. W. (1973). An ultrastructural and radioautographic study of early oogenesis in the toad *Xenopus laevis*. *J. Cell Sci.* **12**, 71–93.
Coleman, A., Morser, J., Lane, C., Besley, J., Wylie, C., and Valle, G. (1981). Fate of secretory proteins trapped in oocytes of *Xenopus laevis* by disruption of the cytoskeleton or by imbalanced subunit synthesis. *J. Cell Biol.* **91**, 770–778.
Colombo, R., Benedusi, P., and Valle, G. (1981). Actin in *Xenopus* development: Indirect immunofluorescence study of actin localization. *Differentiation* **20**, 45–51.
Czolowska, R. (1969). Observations on the origin of the "germinal cytoplasm" in *Xenopus laevis*. *J. Embryol. Exp. Morphol.* **22**, 229–251.
Danilchik, M., and Gerhart, J. (1987). Differentiation of the animal-vegetal axis in *Xenopus laevis* oocytes. I. Polarized intracellular translocation of platelets established the yolk gradient. *Dev. Biol.* **122**, 101–112.
Dent, J. A., and Klymkowsky, M. W. (1989). Whole-mount analysis of cytoskeletal reorganization and function during oogenesis and early embryogenesis in *Xenopus*. *In* "The Cell Biology of Fertilization." (H. Schatten and G. Schatten, eds.), pp. 63–103. Academic Press, Orlando, Florida.
Dent, J. A., Polson, A. G., and Klymkowsky, M. W. (1989). A whole-mount immunocytochemical analysis of the expression of the intermediate filament protein vimentin in *Xenopus*. *Development* **105**, 61–74.
Dent, J. A., Cary, R. B., Bachant, J. B., Domingo, A., and Klymkowsky, M. W. (1992). Host cell factors controlling vimentin organization in the *Xenopus* oocyte. *J. Cell Biol.* **119**, 855–866.
Dictus, W. J. A. G., van Zoelen, E. J. J., Tetteroo, P. A. T., Tertoolen, L. G. J., De Laat, S. W., and Bluemink, J. G. (1984). Lateral mobility of plasma membrane lipids in *Xenopus*

eggs: Regional differences related to animal/vegetal polarity become extreme upon fertilization. *Dev. Biol.* **101,** 201–211.
Dietrich, H. W., Prasad, V., and Luduena, R. F. (1987). Cold-stable MTs from Antarctic fishes contain unique α tubulins. *J. Biol. Chem.* **262,** 8360–8366.
Dumont, J. (1972). Oogenesis in *Xenopus laevis* (Daudin). I. Stages of oocyte development in laboratory maintained animals. *J. Morph.* **136,** 153–180.
Elinson, R., and Rowning, B. (1988). A transient array of parallel microtubules in frog eggs: Potential tracks for a cytoplasmic rotation that specifies the dorso-ventral axis. *Dev. Biol.* **128,** 185–197.
Euteneuer, U., and McIntosh, J. R. (1981a). Polarity of some motility-related microtubules. *Proc. Natl. Acad. Sci. USA* **78,** 372–376.
Euteneuer, U., and McIntosh, J. R. (1981b). Structural polarity of kinetochore microtubules in PtK1 cells. *J. Cell Biol.* **89,** 338–345.
Evans, J. P., Page, B. D., and Kay, B. K. (1990). Talin and vinculin in the oocytes, eggs, and early embryos of *Xenopus laevis:* A developmentally regulated change in distribution. *Dev. Biol.* **137,** 403–413.
Fellous, A., Huchon, D., Thibier, C., and Jessus, C. (1991). Intracellular location of MAP2-related protein (O-map) in prophase I and metaphase II oocytes of *Xenopus*. *Mech. Dev.* **33,** 139–146.
Fouquet, B., Herrmann, H., Franz, J. K., and Franke, W. W. (1988). Expression of intermediate filament proteins during development of *Xenopus laevis* III. Identification of mRNAs encoding cytokeratins typical of complex epithelia. *Development* **104,** 533–548.
Franke, W. W., Rathke, P. C., Seib, E., Trendelenburg, M. F., Osborn, M., and Weber, K. (1976). Distribution and mode of arrangement of microfilamentous structures and actin in the cortex of the amphibian oocyte. *Cytobiologie* **14,** 111–130.
Franz, J. K., Gall, L., Williams, M. A., Picheral, B., and Franke, W. W. (1983). Intermediate-sized filaments in a germ cell: Expression of cytokeratins in oocytes and eggs of the frog *Xenopus*. *Proc. Natl. Acad. Sci. USA* **80,** 6254–6258.
Gard, D. L. (1991a). Organization, nucleation, and acetylation of microtubules in *Xenopus laevis* oocytes: A study by confocal immunofluorescence microscopy. *Dev. Biol.* **143,** 346–362.
Gard, D. L. (1991b). MT organization and spindle assembly during meiotic maturation of *Xenopus* oocytes. *J. Cell Biol.* **115,** 46a.
Gard, D. L. (1992). Microtubule organization during maturation of *Xenopus* oocytes: Assembly and rotation of the meiotic spindles. *Dev. Biol.* **151,** 516–530.
Gard, D. L. (1993). Ectopic spindle assembly during maturation of *Xenopus* oocytes: Evidence for functional polarization of the oocyte cortex. *Dev. Biol.* **159,** 298–310.
Gard, D. L. (1994). γ-Tubulin is asymmetrically distributed in the cortex of *Xenopus* oocytes. *Dev. Biol.* **161,** 131–140.
Gard, D. L., and Kirschner. M. (1987a). Microtubule assembly in cytoplasmic extracts of *Xenopus* oocytes and eggs. *J. Cell Biol.* **105,** 2191–2201.
Gard, D. L., and Kirschner. M. (1987b). A microtubule-associated protein from *Xenopus* eggs that specifically promotes assembly at the plus-end. *J. Cell Biol.* **105,** 2203–2215.
Gard, D. L., Hafezi, S., Zhang, T., and Doxsey, S. J. (1990). Centrosome duplication continues in cycloheximide-treated *Xenopus* blastulae in the absence of a detectable cell cycle. *J. Cell Biol.* **110,** 2033–2042.
Gerhart, J. C. (1980). Mechanisms regulating pattern formation in the amphibian egg and early embryo. *In* "Biological Regulation and Development" (R. F. Goldberger, ed.), pp. 133–316. Plenum Press, New York.
Gerhart, J., Black, S., Gimlich, R., and Scharf, S. (1983). Control of polarity in the amphibian egg. *In* "Time, Space, and Pattern in Embryonic Development" (W. R. Jeffery and R. A. Raff, eds.), pp. 261–286. Liss, New York.

7. Axis Formation during *Xenopus* Oogenesis

Gerhart, J., Danilchik, M., Roberts, J., Rowning, B., and Vincent, J.-P. (1986). Primary and secondary polarity of the amphibian oocyte and egg In "Gametogenesis and the Early Embryo" (J. C. Gall, ed.), pp. 305–319. Liss, New York.

Gimlich, R. L., and Gerhart, J. C. (1980). Early cellular interactions promote embryonic axis formation in *Xenopus laevis*. *Dev. Biol.* **104**, 117–130.

Godsave, S. F., Anderton, B. H., Heasman, J., and Wylie, C. C. (1984a). Oocytes and early embryos of *Xenopus laevis* contain intermediate filaments which react with anti-mammalian vimentin antibodies. *J. Embryol. Exp. Morphol.* **83**, 169–187.

Godsave, S. F., Wylie, C. C., Lane, E. B., and Anderton, B. H. (1984b). Intermediate filaments in the *Xenopus* oocyte: The appearance and distribution of cytokeratin-containing filaments. *J. Embryol. Exp. Morphol.* **83**, 157–167.

Gururajan, R., Perry-O'Keefe, H., Melton, D., and Weeks, D. (1991). The *Xenopus* localized messenger RNA An3 may encode an ATP-dependent RNA helicase. *Nature* **349**, 717–719.

Hausen, P., and Riebesell, M. (1990). "The Early Development of *Xenopus laevis*. An Atlas of Histology." Springer-Verlag, Berlin.

Heasman, J., Quarmby, J., and Wylie, C. C. (1984a). The mitochondrial cloud of *Xenopus* oocytes: The source of germinal granule material. *Dev. Biol.* **105**, 458–469.

Heasman, J., Wylie, C. C., Hausen, P., and Smith, J. C. (1984b). Fates and states of determination of single vegetal pole blastomeres of *Xenopus laevis*. *Cell* **37**, 185–194.

Heidemann, S. R., and Gallas, P. T. (1980). The effect of taxol on living eggs of *Xenopus laevis*. *Dev. Biol.* **80**, 489–494.

Heidemann, S. R., and Kirschner, M. W. (1975). Aster formation in eggs of *Xenopus laevis*. Induction by isolated basal bodies. *J. Cell Biol.* **67**, 105–117.

Heidemann, S. R., Hamborg, M. A., Balasz, J. E., and Lindley, S. (1985). Microtubules in immature oocytes of *Xenopus laevis*. *J. Cell Sci.* **77**, 129–141.

Herman, H., Foquet, B., and Franke, W. (1989). Expression of intermediate filament proteins during development of *Xenopus laevis*. I. cDNA clones of vimentin. *Development* **105**, 279–298.

Huchon, D., Crozet, N., Cantenot, N., and Ozon, R. (1981). Germinal vesicle breakdown in the *Xenopus laevis* oocyte: Description of a transient microtubular structure. *Reprod. Nutr. Develop.* **21**, 135–148.

Huchon, D., Jessus, C., Thibier, C., and Ozon, R. (1988). Presence of microtubules in isolated cortices of prophase I and metaphase II oocytes in *Xenopus laevis*. *Cell Tis. Res.* **154**, 415–420.

Irish, V., Lehmann, R., and Akam, M. (1989). The *Drosophila* posterior gene nanos functions by repressing hunchback activity. *Nature* **338**, 646–648.

Jessus, C., Friederich, E., Francon, J., and Ozon, R. (1984). In vitro inhibition of tubulin assembly by a ribonucleoprotein complex associated with the free ribosomal fraction isolated from *Xenopus laevis* oocytes: Effect at the level of microtubule-associated proteins. *Cell Diff.* **14**, 179–187.

Jessus, C., Huchon, D., and Ozon, R. (1986). Distribution of microtubules during the breakdown of the nuclear envelope of the *Xenopus* oocyte. An immunocytochemical study. *Biol. Cell* **56**, 113–120.

Jessus, C., Thibier, C., and Ozon, R. (1987). Levels of microtubules during meiotic maturation of the *Xenopus* oocyte. *J. Cell Sci.* **87**, 705–712.

Joshi, H. C., Palacios, M. J., McNamera, L., and Cleveland, D. (1992). γ-Tubulin is a centrosomal protein required for cell-cycle-dependent microtubule nucleation. *Nature* **356**, 80–83.

Karsenti, E., Gounon, P., and Bornens, M. (1978). Immunocytochemical study of lampbrush chromosomes: Presence of tubulin and actin. *Biol. Cell* **31**, 219–224.

Keller, R. E. (1975). Vital dye mapping of the gastrula and neural of *Xenopus laevis*. I. Prospective areas and morphogenetic movements of the superficial layer. *Dev. Biol.* **42**, 222–241.

King, M. L., and Barklis, E. (1985). Regional distribution of maternal messenger RNA in the amphibian oocyte. *Dev. Biol.* **112**, 203–212.

King, M. L., Forristall, C., and Zhou, Y. (1993). RNA localization to the vegetal cortex of *Xenopus* oocytes. *Mol. Biol. Cell (Suppl.)* **4**, 5a.

Kirschner, M., and Mitchison, T. (1986). Beyond self-assembly: From microtubules to morphogenesis. *Cell* **45**, 329–342.

Klymkowsky, M. W., Maynell, L. A., and Polson, A. G. (1987). Polar asymmetry in the organization of the cortical cytokeratin system of *Xenopus laevis* oocytes and embryos. *Development* **100**, 543–557.

Klymkowsky, M. W., Maynell, L. A., and Nislow, C. (1991). Cytokeratin phosphorylation, cytokeratin filament severing, and the solubilization of the maternal mRNA Vg1. *J. Cell Biol.* **114**, 787–797.

Lehmann, R., and Nusslein-Volhard, C. (1991). The maternal gene nanos has a central role in posterior pattern formation of the *Drosophila* embryo. *Development* **112**, 679–693.

Lessman, C. A. (1987). Germinal vesicle migration and dissolution in *Rana pipiens* oocytes: Effects of steroids and microtubule poisons. *Cell Differentiation* **20**, 238–251.

Masui, Y. (1972). Distribution of the cytoplasmic activity inducing germinal vesicle breakdown in frog oocytes. *J. Exp. Zool.* **179**, 365–378.

McIntosh, J. R. (1983). The centrosome as an organizer of the cytoskeleton. *Mod. Cell Biol.* **2**, 115–142.

Melton, D. (1987). Translocation of a localized maternal mRNA to the vegetal pole of *Xenopus* oocytes. *Nature* **328**, 80–82.

Merriam, R. W., and Clark, T. G. (1978). Actin in *Xenopus* oocytes. II. Intracellular distribution and polymerizability. *J. Cell Biol.* **77**, 439–447.

Merriam, R. W., Sauterer, R. A., and Christensen, K. (1983). A subcortical pigment-containing structure in *Xenopus* eggs with contractile properties. *Dev. Biol.* **95**, 439–446.

Mitchison, T. J., and Kirschner, M. W. (1984). Microtubule assembly nucleated by isolated centrosomes. *Nature* **312**, 232–236.

Mogensen, M. M., Tucker, J. B., and Stebbings, H. (1989). Microtubule polarities indicate that nucleation and capture of microtubules occurs at cell surfaces in *Drosophila*. *J. Cell Biol.* **108**, 1445–1452.

Mosquera, L., Forristall, C., Zhou, Y., and King, M. L. (1993). A mRNA localized to the vegetal cortex of *Xenopus* oocytes encodes a protein with a nanos-like zinc finger domain. *Development* **117**, 377–386.

Mowry, K. L., and Melton, D. A. (1992). Vegetal messenger RNA localization directed by a 340-nt RNA sequence element in *Xenopus* oocytes. *Science* **255**, 991–994.

Nieuwkoop, P. D., and Sutasurya, L. A. (1979). "Primordial Germ Cells in the Chordates." Cambridge University Press, Cambridge.

Oakley, B. R., Oakley, C. E., Yoon, Y., and Jung, M. K. (1990). γ-Tubulin is a component of the spindle pole body that is essential for microtubule function in *Aspergillus nidulans*. *Cell* **61**, 1289–1301.

Oakley, C. E., and Oakley, B. R. (1989). Identification of γ-tubulin, a new member of the tubulin superfamily encoded by the mipA gene of *Aspergillus nidulans*. *Nature* **338**, 662–664.

Palecek, J., Habrova, V., Nedvidek, J., and Romanovsky, A. (1985). Dynamics of tubulin structures in *Xenopus laevis* oogenesis. *J. Embryol. Exp. Morphol.* **87**, 75–86.

Pestell, R. Q. W. (1975). Microtubule protein synthesis during oogenesis and early embryogenesis in *Xenopus laevis*. *Biochem. J.* **145**, 527–534.

Piperno, G., LeDizet, M., and Chang, X.-J. (1987). Microtubules containing acetylated α-tubulin in mammalian cells in culture. *J. Cell Biol.* **104**, 289–302.

Pondel, M., and King, M. L. (1988). Localized maternal mRNA related to transforming growth factor β mRNA is concentrated in a cytokeratin-enriched fraction from *Xenopus* oocytes. *Proc. Natl. Acad. Sci. USA* **85**, 7612–7616.

7. Axis Formation during *Xenopus* Oogenesis 251

Rebagliati, M., Weeks, D. L., Harvey, R. P., and Melton, D. A. (1985). Identification and cloning of localized maternal RNAs from *Xenopus* eggs. *Cell* **42**, 769–777.

Roeder, A. D., and Gard, D. L. (1994). Confocal microscopy of F-actin distribution in *Xenopus* oocytes. *Zygote* **2**, 111–124.

Rungger, D., Rungger-Brandle, E., Chaponnier, C., and Gabiani, G. (1979). Intranuclear injection of anti-actin antibodies into *Xenopus* oocytes blocks chromosome condensation. *Nature* **282**, 320–321.

Ryabova, L. P., Betina, M. A., and Vassetzky, S. G. (1986). Influence of cytochalasin B on oocyte maturation in *Xenopus laevis*. *Cell Diff.* **19**, 89–96.

Scharf, S., and Gerhart, J. (1983). Axis determination in eggs of *Xenopus laevis:* A critical period before first cleavage, identified by the common effects of cold, pressure, and ultraviolet irradiation. *Dev. Biol.* **99**, 75–87.

Scheer, U., Hinssen, H., Franke, W. W., and Jockusch, B. M. (1984). Microinjection of actin-binding proteins and actin antibodies demonstrates involvement of nuclear actin in transcription of lampbrush chromosomes. *Cell* **39**, 111–122.

Schulze, E., Asai, D. J., Bulinski, J. C., and Kirschner, M. W. (1987). Post-translational modification and microtubule stability. *J. Cell Biol.* **105**, 2167–2177.

Schwartz, S. P., Aisenthal, L., Elisha, Z., Oberman, F., and Yisraeli, J. K. (1992). A 69-kDa RNA-binding protein from *Xenopus* oocytes recognizes a common motif in two vegetally localized maternal mRNAs. *Proc. Natl. Acad. Sci. USA* **89**, 11895–11899.

Shiina, N., Moriguchi, T., Ohta, K., Gotoh, Y., and Nishida, E. (1992). Regulation of a major microtubule-associated protein by MPF and MAP kinase. *EMBO J.* **11**, 3977–3984.

Skoufias, D. A., and Scholey, J. M. (1993). Cytoplasmic microtubule-based motor proteins. *Curr. Opin. Cell Biol.* **5**, 95–104.

Smith, L. D., Michaël, P., and Williams, M. A. (1983). Does a predetermined germ line exist in amphibians? *In* "Current Problems in Germ Cell Differentiation" (A. McLaren and C. C. Wylie, eds.), pp. 19–39. Cambridge University Press, Cambridge.

Snape, A., Wylie, C. C., Smith, J. C., and Heasman, J. (1987). Changes in commitment of single animal pole blastomeres of *Xenopus laevis*. *Dev. Biol.* **119**, 503–510.

Stearns, T., Evans, L., and Kirschner, M. (1991). γ-Tubulin is a highly conserved component of the centrosome. *Cell* **65**, 825–836.

Tang, P., Sharpe, C. R., Mohun, T. J., and Wylie, C. C. (1988). Vimentin expression in oocytes, eggs, and early embryos of *Xenopus laevis*. *Development* **103**, 279–287.

Tannahill, D., and Melton, D. (1989). Localized synthesis of the Vg1 protein during early *Xenopus* development. *Development* **106**, 775–785.

Theurkauf, W. E., Smiley, S., Wong, M. L., and Alberts, B. M. (1992). Reorganization of the cytoskeleton during *Drosophila* oogenesis: Implications for axis specification and intracellular transport. *Development* **115**, 923–936.

Thibier, C., Denoulet, P., Jessus, C., and Ozon, R. (1992). A predominant basic alpha-tubulin isoform present in prophase *Xenopus* oocytes decreases during meiotic maturation. *Biol. Cell* **75**, 173–180.

Thomsen, G. H., and Melton, D. A. (1993). Processed Vg1 protein is an axial mesoderm inducer in *Xenopus*. *Cell* **74**, 433–441.

Torpey, N. P., Heasman, J., and Wylie, C. C. (1990). Identification of vimentin and novel vimentin-related proteins in *Xenopus* oocytes and early embryos. *Development* **10**, 1185–1195.

Torpey, N. P., Heasman, J., and Wylie, C. C. (1992). Distinct distribution of vimentin and cytokeratin in *Xenopus* oocytes and early embryos. *J. Cell Sci.* **101**, 151–160.

Troutt, L. L., and Burnside, B. (1988). The unusual microtubule polarity in teleost retinal pigment epithelial cells. *J. Cell Biol.* **107**, 1461–1464.

Tucker, J. B., Paton, C. C., Richardson, G. P., Mogensen, M. M., and Russell, I. J. (1992). A cell surface-associated centrosomal layer of microtubule-organizing material in the inner pillar cell of the mouse cochlea. *J. Cell Sci.* **102**, 215–226.

Vandre, D. D., and Borisy, G. G. (1989). The centrosome cycle in animal cells. *In* "Mitosis: Molecules and Mechanisms." (J. S. Hyams and B. R. Brinkley, eds.), pp. 39–75. Academic Press, New York.

Verde, F., Berrez, J.-M., Antony, C., and Karsenti, E. (1991). Taxol-induced microtubule asters in mitotic extracts of *Xenopus* eggs: Requirements for phosphorylated factors and cytoplasmic dynein. *J. Cell Biol.* **112**, 1177–1187.

Vincent, J.-P., Oster, G., and Gerhart, J. (1986). Kinematics of gray crescent formation in *Xenopus* eggs: The displacement of subcortical cytoplasm relative to the egg surface. *Dev. Biol.* **113**, 484–500.

Wang, C., and Lehmann, R. (1991). *nanos* is the localized posterior determinant in *Drosophila*. *Cell* **66**, 637–648.

Wang, T., Lessman, C. A., and Gard, D. L. (1993). Developmental regulation of cold-stable microtubules during oogenesis in *Rana pipiens*. *Mol. Biol. Cell (Suppl.)*, **4**, 26a.

Warn, R. M., Gutzeit, H. O., Smith, L., and Warn, A. (1985). F-actin rings are associated with the ring canals of the *Drosophila* egg chamber. *Exp. Cell Res.* **157**, 355–363.

Webster, D. R., and Borisy, G. G. (1989). Microtubules are acetylated in domains that turn over slowly. *J. Cell Sci.* **92**, 57–65.

Weeks, D. L., and Melton, D. A. (1987). A maternal mRNA localized to the vegetal hemisphere in *Xenopus* eggs codes for a growth factor related to TGF-α. *Cell* **51**, 861–867.

Williams, R. C. Jr., Correia, J. J., and deVries, A. L. (1985). Formation of microtubules at low temperature by tubulin from Antarctic fish. *Biochemistry* **24**, 2790–2798.

Wylie, C. C., Brown, D., Godsave, S. F., Quarmby, J., and Heasman, J. (1985). The cytoskeleton of *Xenopus* oocytes and its role in development. *J. Embryol. Exp. Morphol. (Suppl.)* **89**, 1–15.

Wylie, C. C., Snape, A., Heasman, J., and Smith, J. C. (1987). Vegetal pole cells and commitment to form endoderm in *Xenopus laevis*. *Dev. Biol.* **119**, 496–502.

Yisraeli, J. K., Sokol, S., and Melton, D. A. (1990). A two-step model for the localization of maternal mRNA in *Xenopus* oocytes: Involvement of microtubules and microfilaments in the translocation and anchoring of Vg1 mRNA. *Development* **108**, 289–298.

Zhou, Y., Forristall, C., and King, M. L. (1993). XCAT-2 RNA localization in *Xenopus* oocytes. *Mol. Biol. Cell* **4** (suppl), 25a.

8
Specifying the Dorsoanterior Axis in Frogs: 70 Years since Spemann and Mangold

Richard P. Elinson and Tamara Holowacz
Department of Zoology
University of Toronto
Toronto, Ontario, Canada M5S 1A1

I. Introduction
   A. Embryonic Axes in *Xenopus*
   B. Outline of *Xenopus* Axial Development
   C. Key Molecular Players
II. How Does the Organizer Form?
   A. Cortical Rotation
   B. Maternal Dorsal Factor
   C. Location of Dorsal Information in the Egg and Embryo
   D. Axis Induction
   E. Summary: Path to the Organizer
III. How Does the Organizer Act?
   A. Dorsalization
   B. Neuralization
IV. Concluding Remarks
   References

I. Introduction

A landmark in our understanding of vertebrate development occurred in 1924 with the discovery of the organizer by Hans Spemann and Hilde Mangold. Their discovery prompted decades of investigations (Hamburger, 1988) that ran their course by approximately 1960 (Fig. 1). The fiftieth anniversary of the Spemann–Mangold paper was marked by a celebratory volume (Nakamura and Toivonen, 1978), published near the nadir of the paper's impact. Despite a vast amount of work, progress toward understanding the organizer had been painfully slow.

In 1994, on the seventieth anniversary of the Spemann–Mangold paper, this field of research has been rediscovered. The desire to understand the organizer has been released, generating a flood of new information (Fig. 1). Two advances that occurred together in the mid-1980s helped bring about this increased interest. First was the discovery that growth factors can function as the long-sought embryonic inducers; second was the demonstration that molecular techniques can

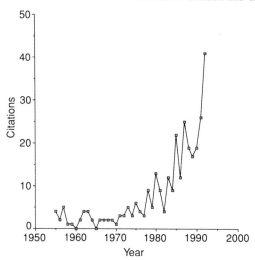

Fig. 1 Yearly citations of Spemann and Mangold (1924).

provide candidates for inducing molecules, as well as critical tests of their function. The organizer finally has begun to yield its secrets.

The organizer is responsible for the development of the dorsal axis, which is the defining feature of vertebrates. The organizer forms some of the dorsal axis and serves as a signaling center for the rest of it. The signals of the organizer dorsalize the mesoderm and neuralize the ectoderm. The organizer is also responsible for the anterior–posterior patterning of the dorsal axis, with a brain at one end and a tail at the other.

Two questions are essential in the analysis of vertebrate development: how does the organizer form and how does it act? The vertebrate that has provided most of the answers to date is the clawed toad, *Xenopus laevis*. Accordingly, we will first describe the basic events of *Xenopus* embryogenesis, and then provide a description of the main molecules implicated in axis development. We will then consider the organizer, emphasizing its formation.

A. Embryonic Axes in *Xenopus*

Insects, all classes of vertebrates, and many other animals have an anterior–posterior axis and a dorsal–ventral axis. The discovery in *Drosophila* of genetic programs to set up these axes maternally demonstrated that the two axes are established independently in the fruit fly (Nüsslein-Volhard, 1991). This independence is not the case in *Xenopus* or other vertebrates. Dorsal and anterior development are linked (Cooke, 1985; Kao and Elinson, 1988; Stewart and

8. Dorsoanterior Axis in Frogs 255

Gerhart, 1990), so we are forced to address dorsoanterior development in the early embryo. The dorsoanterior structures consist of the head at the anterior end and the spinal cord, notochord, and somites of the trunk and tail.

All phenotypes resulting from changes in axial specification in *Xenopus* can be arranged in a continuous series called the dorsoanterior index (DAI; Kao and Elinson, 1988; Fig. 2). Starting from normal (DAI 5), reductions in dorsal development result in progressive losses of anterior dorsal structures, leaving many posterior dorsal structures intact (DAI 1-4). The elimination of all dorsal development leads to a radially symmetric limit form (DAI 0), which lacks all axial structures and has only ventral ones. Conversely, increases in dorsal development result in more anterior development and less posterior development (DAI 6-9). The limit form is again radially symmetric, but with anterior structures such as eye and heart and no posterior structures (DAI 10).

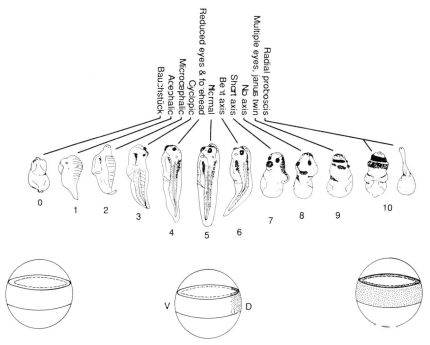

Fig. 2 Relationship between the dorsoanterior index (DAI) and the amount of organizer. The DAI is a continuous series of variation in *Xenopus* dorsoanterior development. Normal embryos are DAI 5, dorsoanterior-reduced embryos are DAI 0-4, and dorsoanterior-enhanced embryos are DAI 6-10. The amount and distribution of organizer activity (stippled) in the early gastrula is represented diagrammatically below the DAI. The organizer in a DAI 5 embryo occupies a 60° arc on the dorsal (D) side. The DAI 0 embryo lacks organizer activity, whereas the DAI 10 embryo has organizer activity in the entire marginal zone. [DAI scale reprinted from Kao and Elinson (1988) with permission.]

The radial extremes of the dorsoanterior series are due to the geometry of the embryo. The organizer arises in the equatorial region of the embryo and occupies an arc of about 60° before the cell movements of gastrulation begin (Fig. 2; Dale and Slack, 1987; Kao and Elinson, 1988; Stewart and Gerhart, 1990). This localization is demonstrated by transplantations, as originally done by Spemann and Mangold (1924). When the equatorial (marginal zone) tissue from the dorsal side is transplanted to the future ventral side, it organizes an ectopic embryonic axis, recruiting host cells into the new dorsal mesoderm and nervous system (Fig. 3). Transplantations of equatorial tissue further from the dorsal side do not generate an ectopic axis.

Variations in the degrees of arc occupied by the organizer generate the dorsoanterior series (Fig. 2; Kao and Elinson, 1988). The organizer arc can be reduced by UV irradiation of fertilized eggs (Grant and Wacaster, 1972; Malacinski et al., 1975; Scharf and Gerhart, 1980), generating dorsoanterior-reduced embryos (DAI 1–4); DAI 0 embryos develop when the organizer arc is eliminated (Fig. 2). Conversely, the organizer arc can be increased by D_2O treatment of fertilized eggs (Scharf et al., 1989) or by lithium treatment of cleaving embryos (Kao and Elinson, 1988). DAI 10 embryos develop when the organizer arc is increased to cover the entire equator (Fig. 2).

These embryological results have molecular correlates. The *goosecoid*, *noggin*, and *pintallavis* genes are expressed in the organizer and serve as a marker for it (Blumberg et al., 1991; Ruiz i Altaba and Jessell, 1992; Smith and Harland, 1992) whereas *Xwnt*-8 is expressed in the ventral mesoderm and not in the organizer (Christian et al., 1991; Smith and Harland, 1991). UV irradiation of the fertilized egg eliminates later *goosecoid*, *noggin*, and *pintallavis* expression

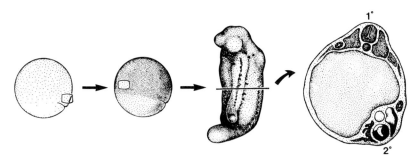

Fig. 3 The Spemann–Mangold experiment. An area above the dorsal lip of a lightly pigmented gastrula of one species was transplanted to the ventral side of a darkly pigmented gastrula of a second species. The resulting embryo had a secondary embryo developing on its flank. Sections revealed the host's primary axis (1°) as well as the induced secondary axis (2°) axis. Donor tissue formed notochord and some somite and neural tube, as indicated by pigmentation, and recruited host cells to form somite, pronephros, and neural tube. [Embryos redrawn from Spemann and Mangold (1924) with permission.]

and enhances *Xwnt*-8 expression, as expected. Conversely, lithium treatment of the cleaving embryo leads to expression of *goosecoid*, *noggin*, and *pintallavis* in a belt pattern, much like the stippling in the DAI 10 case in Fig. 2, and eliminates *Xwnt*-8 expression (Cho *et al.*, 1991; Christian *et al.*, 1991; Smith and Harland, 1991,1992; Ruiz i Altaba and Jessell, 1992).

B. Outline of *Xenopus* Axial Development

The fertilized egg is radially symmetric, with an animal–vegetal axis. Dorsoanterior development is specified about 1 hr after fertilization by a 30° rotation of the egg cortex relative to the cytoplasm (Fig. 4; Gerhart *et al.*, 1989). The cortical rotation produces the gray crescent that marks the later location of the organizer in eggs of many amphibians, although not the lightly pigmented *Xenopus* egg. The cortical rotation leads to information for dorsoanterior development, spread along one side of the early embryo (Fig. 4).

Although dorsoanterior is specified before the egg divides into two cells, mesoderm is specified only after cleavage begins. Signals sent from the yolky vegetal cells induce animal cells to form mesoderm (Nieuwkoop, 1969a; Sudarwati and Nieuwkoop, 1971; Nieuwkoop and Ubbels, 1972; Jones and Woodland, 1987). The induction incorporates the earlier dorsoanterior specification, so dorsal vegetal cells (gray crescent side) induce the formation of dorsal mesoderm, with accompanying organizer activity (Nieuwkoop, 1969b; Dale and Slack, 1987). The term "Nieuwkoop center" has been applied to vegetal cells that can induce dorsal mesoderm but do not form dorsal mesoderm themselves (Gerhart *et al.*, 1989).

As a result of mesoderm induction, much of the equatorial region is fated to become mesoderm. This belt of mesoderm has prospective notochord on the dorsal side, flanked successively by prospective somites, intermediate mesoderm, lateral plate mesoderm, and blood. Organizer activity is found on the dorsal side of the mesodermal belt, but is not limited to a prospective tissue or even to the mesoderm. In *Xenopus*, prospective mesoderm is found deep, yet organizer activity is also found on the surface in prospective endoderm (Shih and Keller, 1992). The formation of axial mesoderm is often considered an indicator of organizer presence, but the two are separable (Stewart and Gerhart, 1991).

Cells begin moving, marking the onset of gastrulation, the first visible sign of which is the dorsal lip of the blastopore. The lip is visible because of the bottle cells, whose change in shape causes an invagination of the embryo's surface. The invagination occurs at the junction between the gray crescent and the vegetal half; the organizer is centered animal to the dorsal lip.

Inside the embryo, cells began moving earlier, with more vigorous migration of cells along the blastocoel roof on the dorsal side than on the ventral side (Fig. 4). Migration of presumptive mesoderm is first vegetal and then animal as the

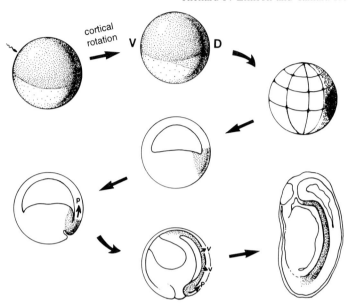

Fig. 4 Outline of frog development. After fertilization, the radially symmetric egg undergoes the cortical rotation that specifies the dorsal (D)–ventral (V) axis. The dorsal side is marked by a gray crescent, as drawn here, although a gray crescent is not visible in *Xenopus*. Transplantation experiments show that, at the 32-cell stage, information for dorsoanterior development (hatched) is found in all cells along the dorsal side. A section of a late blastula shows the location of the organizer (stippled) at this stage. With gastrulation, the organizer (stippled) involutes and migrates under the ectoderm, which is induced to form the central nervous system. Two routes have been proposed for neural induction. In one, signals from the organizer spread in the plane of the tissue (→P) into the adjacent ectoderm. In the other, signals from the involuted organizer spread vertically (→V) to the overlying ectoderm.

presumptive mesoderm involutes. In essence, the mesodermal belt turns outside-in and migrates toward the animal pole. Movement on the dorsal side is far more extensive as the dorsal part of the mesodermal belt elongates (Fig. 4). The elongation is powered by convergent extension, as dorsal mesodermal cells crowd toward the dorsal midline (Keller *et al.*, 1992). The dorsal mesoderm of the trunk stretches from anterior to posterior and underlies the developing central nervous system.

C. Key Molecular Players

The *Xenopus* embryo has been instrumental in identifying molecules involved in vertebrate axis formation. The virtual unavailability of genetic approaches in *Xenopus* is compensated for by the large size and ease of manipulation of the

8. Dorsoanterior Axis in Frogs 259

early embryo. These features permit simple functional assays for candidate molecules. The two main assays of function are mesoderm induction and ectopic axis formation.

1. Mesoderm Induction Assay

The mesoderm induction assay takes advantage of the fact that the animal cells of the blastula can be induced by vegetal cells to form mesoderm (Fig. 5). The type of mesoderm induced depends on the type of vegetal cells used. Ventral vegetal cells induce the formation of mesothelium, mesenchyme, and blood, characteristic of ventral mesoderm, whereas dorsal vegetal cells induce the formation of notochord and muscle, characteristic of dorsal mesoderm (Nieuwkoop, 1969b; Dale and Slack, 1987). To assay potential mesoderm inducers, isolated animal caps are incubated with proteins or extracts, and the caps are examined for the

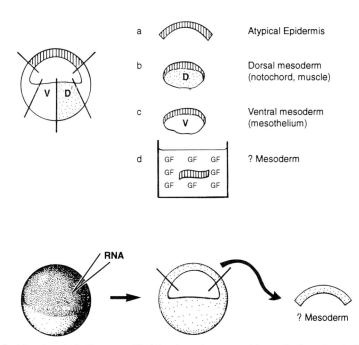

Fig. 5 Mesoderm induction assay. The blastula can be separated into animal cap (hatched), marginal zone, dorsal vegetal mass (D), and ventral vegetal mass (V). Animal cap cultured alone (a) produces atypical epidermis, whereas animal cap cocultured with vegetal masses (b,c) produces mesoderm. The type of mesoderm depends on the type of yolk mass. To assay a protein such as a growth factor (GF), animal cap is cultured in solution (d). To assay an RNA, RNA is injected into eggs and allowed to diffuse and be translated. Animal caps are isolated from the resulting blastulae, cultured to the appropriate stage, and examined for mesodermal derivatives.

presence of mesodermal tissues (Fig. 5). The assay can be adapted for RNA by injecting RNA into a fertilized egg, isolating animal caps, and determining whether the caps form mesoderm in response to proteins synthesized from the injected RNA.

Five types of molecule have activity in the mesoderm induction assay:

1. Several members of the transforming growth factor β (TGF-β) family induce dorsal mesoderm. Activin is the most potent of these (Rosa *et al.*, 1988; Albano *et al.*, 1990; Asashima *et al.*, 1990; Smith *et al.*, 1990; Thomsen *et al.*, 1990; van den Eijnden-van Raaij *et al.*, 1990). Vg1, another member of the family, is represented by a localized mRNA in the oocyte (Weeks and Melton, 1987). Vg1 RNA is not active in the mesoderm induction assay but, when present in a construct designed to ensure production of the putative secreted portion, Vg1 produces dorsal mesoderm (Dale *et al.*, 1993; Thomsen and Melton, 1993).
2. Members of the fibroblast growth factor (FGF) family induce ventral mesoderm (Slack *et al.*, 1987; Paterno *et al.*, 1989; Isaacs *et al.*, 1992; Kimelman and Maas, 1992) and can enhance the activity of TGF-β-like molecules (Kimelman and Kirschner, 1987; Green *et al.*, 1992).
3. Bone morphogenetic protein (BMP-4), a TGF-β-like molecule, induces ventral mesoderm but, unlike FGF, BMP-4 prevents formation of dorsal mesoderm in response to activin (Dale *et al.*, 1992; Jones *et al.*, 1992).
4. Xwnt-8, a member of a family of secreted proteins homologous to *Drosophila wingless* gene and the mouse mammary tumor gene *int*, probably does not induce mesoderm (Christian *et al.*, 1991,1992; Sokol and Melton, 1992), but this point has been controversial (Sokol *et al.*, 1991; Chakrabarti *et al.*, 1992; Sokol, 1993). Xwnt-8, however, synergizes with FGF or activin in producing dorsal mesoderm (Chakrabarti *et al.*, 1992; Christian *et al.*, 1992; Sokol and Melton, 1992). Molecules like Xwnt-8 that enhance induction have been called competence modifiers because they make the cells more competent, that is, more able to respond to inducing signals (Kimelman *et al.*, 1992; Moon and Christian, 1992).
5. Noggin, a novel secreted protein, does not induce mesoderm but can convert ventral mesoderm to dorsal mesoderm (Lamb *et al.*, 1993; Smith *et al.*, 1993).

2. Ectopic Axis and UV Rescue Assays

The ectopic axis assay is a molecular analog of the Spemann–Mangold experiment. Instead of transplanting tissues to the ventral marginal zone, RNAs or other molecules are injected into a cell whose descendants will populate the ventral marginal zone (Fig. 6A). Injections are usually made at the 4- to 32-cell

8. Dorsoanterior Axis in Frogs 261

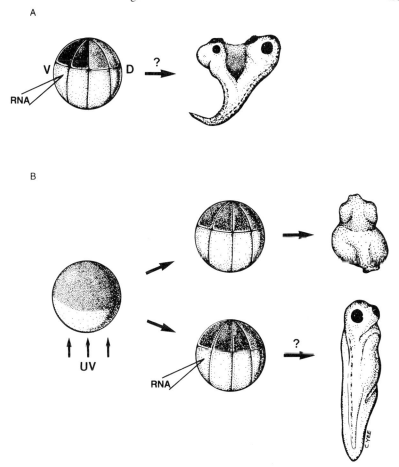

Fig. 6 (A) Ectopic axis assay. A normal embryo has a dorsal (D) and a ventral side and develops into a tadpole with an axis. To assay for ectopic axis activity, an RNA is injected into a cell on the ventral side, and embryos are scored for extra axial structures. (B) A variant of this assay is UV rescue. When eggs are UV irradiated, they develop without an axis (DAI 0). RNA is injected into a cell in an attempt to produce axial structures and rescue these embryos.

stage, and the embryos are examined for ectopic axes. In the most extreme case, a complete secondary axis is produced so a Siamese twin is formed. With less activity, fewer dorsoanterior structures are found, in accordance with the DAI. A variant of the ectopic axis assay is the UV rescue assay. Instead of being injected into a normal embryo to produce an extra axis, molecules are injected into an embryo developing from a UV-irradiated egg. Such embryos lack a dorsoanterior axis (DAI 0; Fig. 2), so the measure of a molecule's activity is its ability to restore an axis in these embryos (Fig. 6B).

The rapid pace of identifying molecules involved in axis formation is illustrated by the fact that, until 1990, lithium was the only substance known to cause ectopic axis formation when injected into *Xenopus* embryos (Kao *et al.*, 1986). Obviously, lithium is not the natural inducer. Six mRNAs that have been found in the embryo are now known to cause ectopic axis formation. Some of these may be RNA equivalents of lithium that do not play a natural role in the formation of the embryo axis, whereas others may represent different steps along the path leading to axis formation.

Activin and secreted Vg1, which generate dorsal mesoderm in the mesoderm induction assay, are also active as RNA in the ectopic axis assay (Thomsen *et al.*, 1990; Sokol *et al.*, 1991; Dale *et al.*, 1993; Thomsen and Melton, 1993). Other active RNAs are activin receptor (Kondo *et al.*, 1991; Hemmati-Brivanlou and Melton, 1992; Mathews *et al.*, 1992), Xwnt-8 (Christian *et al.*, 1991; Smith and Harland, 1991; Sokol *et al.*, 1991), noggin (Smith and Harland, 1992), and goosecoid, a homeobox-containing gene expressed in the organizer (Cho *et al.*, 1991; Steinbesser *et al.*, 1993). Secreted Vg1, noggin, and Xwnt-8 produce a rather complete axis, including a head (DAI 5), whereas activin, activin receptor, and goosecoid usually produce posterior dorsal structures without heads (DAI 1,2). One unexpected recent finding is that an antibody against β-catenin can cause ectopic axis formation when injected into the embryo (McCrea *et al.*, 1993).

II. How Does the Organizer Form?

A. Cortical Rotation

Many steps in oogenesis and early embryogenesis lead to the formation of Spemann's organizer and, subsequently, to the development of dorsoanterior structures. These steps involve maternally established cytoplasmic localizations in the oocyte, their rearrangement following fertilization of the egg, and intercellular signaling events in the blastula and gastrula.

The unfertilized *Xenopus* egg displays an obvious polarity about the animal–vegetal axis. Visible pigment granules are distributed primarily in the animal hemisphere. The yolky vegetal hemisphere is far less pigmented, and an equatorial zone often appears as a white band between the two hemispheres. A less obvious, but equally important, inside–outside polarity also exists. The outer layer of cytoplasm, or "cortex", is about 10 μm thick and behaves as a coherent structure. This cortex can be manually separated from the mass of the inner cytoplasm from oocytes (Elinson *et al.*, 1993) and unfertilized (Goldenberg and Elinson, 1980) or fertilized (Houliston and Elinson, 1991a) eggs.

Fertilization creates a further polarity that specifies dorsoanterior development of the embryo (Fig. 4). A single sperm enters the egg anywhere in the animal hemisphere; this site corresponds to the future ventral side of the embryo (Roux,

8. Dorsoanterior Axis in Frogs 263

1887). (The year 1994 is not only the 70th anniversary of the Spemann–Mangold paper but also the 100th anniversary of the founding by Roux of the journal in which it was published.) During the first embryonic cell cycle, the cortex rotates about 30° relative to the inner cytoplasm, translocating vegetal cortical cytoplasm toward the equatorial zone on the dorsal side (Ancel and Vintemberger, 1948; Vincent et al., 1986). The organizer eventually forms in the region where vegetal cortex is juxtaposed to animal cytoplasm.

The mechanism of cortical rotation relies on a parallel array of microtubules within the vegetal cortex (Elinson and Rowning, 1988; Houliston and Elinson, 1992). The plus ends of the microtubules point in the direction of cortical rotation, away from the sperm entry side (Houliston and Elinson, 1991b). Microtubules also radiate out through the egg cytoplasm from the sperm aster in the animal hemisphere. Those that reach the vegetal pole become reoriented and contribute to the parallel array (Houliston and Elinson, 1991b; Schroeder and Gard, 1992). Because the array of parallel microtubules is continuous with deeper cytoplasmic ones, it is considered part of the cytoplasmic component of the fertilized egg. During cortical rotation, the array may serve as a track along which the layer of cortical endoplasmic reticulum moves (Houliston and Elinson, 1991a). Several findings suggest that kinesin is the molecular motor responsible for generating the force for cortical movement. Using antibody staining, investigators have shown that a kinesin-like molecule is associated with microtubules of the array. Also, since the direction of cortical rotation is toward the plus ends of the microtubules, a plus-end-directed motor such as kinesin is necessary (Houliston and Elinson, 1991a,1992).

The microtubules of the sperm aster provide a cue for the orientation of the parallel array and the subsequent direction of the cortical rotation, but they are not necessary for either. A sperm aster is not present in enucleated, bisected, or artificially activated eggs (Houliston and Elinson, 1991b; Schroeder and Gard, 1992; Elinson and Paleček, 1993), but in these cases microtubules extend from the animal hemisphere toward the vegetal surface, and a parallel array forms. Cortical rotation occurs in activated eggs or vegetal fragments, although the direction of rotation is not predictable (Vincent et al., 1986,1987). Polyspermy produces two or more sperm asters from which the microtubules can originate. Once these microtubules reach the vegetal pole, the cortical array must orient in a parallel pattern, since a normal cortical rotation follows (Vincent and Gerhart, 1987). One dorsoanterior axis forms in the embryo and is oriented with the direction of the rotation movement. In these polyspermic embryos, axis orientation cannot be predicted on the basis of the sperm entry points (Render and Elinson, 1986).

In addition to the influence of the sperm aster, gravity can have a strong effect on cytoplasmic movement in the egg and on the location of the future dorsal side (Ancel and Vintemberger, 1948; Gerhart et al., 1981). Brief tilting of the egg before the array of parallel microtubules forms can orient the array (Zisckind and

Elinson, 1990); stronger gravity treatments can move the cytoplasm relative to the cortex, overriding the normal cortical rotation (Gerhart et al., 1981; Black and Gerhart, 1985). When the egg is tilted 90° so that the animal–vegetal axis is horizontal instead of vertical, the dense vegetal cytoplasm shifts downward relative to the cortex (Vincent and Gerhart, 1987). The dorsal side of the embryo arises on the uppermost side of the egg.

The cortical rotation can be inhibited by UV irradiation of the vegetal hemisphere shortly before cortical rotation begins (Manes and Elinson, 1980; Vincent and Gerhart, 1987). The resulting embryos lack dorsoanterior structures, including the head, spinal cord, notochord, and muscles (DAI 0; Malacinski et al., 1975; Scharf and Gerhart, 1980). UV light prevents the formation of the array of parallel microtubules so that cortical rotation does not occur (Elinson and Rowning, 1988). Tilting against gravity during the first cell cycle makes it possible to obtain complete rescue of dorsoanterior structures in UV-irradiated eggs (Chung and Malacinski, 1980; Scharf and Gerhart, 1980).

Concomitant with the cortical rotation, animal cytoplasm undergoes dramatic rearrangements (Danilchik and Denegre, 1991). These cytoplasmic changes are due to the shifting of the vegetal yolk mass and the growth of the animal sperm aster (Brown et al., 1993; Denegre and Danilchik, 1993). Although the cytoplasmic patterns have dorsal/ventral differences, researchers have not determined whether these differences are simply consequences of the other changes or whether they are important for dorsoanterior specification.

B. Maternal Dorsal Factor

Cortical rotation of the fertilized egg specifies dorsoanterior development, indicating that factors present before rotation are activated by this event. Such a dorsal factor has been detected in fully grown oocytes. Nothing is known about the biochemical identity of this factor; however, the location and properties of the dorsal factor during oocyte maturation and early development have been followed using embryological methods. These methods include UV irradiation of the oocyte to alter the behavior of the dorsal factor and microinjection of cytoplasm from regions of the egg and embryo to test for axis-inducing activity.

UV irradiation of the vegetal surface of prophase I oocytes leads to the development of ventralized embryos (DAI 0) identical to those produced by UV irradiation of fertilized eggs (Holwill et al., 1987). Since UV light has limited penetrance into the oocyte, its target must be near the vegetal surface. Unlike UV-irradiated fertilized eggs, embryos derived from UV-irradiated oocytes are capable of forming an array of parallel microtubules, and can undergo cortical rotation. These oocytes cannot be rescued by tilting against gravity (Elinson and Pasceri, 1989).

The inability of gravity to rescue UV-irradiated oocytes suggests that the UV

8. Dorsoanterior Axis in Frogs 265

target is different in oocytes than in eggs. We hypothesize that a dorsal factor (D) in oocytes changes its form during oocyte maturation and loses UV sensitivity (d) (Fig. 7). As a consequence of the cortical rotation, the altered factor (d) mixes with animal cytoplasm and is activated (d*). This activation is necessary for subsequent dorsoanterior development. UV irradiation of oocytes inactivates the dorsal factor (Fig. 7). Such oocytes cannot be rescued by gravity rearrangement of the cytoplasm, since they lack a necessary dorsal factor. UV irradiation of fertilized eggs blocks cortical rotation, but the dorsal factor, which can be activated by gravity rearrangement, is retained.

Cytoplasmic transfers provide further evidence in support of this model. Vegetal cortical cytoplasm from the egg induces secondary axis formation when microinjected into the ventral vegetal cells of 16-cell recipient embryos (Fujisue et al., 1993; Holowacz and Elinson, 1993). Animal cortical, equatorial cortical,

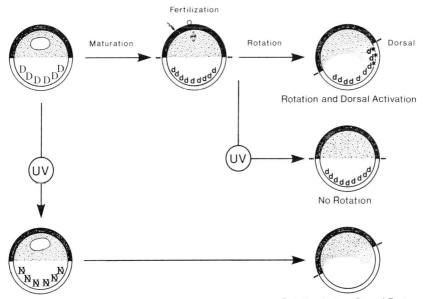

Fig. 7 Model for a maternal dorsal factor. The oocyte (*top left*) has a hypothetical dorsal factor (D) localized near the vegetal surface. With oocyte maturation, the factor changes in some way (d). After fertilization, the cortex rotates relative to the cytoplasm; the mixing of vegetal cortex and animal cytoplasm activates the dorsal factor (d*). Activated factor (d*) is required for further dorsoanterior development. UV irradiation of the oocyte destroys the original factor (N). Although rotation occurs later, there is no factor to be activated so there is no dorsoanterior development (DAI 0). Such eggs cannot be rescued by gravity. UV irradiation of the fertilized egg blocks rotation so "d*" does not form, and there is no dorsoanterior development (DAI 0). Such eggs can be rescued by gravity, presumably because cytoplasmic mixing converts "d" to "d*". [Modified from Elinson (1990) with permission.]

and deep vegetal cytoplasm do not possess any axis-inducing activity. Axis-inducing activity is retained in vegetal cortical cytoplasm obtained from UV-irradiated fertilized eggs, so UV irradiation after fertilization does not inactivate the factor (Fig. 8). On the other hand, if eggs are irradiated as oocytes, axis-inducing activity is not found in their vegetal cortical cytoplasm (Holowacz and Elinson, 1993). This difference between UV-irradiated oocytes and eggs in transferable axis-inducing activity is expected from the model in Fig. 7.

Although the identity of the dorsal factor is unknown, the vegetal cortex of the stage VI oocyte represents an important cytoplasmic domain. In addition to the UV-sensitive factor(s) (Holwill *et al.*, 1987; Elinson and Pasceri, 1989), RNAs are specifically localized to the vegetal cortex; these include Vg1, TGF-β-5,

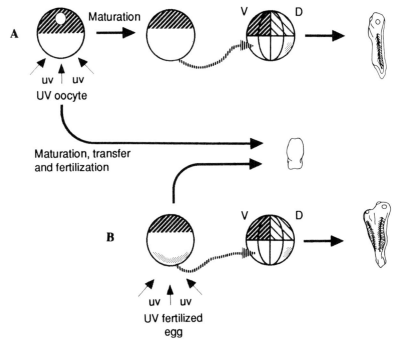

Fig. 8 UV irradiation and axis-inducing activity of vegetal cortical cytoplasm. Stippled regions indicate the location of putative cytoplasmic dorsal activity. The effect of UV light depends on whether oocytes or fertilized eggs are irradiated. (A) Oocytes were UV irradiated, matured, and used as donors of vegetal cytoplasm. This cytoplasm did not induce a secondary axis when transferred into ventral (V) vegetal cells of a normal embryo (hatched arrow). When a subset of the oocytes was allowed to develop into embryos (curved arrow), they lacked all dorsal structures (DAI 0). (B) Fertilized eggs were UV irradiated and used as donors of vegetal cortical cytoplasm. This cytoplasm induced a secondary axis when injected into ventral (V) vegetal cells of a normal embryo (hatched arrow). Sibling UV-treated embryos (curved arrow) lacked an axis (DAI 0). [Reprinted from Holowacz and Elinson (1993) with permission.]

8. Dorsoanterior Axis in Frogs 267

Xcat-2, and Xcat-3 (Rebagliati *et al.*, 1985; Melton, 1987; Schwartz *et al.*, 1992; Mosquera *et al.*, 1993; Elinson *et al.*, 1993). This localization is discrete, unlike that seen for animally localized RNAs, which tend to be present throughout the animal half. The vegetal cortex has a striking network of cytokeratin (Klymkowsky *et al.*, 1987); a cortical fraction can be isolated that retains the RNAs and the peculiar cytoskeleton (Elinson *et al.*, 1993).

Researchers should be able to link some property of the oocyte vegetal cortex to the axis-inducing activity; the best current hypothesis is that Vg1 is important (Dale *et al.*, 1993; Thomsen and Melton, 1993). We will consider this and other possibilities in Sections II,D and II,E.

C. Location of Dorsal Information in the Egg and Embryo

During the first cell cycle after fertilization, axis-inducing activity, as assayed by microinjection of cytoplasm, gradually declines from the vegetal cortex and rises in the marginal zone on the dorsal side of the embryo, coinciding with cortical rotation (Fujisue *et al.*, 1993). Cortical axis-inducing activity remains in this location, at least until the 16-cell stage (Yuge *et al.*, 1990). Some activity is also present in dorsal animal blastomeres, but this activity is significantly less effective (Fujisue *et al.*, 1993; Holowacz and Elinson, 1993). Note that Curtis (1960, 1962) argued for a similar localization of axis-inducing activity based on transplantations of gray crescent cortex. His results were generally discredited once investigators found that gravity alone could produce the Siamese twin phenotype (Gerhart *et al.*, 1981). Unfortunately, the hypothesis of an axis-inducing cytoplasmic localization near the cortex was discredited along with Curtis's results. Results from microinjection of cortical cytoplasm renew support for a cortical axis inducing activity (Yuge *et al.*, 1990; Fujisue *et al.*, 1993; Holowacz and Elinson, 1993).

Attempts to isolate this axis-inducing activity have not yet been successful. Isolated cytosols from 16-cell animal dorsal or vegetal dorsal blastomeres are incapable of inducing axis formation, although they have some activity in changing the fate of injected cells (Hainski and Moody, 1992). Since the axis-inducing activity of whole cytoplasm is concentrated in the cortex of vegetal blastomeres, isolation of cytosol from intact blastomeres may lead to dilution of the active component. RNA from animal dorsal blastomeres of the 16-cell stage induces an axis, whereas RNA from vegetal dorsal blastomeres does not (Hainski and Moody, 1992). This result contrasts with the presence of more axis-inducing activity in vegetal cortical cytoplasm than in animal cortical cytoplasm. Dorsal information in the cortical cytoplasm of dorsal vegetal cells may be present in a form other than RNA.

The location of axis-inducing activity during early cleavage stages is demonstrated by numerous cell transplantation studies (summarized in Fig. 4). In 16- and 32-cell embryos, dorsal blastomeres can induce the formation of a secondary axis when transplanted into the ventral side of a host embryo. The ability to induce axis formation tends to predominate in the more vegetal and equatorial dorsal blastomeres (Gimlich and Gerhart, 1984; Gimlich, 1986; Takasaki and Konishi, 1989; Kageura, 1990), with some axis-inducing activity present in animal dorsal blastomeres (Kageura, 1990; Gallagher et al., 1991). Similarly, transplantation of dorsal cells into UV-ventralized embryos restores the host axis (Gimlich and Gerhart, 1984; Gimlich, 1986). Lineage tracing of transplanted cells shows that their progeny do not always contribute to the axial structures of the host embryo. Transplanted cells derived from the dorsal vegetal region form endoderm, while they induce neighboring equatorial cells to form dorsal mesodermal structures such as notochord and muscle (Gimlich and Gerhart, 1984). Transplanted cells from the dorsal equatorial region both contribute and induce additional axial structures (Gimlich, 1986). The term "Nieuwkoop center" is used to define the region of the cleavage-stage embryo that has the ability to induce surrounding cells to form axial structures (Gerhart et al., 1989).

Differences between dorsal and ventral blastomeres have been detected by means other than cell transplantation. At the 8-cell stage, animal dorsal blastomeres develop differently than animal ventral blastomeres. First, isolated ventral blastomeres express more of an epidermal marker (Epi 1) than do isolated dorsal blastomeres (London et al., 1988). The marker is expressed 1–2 days after blastomere isolation. Second, isolated dorsal blastomeres respond to a brief activin treatment by forming muscle and notochord 1–2 days later. Isolated ventral blastomeres are less likely to generate dorsal mesoderm (Kinoshita et al., 1993). At the 16- to 64-cell stage, gap junctional communication between animal half cells is more prevalent on the dorsal side than on the ventral side (Guthrie, 1984; Guthrie et al., 1988). Agents that respecify ventral cells to dorsal ones, such as lithium or Xwnt-8, also increase gap junctional communication between ventral cells (Nagajski et al., 1989; Olson et al., 1991; Olson and Moon, 1992).

All these differences appear before zygotic transcription begins at the midblastula stage (Newport and Kirschner, 1982); they also are likely to appear before mesoderm induction starts. The timing of mesoderm induction has been determined by culturing animal and vegetal cells of different stages together (Jones and Woodland, 1987). Animal cells are responsive to signals from the vegetal cells from stage 6.5 (64-cell) to stage 10.5 (early gastrula). Vegetal cells are capable of inducing mesoderm from stages 6 to 6.5, and perhaps earlier. This capability increases markedly at stages 7.5 to 8, and then disappears by stage 11 (late gastrula). The presence of dorsal/ventral differences in animal blastomeres before the 64-cell stage suggests that they arise directly in response to the cortical rotation and the associated rearrangements of deep cytoplasm.

At the beginning of gastrulation, the organizer occupies an arc of 60°, centered on the dorsal midline (Stewart and Gerhart, 1990). This small region seems to be achieved by focusing previous dorsoanterior activities. Vegetal dorsal cells begin

8. Dorsoanterior Axis in Frogs 269

to lose their activity by the late blastula stage (stage 9), whereas the dorsal equatorial zone cells gradually gain this activity (Gimlich, 1986; Stewart and Gerhart, 1991).

At the early gastrula stage (stage 10), involuting cells are visible in the organizer region. These cells can be transplanted to the ventral midline to induce a secondary axis (Spemann and Mangold, 1924; Gimlich and Cooke, 1983; Smith and Slack, 1983). More recent work has shown that trunk and tail organizer activity is contained in the epithelial layer of the involuting organizer tissue, whereas a somewhat different inducing activity is attributable to deeper cells (Shih and Keller, 1992). In normal development, the epithelial layer is fated to form the roof of the archenteron, an endodermal derivative. Even at this advanced stage, cells with organizer activity can be distinguished from those forming dorsal mesodermal structures.

D. Axis Induction

The molecules active in the ectopic axis assay (Section I,C,2) can be divided into two groups based on the mesoderm induction assay (Section I,C,1). Activin and FGF are mesoderm inducers and should be released from signaling cells, whereas Xwnt-8 and noggin are competence modifiers and affect responding cells. For any of these molecules to be considered a candidate for the initial axis-inducing factor, it must be found in the egg and must become active only in the dorsal blastomeres of the early cleavage-stage embryo. Several molecules satisfy the former criterion, but none have yet satisfied the latter.

Vg1, like activin, is a member of the TGF-β family. Its RNA is localized to the vegetal cortex of the oocyte and is distributed throughout the vegetal cytoplasm of the cleaving embryo (Melton, 1987; Weeks and Melton, 1987). Previous attempts to show that Vg1 could induce mesoderm or an axis were not successful, perhaps because of the absence of a secretion signal on the Vg1 protein. The engineering of fusion proteins that consist of the secretory signal from BMP-2 or BMP-4 and the C-terminal domain of Vg1 with TGF-β homology has renewed interest in the possible role of Vg1 (Dale et al., 1993; Thomsen and Melton, 1993). These altered versions of Vg1 can be cleaved into a secreted form, and have potent effects in the embryo. When Vg1 is fused with the secretory region of BMP-4, it acts similarly to activin; in other words, it can induce a partial axis and has mesoderm-inducing activity in isolated animal caps (Dale et al., 1993). When Vg1 is fused with BMP-2, it can induce a full axis (DAI 5) superior to that induced by activin mRNA, and it can induce dorsal mesoderm in isolated animal caps (Thomsen and Melton, 1993). Such powerful effects, in addition to its vegetal localization, provide strong support for the proposal that secreted Vg1 is important in axis formation. Nevertheless, detecting a secreted form of Vg1 in dorsal blastomeres has not been possible to date. Identification of a receptor for Vg1 and of a protease that cleaves Vg1 into a secreted form would also help our understanding of its role.

Activin is another candidate for the role of axis inducer. Synthesis of activin mRNA begins in late blastula, after mesoderm induction has begun (Thomsen *et al.*, 1990; Dohrmann *et al.*, 1993), but activin activity is present in the unfertilized egg and blastula (Asashima *et al.*, 1991). Since activin mRNA is present in the follicle cells surrounding developing oocytes, activin polypeptides may be synthesized by the follicle cells and then transported to the oocyte (Rebagliati and Dawid, 1993; Dohrmann *et al.*, 1993). The specific distribution of activin protein in the early embryo is unknown. The effects of exogenous activin on animal caps and whole embryos imply that an activin receptor is present in the early embryo. Indeed, maternal accumulation of activin receptor mRNA is found in the egg (Kondo *et al.*, 1991; Hemmati-Brivanlou *et al.*, 1992; Mathews *et al.*, 1992).

A compelling argument for activin's role in mesoderm induction is provided by the utilization of a dominant negative mutant for the activin receptor (Hemmati-Brivanlou and Melton, 1992). The mutated receptor competes with the endogenous one for ligand, but cannot activate intracellular signaling. Injection of the mRNA encoding this mutant activin receptor (Δ1XAR1) blocks activin-mediated mesoderm induction in animal caps. Whole embryos expressing the mutant receptor often fail to gastrulate and do not express mesodermal markers. The absence of mesodermal derivatives in these embryos is paradoxical. FGF is present maternally in embryos, and the mutant activin receptor does not block mesoderm induction by FGF in isolated animal caps. Perhaps an activin-mediated event is required for FGF secretion in the embryo.

A similar inhibition of mesoderm formation was found using a dominant negative mutant for $p21^{ras}$ (Whitman and Melton, 1992). Ras is a possible component of the signal transduction pathway for both activin and FGF. In animal caps, the dominant negative Ras mutant inhibits mesoderm induction by both activin and FGF, as well as by the natural mesoderm inducer released by vegetal cells. Both the mutant activin receptor and the mutant Ras disrupt development, generating abnormal gastrulae or preventing any gastrulation. As a result, these experiments do not allow assessment of activin's role in axis formation, since the embryos are blocked at a stage before axial elements appear. The dominant negative mutants are injected into early cleaving embryos, so the observed effects could be due to inhibition of action of maternally derived activin in mesoderm induction. Inhibiting only the later zygotic expression of activin might reveal whether gastrulation or axis formation is affected.

Use of the dominant negative mutants strongly implicates activin in mesoderm induction. Experiments by Slack (1992) are not in agreement, however, and suggest that other secreted factors are involved. When a millipore filter is placed between vegetal and animal cells, the two tissues cannot adhere to each other, so secreted factors must traverse a liquid gap to reach the target tissue. Vegetal cells can induce animal cells to form muscle and mesothelium when the two tissues are cultured together with an intervening filter. In this "transfilter recombinant," inhibitors can be added to the culture medium and can have access to the secreted factors. Follistatin is a peptide that binds to activin and inhibits its biological activity, but neither follistatin nor inhibitory antibodies against basic FGF block mesoderm induction in transfilter recombinants. These transfilter recombinants indicate that other unknown growth factors may play important roles in axis formation.

The suggestion that activin alone is insufficient for axis formation is supported by other observations. First, activin does not induce a full axis when it is injected into ventral cells at the 32-cell stage (Thomsen et al., 1990; Sokol et al., 1991). Second, communication via gap junctions is not enhanced by activin mRNA to the extent found between normal dorsal blastomeres or to the extent that communication is enhanced by Xwnt-8 mRNA or lithium (Guthrie et al., 1988; Nagajski et al., 1989; Olson et al., 1991; Olson and Moon, 1992). Third, when the ventral half of an animal cap is exposed to activin, muscle develops but notochord rarely does (Sokol and Melton, 1991). Induction of notochord by activin only occurs in the dorsal half of the animal cap. These dorsal animal caps can also undergo secondary inductions yielding "embryoids" that possess anterior structures such as eyes, cement glands, and neural tissue. As mentioned, the dorsal/ventral difference in response to activin exists as early as the 8-cell stage (Kinoshita et al., 1993). These results indicate that an activity localized to dorsal cells is required for the induction of notochord and anterior head structures by activin. Such an activity may originate from the egg vegetal cortex that moved to the dorsal side during cortical rotation.

Xwnt-8 RNA and lithium do not themselves induce mesoderm, but they can provide an activity that enhances responses to activin (Cooke et al., 1989; Sokol and Melton, 1992; Sokol, 1993). When treated with Xwnt-8, ventral animal caps produce notochord and other dorsoanterior structures in response to activin. A similar ventral-to-dorsal conversion occurs when animal caps from UV-ventralized embryos are used (Sokol and Melton, 1992). When Xwnt-8- or lithium-treated animal caps are exposed to the ventral mesoderm inducer FGF, more abundant muscle and notochord is induced than when animal caps are treated with FGF alone (Slack et al., 1988; Kao and Elinson, 1989; Chakrabarti et al., 1992; Christian et al., 1992; T. Holowacz, unpublished results). All these experiments indicate that Xwnt-8 RNA and lithium can affect cell response to growth factors in such a way that dorsal mesoderm is produced.

The finding that FGF in combination with Xwnt-8 produces dorsal mesoderm (Christian et al., 1992) raises the question of FGF's role in axis formation. A dominant negative mutant of the FGF receptor provides an answer. Expression of the dominant negative mutant causes incomplete blastopore closure and deficiencies in notochord and trunk muscle (Amaya et al., 1991,1993). Although trunk development is severely affected, the head appears normal, so the embryo looks like a Norse boat. The Norse boat syndrome frequently arises from gastrulation failures in which the blastopore lips do not close properly over the yolk. Before the development of Vogt's (1929) fate map, this developmental failure led to a mistaken view about where the nervous system originated in the embryo (Sander, 1991). Great pains were taken by Amaya and co-workers (1991,1993) in their experiments with dominant negative FGF receptor mutants to rule out nonspecific inhibition.

Similar results were obtained with a dominant negative mutant of Raf-1 kinase, which is involved in the FGF intracellular signaling pathway (MacNicol et al., 1993). Both FGF receptor and Raf-1 kinase dominant negative mutants block muscle-specific actin expression in animal caps in response to FGF but not activin. These inhibitions show that FGF and Raf-1 kinase are not involved in head organizer formation, but play a role in trunk development.

The effect of Xwnt-8 in promoting dorsal mesoderm development is identical to the effect of lithium, suggesting that these two competence modifiers may operate through the same pathway (Christian et al., 1992; Sokol and Melton, 1992; Sokol, 1993). Some understanding of the mechanism of lithium action has been achieved based on its well-known ability to inhibit the phosphoinositide (PI) signal transduction pathway (Berridge et al., 1989). Co-injection of myo-inositol prevents the dorsal enhancement of injected lithium, whereas co-injection of an inactive isomer, epi-inositol, has no effect (Busa and Gimlich, 1989). Myo-inositol restores PI cycling blocked by lithium. More direct involvement of the PI cycle is indicated from the measurement of inositol phosphate levels within the embryo (Maslanski et al., 1992). In normal embryos, the onset of mesoderm induction at the 32- to 64-cell stage is correlated with a dramatic increase in inositol trisphosphate (IP_3) levels, which remain high through the mid-blastula stage. Doses of lithium enhance dorsoanterior development, increase total inositol phosphate, and reduce myo-inositol and IP_3. These correlations are consistent with the hypothesis that an inhibition of the PI cycle in responding cells leads to their development as dorsal mesoderm.

Immunocytochemical staining shows that the receptor for IP_3 is found throughout the egg and is concentrated in the vegetal cortex (Kume et al., 1993). Since lithium, an inhibitor of inositol cycling, is a powerful axis inducer, the IP_3 receptor may be directly involved in axis induction. Investigation of this possibility would be interesting.

Yet another molecule that may have a role in axis induction is β-catenin. β-Catenin binds to the extracellular matrix molecule C-cadherin, and may act as a link to the intracellular actin cytoskeleton. Both β-catenin and C-cadherin are present in the egg and early embryo (Angres et al., 1991; McCrea et al., 1993; Schneider et al., 1993). Injection of an antibody against β-catenin into a ventral cell of a 4-cell embryo or into a UV-ventralized embryo induces an axis (McCrea et al., 1993). Xenopus β-catenin shares sequence homology with the *Drosophila* segment polarity gene *armadillo* (McCrea et al., 1991); expression of *armadillo* depends on the secretion of wingless protein by neighboring cells (Riggleman et al., 1990). Wingless is similar to Xwnt-8 and, like Xwnt-8, is capable of inducing an axis in *Xenopus* (McMahon and Moon, 1989; Chakrabarti et al., 1992). McCrea and co-workers (1993) have suggested that β-catenin and Xwnt-8 function in the same developmental pathway.

E. Summary: Path to the Organizer

Dorsoanterior development is specified by the cortical rotation in the first cell cycle. From that point on, an activity must be localized to one side of the embryo that culminates in the appearance of the organizer at the start of gastrulation. The activity could be represented, through this stage, by one molecular type but, given the number of interesting activities found in embryos to date, a sequence of molecules is likely to be involved. The sequence that leads to the organizer can be stimulated in various ways and at various times, as demonstrated by generation of an ectopic axis. However, this malleability of axis specification does not negate the requirement for localization of specific activities on the dorsal side through normal development.

The simplest view of the cortical rotation is that it creates an interaction between a vegetal cortical and an animal cytoplasmic component, activating one of them (Fig. 7). Activation could involve any number of post-transcriptional processes, including selective translation of an RNA or phosphorylation of a protein. The uniqueness of the vegetal cortical domain (Elinson et al., 1993) and the activity of vegetal cortical cytoplasm in ectopic axis formation (Fujisue et al., 1993; Holowacz and Elinson, 1993) suggest that the activated component comes from the vegetal cortex.

Vg1 secreted protein is an excellent candidate for a molecule in the localized chain of events leading to the organizer. Vg1 RNA and protein are present in the egg, and Vg1 secreted protein has axis-inducing activity (Dale et al., 1993; Thomsen and Melton, 1993). One model for dorsoanterior specification would be for the cortical rotation to activate a protease, which then specifically generates the Vg1 secreted form from Vg1 protein on the dorsal side. The activated protease would then be the first localized activity in dorsoanterior specification (d*; Fig. 7) and Vg1 secreted protein would be the second.

This model requires the demonstration of localized Vg1 secreted protein; until this molecule is found, other models are possible. For instance, a Xwnt localized to the vegetal cortex could be moved animally during the cortical rotation and could be activated by an animal cytoplasmic component. The activated Xwnt would combine on the dorsal side with a generally distributed mesoderm inducer, such as FGF, to produce dorsal mesoderm and organizer on that side. Noggin may also be able to serve this function. This protein is present maternally and can convert ventral mesoderm to dorsal mesoderm (Smith and Harland, 1992; Smith et al., 1993). Such a model requires a localized Xwnt or noggin activity which, like localized Vg1 secreted protein, has not been reported.

A similar ambiguity surrounds the final steps in setting up the organizer. Activin is a direct activator of organizer-specific gene expression and requires no intervening steps of protein synthesis (Cho et al., 1991; Dirksen and Jamrich, 1992; Ruiz i Altaba and Jessell, 1992; Taira et al., 1992; van Dassow et al., 1993). The appearance of organizer markers could be interpreted as a bioassay for activin activity, but a direct demonstration of activin activity localized in the dorsal, equatorial region of the blastula has not been achieved. This leaves room to consider a novel Xwnt on the dorsal side but, again, such a molecule has not been described.

Various models have been proposed, each of which presents various combinations of molecular activities that give the dorsal side its unique features (Kimelman et al., 1992; Sive, 1993; Slack, 1993; Smith, 1993). Despite the availability of many molecular candidates, no demonstration has been made of any molecular activity that is restricted to the dorsal side. Such a demonstration will clarify how the organizer forms.

III. How Does the Organizer Act?

A. Dorsalization

When Spemann and Mangold transplanted the organizer from one embryo to another, an ectopic axis was induced to form. The transplant produced notochord and parts of the somite and neural tube; however, host tissues formed much of the neural tube and somite, as well as the pronephros, the mesoderm lateral to the somite (Fig. 3). Two functions of the organizer were clear: it induced ectoderm to form neural tube and it recruited more lateral mesoderm to form dorsal mesoderm. The latter activity is called dorsalization.

Mesoderm induction, discussed earlier, differs from dorsalization in several ways. In mesoderm induction, presumptive ectoderm is induced to form mesoderm, including notochord and muscle. The ability of a tissue to respond to an inducing signal is known as competence. The competence of ectoderm for mesoderm induction is high in the blastula but declines by early gastrula (Jones and Woodland, 1987; Lettice and Slack, 1993; Smith et al., 1993). In dorsalization, ventral mesoderm is converted to more dorsal mesoderm. Rather than forming mesothelium and mesenchyme, the dorsalized mesoderm forms muscle and pronephros, with a limited ability to form notochord (Slack and Forman, 1980; Lettice and Slack, 1993; Smith et al., 1993). The competence of ventral mesoderm for dorsalization continues through gastrulation, declining at the end. Thus, blastular presumptive ectoderm is induced to form mesoderm, whereas gastrular presumptive ventral mesoderm is dorsalized to dorsal mesoderm.

Dorsalizing activity is diffusible (Lettice and Slack, 1993). The best current candidate molecule for this activity is noggin protein (Smith *et al.*, 1993). Noggin was identified in a screen for RNAs active in the UV rescue assay; its RNA is transcribed in the organizer region at the beginning of gastrulation (Smith and Harland, 1992). The effects of noggin on the embryo differ from those of the dorsal mesoderm inducer activin and the competence modifier Xwnt-8 (Lamb *et al.*, 1993; Smith *et al.*, 1993). Activin induces blastula animal cap, but not gastrula ventral marginal zone, to form muscle, whereas noggin induces ventral marginal zone but not animal cap to form muscle. Xwnt-8 causes ectopic axis formation when injected into blastulae, but inhibits axis formation in the gastrula (Christian and Moon, 1993). Noggin, on the other hand, causes axis formation either when present in the blastula or when expressed in the gastrula.

Noggin is present at the right time and place and has the activity expected of a dorsalizer. Inhibiting noggin function in the embryo might provide further evidence for its role in the organizer. The expectation is that the amount of mesoderm contributing to notochord, muscle, and pronephros would be reduced, leading to corresponding reductions in dorsoanterior development (DAI 1–3). This prediction is confounded by the possibility that noggin might also function in neuralization (Lamb *et al.*, 1993), as will be discussed.

B. Neuralization

The second function of the organizer, and the one for which it is famous, is neuralization. Spemann called the conversion of ectoderm to neural tissue primary neural induction. In the last several years, a revolution in thinking has occurred with respect to the routes of signaling involved in neural induction. Subsequently, the first bona fide candidate for a neural inducer has emerged. Neural induction has been reviewed authoritatively (Doniach, 1993; Ruiz i Altaba, 1993), so we will highlight only a few issues.

1. Planar or Vertical Induction?

Spemann (1938) believed that the neural induction signal could either spread from the organizer into the adjacent ectoderm ("progressive determination" or planar induction) or move from the involuted organizer vertically into the overlying ectoderm (Fig. 4). Two experiments yielded evidence that strongly favored vertical induction by the involuted archenteric roof. First, Holtfreter (1933) examined urodele exogastrulae, embryos in which tissues turn outward instead of inward during gastrulation. The exogastrulae consist of a crumpled ectoderm attached at one end to an elongated body, with axial mesoderm surrounded by endoderm. The organizer or archenteric roof never underlies the ectoderm, so vertical induction is precluded whereas planar induction is possible. The exogastrulae form no nerve cells or other recognizable neural tissue, so Holtfreter concluded that vertical transmission is the only route for neural induction. Second, Otto Mangold (1933) showed that pieces of archenteric roof had neural-inducing activity and that the activity had anteroposterior differences. Anterior roof induced brain and head structures, whereas posterior roof induced tail structures. The Holtfreter and Mangold results argued strongly for vertical induction.

According to Victor Hamburger, Spemann never abandoned planar induction, but by 1988, Hamburger concluded that no evidence was ever forthcoming to support Spemann's view. Nonetheless, in 1993, Doniach and Ruiz i Altaba each marshalled the now considerable evidence in favor of planar induction. Kintner and Melton (1987) had re-examined exogastrulae in *Xenopus* with neural molecular markers and had found neural cell adhesion molecule (NCAM) expression. *Xenopus* exogastrulae also produced neurons and expressed markers of anterior neural development (Ruiz i Altaba, 1990,1992; but see Hemmati-Brivanlou and Harland, 1989). Similar expressions of neural markers were found when strips of ectoderm plus organizer were cultured in a planar arrangement (Dixon and Kintner, 1989; Savage and Phillips, 1989; Papalopulu and Kintner, 1993; Zimmerman *et al.*, 1993). In a convincing demonstration, Doniach and co-workers (1992) published a photograph showing the correct antero-posterior arrangement of three neural markers, expressed as a result of planar induction. Apparently both planar and vertical induction contribute to the formation of the nervous system. Spemann was right in clinging to his point of view.

Note that all the recent results come from *Xenopus*, whereas Spemann, Holtfreter, and Mangold worked with urodeles. Compared with *Xenopus*, urodele embryos are larger, their development is much slower, and the fate map of the early gastrula is somewhat different (Brun and Garson, 1984; Lundmark, 1986; Shi *et al.*, 1987; Imoh, 1988; Dettlaff, 1993). Obviously, the possibility exists that the relative roles of vertical and planar induction differ in urodeles and *Xenopus*. Although urodele embryos should be re-examined, the realization that planar induction exists will have a big impact on views of avian and mammalian development. In these amniotes, Hensen's node is considered to be the organizer; the spreading of signals from Hensen's node into the adjacent epiblast tissue by planar induction is easy to envision.

2. Neural Inducers

A parallel exists between the analyses of egg activation at fertilization and neuralization. Butyric acid, needle pricks, and many other events activate eggs, whereas guinea pig liver, methylene blue, and many other materials neuralize ectoderm. In the past, the information about normal events yielded by any of these treatments was unclear. In egg activation, Ca^{2+} was found to be a key component of the signal transduction pathway; most of the known activators of eggs cause an increase in free Ca^{2+}. The importance of Ca^{2+} provided a first step in the investigation of other events in the signal transduction pathway (Nuccitelli, 1991). Very recently, a candidate for a sperm receptor has been identified that may initiate the signaling cascade of activation (Foltz et al., 1993).

Considering the history of Ca^{2+} in egg activation, researchers expect that many of the artificial neuralizers discovered previously will be found to act through a common signaling cascade. Signal transduction mechanisms involving protein kinase C have been implicated in neural induction (Otte et al., 1988,1991; Otte and Moon, 1992), although exploring these events is more difficult in ectoderm than in eggs. Also, the strange finding that neural genes are expressed in embryos poisoned with the mutant activin receptor must be considered (Hemmati-Brivanlou and Melton, 1992), because of the possibility that activin signaling inhibits neuralization.

Noggin has emerged as a candidate for a neural inducer (Lamb et al., 1993). Noggin causes ectoderm to express neural markers in the absence of a mesodermal intermediate. Gastrula ectoderm, which is no longer competent to respond to mesoderm induction, is neuralized by noggin. The expression of noggin in the organizer places it at the right place and time to be a neural inducer. The 70-yr search begun by Spemann and Mangold's demonstration of neural induction may finally have uncovered a neural inducer.

IV. Concluding Remarks

Currently, more than enough molecules have been identified to generate mesoderm and axial structures in the *Xenopus* embryo, but that does not guarantee that all important molecules have been discovered. For instance, if Vg1 is important, then its receptor and the molecule involved in cleaving the secreted portion must play major roles. In addition, other ligand–receptor combinations and molecules that are yet to be uncovered are probably involved in signal transduction. Finally, more molecules necessary for neural induction will certainly be found. The absence of genetics makes it difficult to know when most of the major molecular players have been detected.

A further problem presented by the absence of genetics is that, as more and more molecules are found, assessing molecular interactions and causal relationships in axis formation becomes increasingly difficult. The assays described earlier explain what a molecule can do but not necessarily what it normally does in the embryo. The assays lack the precision of genetic mutants, since time and place of expression relative to normal events are not well controlled. Careful determination of temporal and spatial relationships, particularly of various signal transduction events, will be necessary to describe the molecular pathway leading to the axis.

Nonetheless, it is remarkable that amphibian embryos, first urodeles and now *Xenopus*, have remained the embryos of choice for investigations of axis specification in vertebrates. Although many developmental biologists derive pleasure from links to the past, the continued use of amphibians to analyze the organizer makes these links particularly strong. The discovery of the organizer 70 years ago by Spemann and Mangold retains its impact today.

Acknowledgments

We thank Cynthia Yee for the excellent drawings. Support for our work was provided by the Natural Sciences and Engineering Research Council and the Medical Research Council of Canada.

References

Albano, R. M., Godsave, S. F., Huylebroeck, D., van Nimmen, K., Isaacs, H. V., Slack, J. M. W., and Smith, J. C. (1990). A mesoderm-inducing factor produced by WEHI-3 murine myelomonocytic leukemia cells is activin A. *Development* **110**, 435–443.

Amaya, E., Musci, T. J., and Kirschner, M. W. (1991). Expression of a dominant negative mutant of the FGF receptor disrupts mesoderm formation in *Xenopus* embryos. *Cell* **66**, 257–270.

Amaya, E., Stein, P. A., Musci, T. J., and Kirschner, M. W. (1993). FGF signalling in the early specification of mesoderm in *Xenopus*. *Development* **118**, 477–487.

Ancel, P., and Vintemberger, P. (1948). Recherches sur le déterminisme de la symétrie bilatérale dans l'oeuf des amphibiens. *Bull. Biol. France Belg. Suppl.* **31**, 1–182.

Angres, B., Muller, A. H. J., Kellermann, J., and Hausen, P. (1991). Differential expression of two cadherins in *Xenopus laevis*. *Development* **111**, 829–844.

Asashima, M., Nakano, H., Shimada, K., Kinoshita, K., Ishii, K., Shibai, H., and Ueno, N. (1990). Mesodermal induction in early amphibian embryos by activin A erythroid differentiation factor. *Roux's Arch. Dev. Biol.* **198**, 330–335.

Asashima, M., Nakano, H., Uchiyama, H., Sugino, H., Nakamura, T., Eto, Y., Ejima, D., Nishimatsu, S.-I., Ueno, N., and Kinoshita, K. (1991). Presence of activin erythroid differentiation factor in fertilized eggs and blastulae of *Xenopus laevis*. *Proc. Natl. Acad. Sci. USA* **88**, 6511–6514.

Berridge, M. J., Downes, C. P., and Hanley, M. R. (1989). Neural and developmental actions of lithium—A unifying hypothesis. *Cell* **59**, 411–419.

Black, S. D., and Gerhart, J. C. (1985). Experimental control of the site of embryonic axis formation in *Xenopus laevis* eggs centrifuged before first cleavage. *Dev. Biol.* **108**, 310–324.

Blumberg, B., Wright, C. V. E., De Robertis, E. M., and Cho, K. W. Y. (1991). Organizer-specific homeobox genes in *Xenopus laevis* embryos. *Science* **253**, 194–196.

Brown, E. E., Denegre, J. M., and Danilchik, M. V. (1993). Deep cytoplasmic rearrangements in ventralized *Xenopus* embryos. *Dev. Biol.* **160**, 148–156.

Brun, R. B., and Garson, J. A. (1984). Notochord formation in the Mexican salamander (*Ambys-*

8. Dorsoanterior Axis in Frogs

toma mexicanum) is different from notochord formation in *Xenopus laevis*. *J. Exp. Zool.* **229**, 235–240.

Busa, W. B., and Gimlich, R. L. (1989). Lithium-induced teratogenesis in frog embryos prevented by a polyphosphoinositide cycle intermediate or a diacylglycerol analog. *Dev. Biol.* **132**, 315–324.

Chakrabarti, A., Matthews, G., Colman, A., and Dale, L. (1992). Secretory and inductive properties of *Drosophila wingless* protein in *Xenopus* oocytes and embryos. *Development* **115**, 355–369.

Cho, K. W. Y., Blumberg, B., Steinbesser, H., and De Robertis, E. M. (1991). Molecular nature of Spemann's organizer: The role of the *Xenopus* homeobox gene *goosecoid*. *Cell* **67**, 1111–1120.

Christian, J. L., and Moon, R. T. (1993). Interactions between *Xwnt-8* and Spemann organizer signalling pathways generate dorsoventral pattern in the embryonic mesoderm of *Xenopus*. *Genes Dev.* **7**, 13–28.

Christian, J. L., McMahon, J. A., McMahon, A. P., and Moon, R. T. (1991). *Xwnt-8*, a *Xenopus Wnt-1/int-1*-related gene responsive to mesoderm-inducing growth factors, may play a role in ventral mesodermal patterning during embryogenesis. *Development* **111**, 1045–1055.

Christian, J. L., Olson, D. J., and Moon, R. T. (1992). *Xwnt-8* modifies the character of mesoderm induced by bFGF in isolated *Xenopus* mesoderm. *EMBO J.* **11**, 33–41.

Chung, H. M., and Malacinski, G. M. (1980). Establishment of the dorsal/ventral polarity of the amphibian embryo: Use of ultraviolet irradiation and egg rotation as probes. *Dev. Biol.* **80**, 120–133.

Cooke, J. (1985). Early specification for body position in mesodermal regions of an amphibian embryo. *Cell Diff.* **17**, 1–12.

Cooke, J., Symes, K., and Smith, E. J. (1989). Potentiation by the lithium ion of morphogenetic responses to a *Xenopus* inducing factor. *Development* **105**, 549–558.

Curtis, A. S. G. (1960). Cortical grafting in *Xenopus laevis*. *J. Embryol. Exp. Morphol.* **8**, 163–173.

Curtis, A. S. G. (1962). Morphogenetic interactions before gastrulation in the amphibian *Xenopus laevis*—The cortical field. *J. Embryol. Exp. Morph.* **10**, 410–422.

Dale, L., and Slack, J. M. W. (1987). Regional specification within the mesoderm of early embryos of *Xenopus laevis*. *Development* **100**, 279–295.

Dale, L., Howes, G., Price, B. M. J., and Smith J. C. (1992). Bone morphogenetic protein 4: A ventralising factor in early *Xenopus* development. *Development* **115**, 573–585.

Dale, L., Matthews, G., and Colman, A. (1993). Secretion and mesoderm-inducing activity of the TGF-β-related domain of *Xenopus* Vg1. *EMBO J.* **12**, 4471–4480.

Danilchik, M. V., and Denegre, J. M. (1991). Deep cytoplasmic rearrangements during early development in *Xenopus laevis*. *Development* **111**, 845–856.

Denegre, J. M., and Danilchik, M. V. (1993). Deep cytoplasmic rearrangements in axis-respecified *Xenopus* embryos. *Dev. Biol.* **160**, 157–164.

Dettlaff, T. A. (1993). Evolution of the histological and functional structure of ectoderm, chordamesoderm and their derivatives in Anamnia. *Roux's Arch. Dev. Biol.* **203**, 3–9.

Dirksen, M. L., and Jamrich, M. (1992). A novel, activin-inducible, blastopore lip-specific gene of *Xenopus laevis* contains a *fork head* DNA-binding domain. *Genes Dev.* **6**, 599–608.

Dixon, J. E., and Kintner, C. R. (1989). Cellular contacts required for neural induction in *Xenopus* embryos: Evidence for two signals. *Development* **106**, 749–757.

Dohrmann, C. E., Hemmati-Brivanlou, A., Thomsen, G. H., Fields, A., Woolf, T. M., and Melton, D. A. (1993). Expression of activin mRNA during early development in *Xenopus laevis*. *Dev. Biol.* **157**, 474–483.

Doniach, T. (1993). Planar and vertical induction of anteroposterior pattern during the development of the amphibian central nervous system. *J. Neurobiol.* **24**, 1256–1275.

Doniach, T., Phillips, C. R., and Gerhart, J. C. (1992). Planar induction of anteroposterior pattern in the developing central nervous system of *Xenopus laevis*. *Science* **257**, 542–545.
Elinson, R. P. (1990). Dorsal specification in the fertilized egg. In "Mechanism of Fertilization: Plants to Humans" (B. Dale, ed.), Vol. H45, pp. 663–671. NATO ASI Series, Springer-Verlag.
Elinson, R. P., and Paleček, J. (1993). Independence of two microtubule systems in fertilized frog eggs: The sperm aster and the vegetal parallel array. *Roux's Arch. Dev. Biol.* **202**, 224–232.
Elinson, R. P., and Pasceri, P. (1989). Two UV-sensitive targets in dorsoanterior specification of frog embryos. *Development* **106**, 511–518.
Elinson, R. P., and Rowning, B. (1988). A transient array of parallel microtubules in frog eggs: Potential tracks for a cytoplasmic rotation that specifies the dorso-ventral axis. *Dev. Biol.* **128**, 185–197.
Elinson, R. P., King, M. L., and Forristall, C. (1993). Isolated vegetal cortex from *Xenopus* oocytes selectively retains localized mRNAs. *Dev. Biol.* **160**, 554–562.
Foltz, K. R., Partin, J. S., and Lennarz, W. J. (1993). Sea urchin egg receptor for sperm: Sequence similarity of binding domain and hsp70. *Science* **259**, 1421–1425.
Fujisue, M., Kobayakawa, Y., and Yamana, K. (1993). Occurrence of dorsal-axis inducing activity around the vegetal pole of an uncleaved *Xenopus* egg and displacement to the equatorial region by cortical rotation. *Development* **118**, 163–170.
Gallagher, B. C., Hainski, A. M., and Moody, S. A. (1991). Autonomous differentiation of dorsal axial structures from an animal cap cleavage stage blastomere in *Xenopus*. *Development* **112**, 1103–1114.
Gerhart, J., Ubbels, G., Black, S., Hara, K., and Kirschner, M. (1981). A reinvestigation of the role of the grey crescent in axis formation in *Xenopus laevis*. *Nature* **292**, 511–516.
Gerhart, J. C., Danilchik, M., Doniach, T., Roberts, S., Rowning, B., and Stewart, R. (1989). Cortical rotation of the *Xenopus* egg: Consequences for the anteroposterior pattern of embryonic dorsal development. *Development Suppl.* **107**, 37–51.
Gimlich, R. L. (1986). Acquisition of developmental autonomy in the equatorial region of the *Xenopus* embryo. *Dev. Biol.* **115**, 340–352.
Gimlich, R. L., and Cooke, J. (1983). Cell lineage and the induction of second nervous systems in amphibian development. *Nature* **306**, 471–473.
Gimlich, R. L., and Gerhart, J. C. (1984). Early cellular interactions promote embryonic axis formation in *Xenopus laevis*. *Dev. Biol.* **104**, 117–130.
Goldenberg, M., and Elinson, R. P. (1980). Animal/vegetal differences in cortical granule exocytosis during activation of the frog egg. *Dev. Growth Diff.* **22**, 345–356.
Grant, P., and Wacaster, J. F. (1972). The amphibian gray crescent region—A site of developmental information? *Dev. Biol.* **28**, 454–471.
Green, J. B. A., New, H. V., and Smith, J. C. (1992). Responses of embryonic *Xenopus* cells to activin and FGF are separated by multiple dose thresholds and correspond to distinct axes of the mesoderm. *Cell* **71**, 731–739.
Guthrie, S. C. (1984). Patterns of junctional communication in the early amphibian embryo. *Nature* **311**, 149–151.
Guthrie, S., Turin, L., and Warner, A. (1988). Patterns of junctional communication during development of the early amphibian embryo. *Development* **103**, 769–783.
Hainski, A. M., and Moody, S. A. (1992). *Xenopus* maternal RNAs from a dorsal animal blastomere induce a secondary axis in host embryos. *Development* **116**, 347–355.
Hamburger, V. (1988). "The Heritage of Experimental Embryology: Hans Spemann and the Organizer." Oxford University Press, Oxford.
Hemmati-Brivanlou, A. H., and Harland, R. M. (1989). Expression of an engrailed-related protein is induced in the anterior neural ectoderm of early *Xenopus* embryos. *Development* **106**, 611–617.

8. Dorsoanterior Axis in Frogs

Hemmati-Brivanlou, A., and Melton, D. A. (1992). A truncated activin receptor inhibits mesoderm induction and formation of axial structures in *Xenopus* embryos. *Nature* **359**, 609–614.

Hemmati-Brivanlou, A., Wright, D. A., and Melton, D. A. (1992). Embryonic expression and functional analysis of a *Xenopus* activin receptor. *Dev. Dynam.* **194**, 1–11.

Holowacz, T., and Elinson, R. P. (1993). Cortical cytoplasm, which induces dorsal axis formation in *Xenopus*, is inactivated by UV-irradiation of the oocyte. *Development* **119**, 277–285.

Holtfreter, J. (1993). Die totale Exogastrulation, eine Selbstablösung des Ektoderms vom Entomesoderm. *Roux Arch. Entwicklungsmech. Org.* **129**, 669–793.

Holwill, S., Heasman, J., Crawley, C. R., and Wylie, C. C. (1987). Axis and germ line deficiencies caused by UV irradiation of *Xenopus* oocytes cultured in vitro. *Development* **100**, 735–743.

Houliston, E., and Elinson, R. P. (1991a). Evidence for the involvement of microtubules, endoplasmic reticulum, and kinesin in the cortical rotation of fertilized frog eggs. *J. Cell Biol.* **114**, 1017–1028.

Houliston, E., and Elinson, R. P. (1991b). Patterns of microtubule polymerization related to cortical rotation in *Xenopus laevis* eggs. *Development* **112**, 107–117.

Houliston, E., and Elinson, R. P. (1992). Microtubules and cytoplasmic reorganization in the frog egg. *Curr. Top. Dev. Biol.* **26**, 53–70.

Imoh, H. (1988). Formation of germ layers and roles of the dorsal lip of the blastopore in normally developing embryos of the newt *Cynops pyrrhogaster*. *J. Exp. Zool.* **246**, 258–270.

Isaacs, H. V., Tannahill, D., and Slack, J. M. W. (1992). Expression of a novel FGF in the *Xenopus* embryo. A new candidate inducing factor for mesoderm formation and anteroposterior specification. *Development* **114**, 711–720.

Jones, C. M., Lyons, K. M., Lapan, P. M., Wright, C. V. E., and Hogan, B. L. M. (1992). DVR-4 (Bone Morphogenetic Protein-4) as a posterior-ventralizing factor in *Xenopus* mesoderm induction. *Development* **115**, 639–647.

Jones, E. A., and Woodland, H. R. (1987). The development of animal caps in *Xenopus*: A measure of the start of animal cap competence to form mesoderm. *Development* **101**, 557–563.

Kageura, H. (1990). Spatial distribution of the capacity to initiate a secondary embryo in the 32-cell embryo of *Xenopus laevis*. *Dev. Biol.* **142**, 432–438.

Kao, K. R., and Elinson, R. P. (1988). The entire mesodermal mantle behaves as Spemann's organizer in dorsoanterior enhanced *Xenopus laevis* embryos. *Dev. Biol.* **127**, 64–77.

Kao, K. R., and Elinson, R. P. (1989). Dorsalization of mesoderm induction by lithium. *Dev Biol.* **132**, 81–90.

Kao, K. R., Masui, Y., and Elinson, R. P. (1986) Lithium-induced respecification of pattern in *Xenopus laevis* embryos. *Nature* **322**, 371–373.

Keller, R., Shih, J., and Domingo, C. (1992). The patterning and functioning of protrusive activity during convergence and extension of the *Xenopus* organizer. *Development (1992 Suppl.)*, 81–91.

Kimelman, D., and Kirschner, M. (1987). Synergistic induction of mesoderm by FGF and TGF-β and the identification of an mRNA coding for FGF in the early *Xenopus* embryo. *Cell* **51**, 869–877.

Kimelman, D., and Maas, A. (1992). Induction of dorsal and ventral mesoderm by ectopically expressed *Xenopus* basic fibroblast growth factor. *Development* **114**, 261–269.

Kimelman, D., Christian, J. L., and Moon, R. T. (1992). Synergistic principles of development: Overlapping patterning system in *Xenopus* mesoderm induction. *Development* **116**, 1–9.

Kinoshita, K., Bessho, T., and Asashima, M. (1993). Competence prepattern in the animal hemisphere of the 8-cell-stage *Xenopus* embryo. *Dev. Biol.* **160**, 276–284.

Kintner, C. R., and Melton, D. A. (1987). Expression of *Xenopus* N-CAM RNA in ectoderm is an early response to neural induction. *Development* **99**, 311–325.

Klymkowsky, M. W., Maynell, L. A., and Polson, A. G. (1987). Polar asymmetry in the organization of the cortical cytokeratin system of *Xenopus laevis* oocytes and embryos. *Development* **100**, 543–557.

Kondo, M., Tashiro, K., Fujii, G., Asano, M., Miyoshi, R., Yamada, R., Muramatsu, M., and Shiokawa, K. (1991). Activin receptor mRNA is expressed early in *Xenopus* embryogenesis and the level of expression affects the body axis formation. *Biochem. Biophys. Res. Comm.* **181**, 684–690.

Kume, S., Muto, A., Aruga, J., Nakagawa, T., Michikawa, T., Furichi, T., Nakade, S., Okano, H., and Mikoshiba, K. (1993). The *Xenopus* IP$_3$ receptor: Structure, function, and localization in oocytes and eggs. *Cell* **73**, 555–570.

Lamb, T. M., Knecht, A. K., Smith, W. C., Stachel, S. E., Economides, A. N., Stahl, N., Yancopolous, G. D., and Harland, R. M. (1993). Neural induction by the secreted polypeptide noggin. *Science* **262**, 713–718.

Lettice, L. A., and Slack, J. M. W. (1993). Properties of the dorsalizing signal in gastrulae of *Xenopus laevis*. *Development* **117**, 263–271.

London, C., Akers, R., and Phillips, C. (1988). Expression of Epi 1, an epidermis-specific marker in *Xenopus laevis* embryos, is specified prior to gastrulation. *Dev. Biol.* **129**, 380–389.

Lundmark, C. (1986). Role of bilateral zones of ingressing superficial cells during gastrulation of *Ambystoma mexicanum*. *J. Embryol. Exp. Morphol.* **97**, 47–62.

MacNicol, A. M., Muslin, A. J., and Williams, L. T. (1993). Raf-1 kinase is essential for early *Xenopus* development and mediates induction of mesoderm by FGF. *Cell* **73**, 571–583.

Malacinski, G. M., Benford, H., and Chung, H. M. (1975). Association of an ultraviolet irradiation sensitive cytoplasmic localization with the future dorsal side of the amphibian egg. *J. Exp. Zool.* **191**, 97–110.

Manes, M. E., and Elinson, R. P. (1980). Ultraviolet light inhibits grey crescent formation on the frog egg. *Roux's Arch. Dev. Biol.* **189**, 73–76.

Mangold, O. (1933). Über die Induktionsfähigkeit der verschiedenen Bezirke der Neurula von Urodelen. *Naturwissenschaften* **21**, 761–766.

Maslanski, J. A., Leshko, L., and Busa, W. B. (1992). Lithium-sensitive production of inositol phosphates during amphibian embryonic mesoderm induction. *Science* **256**, 243–245.

Mathews, L. S., Vale, W. W., and Kintner, C. R. (1992). Cloning of a second type of activin receptor and functional characterization in *Xenopus* embryos. *Science* **255**, 1702–1703.

McCrea, P. D., Turck, C. W., and Gumbiner, B. (1991). A homolog of the *armadillo* protein in *Drosophila* (plakoglobin) associated with β-catenin. *Science* **254**, 1359–1361.

McCrea, P. D., Brieher, W. M., and Gumbiner, B. M. (1993). Induction of a secondary body axis in *Xenopus* by antibodies to β-catenin. *J. Cell Biol.* **123**, 477–484.

McMahon, A. P., and Moon, R. T. (1989). Ectopic expression of the proto-oncogene *int-1* in *Xenopus* embryos leads to duplication of the embryonic axis. *Cell* **58**, 1075–1084.

Melton, D. A. (1987). Translocation of a localized maternal mRNA to the vegetal pole of *Xenopus* oocytes. *Nature* **328**, 80–82.

Moon, R. T., and Christian, J. L. (1992). Competence modifiers synergize with growth factors during mesoderm induction and patterning in *Xenopus*. *Cell* **71**, 709–712.

Mosquera, L., Forristall, C., Zhou, Y., and King, M. L. (1993). A mRNA localized to the vegetal cortex of *Xenopus* oocytes encodes a protein with a *nanos*-like zinc finger domain. *Development* **117**, 377–386.

Nagajski, D. J., Guthrie, S. C., Ford, C. C., and Warner, A. E. (1989). The correlation between patterns of dye transfer through gap junctions and future developmental fate in *Xenopus*: The consequences of uv irradiation and lithium treatment. *Development* **105**, 747–752.

Nakamura, O., and Toivonen, S., eds. (1978). "Organizer—A Milestone of a Half-Century from Spemann." Elsevier/North Holland Biomedical Press, Amsterdam.

Newport, J., and Kirschner, M. (1982). A major developmental transition in early *Xenopus* embryos. I. Characterization and timing of cellular changes at the midblastula transition. *Cell* **30**, 675–686.

8. Dorsoanterior Axis in Frogs

Nieuwkoop, P. D. (1969a). The formation of the mesoderm in urodelean amphibians. I. Induction by the endoderm. *Roux Arch. Entwicklungsmech. Org.* **162**, 341–373.

Nieuwkoop, P. D. (1969b). The formation of the mesoderm in urodelean amphibians. II. The origin of the dorso-ventral polarity of the mesoderm. *Roux Arch. Entwicklungsmech. Org.* **163**, 298–315.

Nieuwkoop, P. D., and Ubbels, G. A. (1972). The formation of the mesoderm in urodelean amphibians. IV. Qualitative evidence for the purely "ectodermal" origin of the entire mesoderm and of the pharyngeal endoderm. *Roux Arch. Entwicklungsmech. Org.* **169**, 185–199.

Nuccitelli, R. (1991). How do sperm activate eggs? *Curr. Top. Dev. Biol.* **25**, 1–16.

Nüsslein-Volhard, C. (1991). Determination of the embryonic axes of *Drosophila*. *Development Suppl.* **1**, 1–10.

Olson, D. J., and Moon, R. T. (1992). Distinct effects of ectopic expression of *Wnt-1*, activin B, and bFGF on gap junctional permeability in 32-cell *Xenopus* embryos. *Dev. Biol.* **151**, 204–212.

Olson, D. J., Christian, J. L., and Moon, R. T. (1991). Effect of *Wnt-1* and related proteins on gap junctional communication in *Xenopus* embryos. *Science* **252**, 1173–1176.

Otte, A. P., and Moon, R. T. (1992). Protein kinase C isozymes have distinct roles in neural induction and competence in *Xenopus*. *Cell* **68**, 1021–1029.

Otte, A. P., Koster, C. H., Snoek, G. T., and Durston, A. J. (1988). Protein kinase C mediates neural induction in *Xenopus laevis*. *Nature* **334**, 618–620.

Otte, A. P., Kramer, I. M., and Durston, A. J. (1991). Protein kinase C and regulation of the local competence of *Xenopus* ectoderm. *Science* **251**, 570–573.

Papalopulu, N., and Kintner, C. (1993). *Xenopus distal-less* related homeobox genes are expressed in the developing forebrain and are induced by planar signals. *Development* **117**, 961–975.

Paterno, G. D., Gillespie, L. L., Dixon, M. S., Slack, J. M. W., and Heath, J. K. (1989). Mesoderm inducing properties of INT-2 and kFGF: Two oncogene-encoded growth factors related to FGF. *Development* **106**, 79–83.

Rebagliati, M. R., and Dawid, I. B. (1993). Expression of activin transcripts in follicle cells and oocytes of *Xenopus laevis*. *Dev. Biol.* **159**, 574–580.

Rebagliati, M. R., Weeks, D. L., Harvey, R. P., and Melton, D. A. (1985). Identification and cloning of localized maternal RNAs from *Xenopus* eggs. *Cell* **42**, 769–777.

Render, J., and Elinson, R. P. (1986). Axis determination in polyspermic *Xenopus laevis* eggs. *Dev. Biol.* **115**, 425–433.

Riggleman, B., Schedl, P., and Wieschaus, E. (1990). Spatial expression of the *Drosophila* segment polarity gene *armadillo* is posttranscriptionally regulated by *wingless*. *Cell* **63**, 549–560.

Rosa, F., Roberts, A. B., Danielpour, D., Dart, L. L., Sporn, M. B., and Dawid, I. B. (1988). Mesoderm induction in amphibians: The role of TGF-β2-like factors. *Science* **239**, 783–785.

Roux, W. (1887). Beitrage zur Entwicklungsmechanik des Embryo. 4. Die Richtungsbestimmung der Medianebene des Froschembryo durch die Copulationsrichtung des Eikernes und des Spermakernes. *Arch. Mikrosk. Anat.* **29**, 157–212.

Ruiz i Altaba, A. (1990). Neural expression of the *Xenopus* homeobox gene *Xhox3*: Evidence for a patterning neural signal that spreads through the ectoderm. *Development* **108**, 595–604.

Ruiz i Altaba, A. (1992). Planar and vertical signals in the induction and patterning of the *Xenopus* nervous system. *Development* **116**, 67–80.

Ruiz i Altaba, A. (1993). Induction and axial patterning of the neural plate: Planar and vertical signals. *J. Neurobiol.* **24**, 1276–1304.

Ruiz i Altaba, A., and Jessell, T. M. (1992). *Pintallavis*, a gene expressed in the organizer and midline cells of frog embryos: Involvement in the development of the neural axis. *Development* **116**, 81–93.

Sander, K. (1991). When seeing is believing: Wilhelm Roux's misconceived fate map. *Roux's Arch. Dev. Biol.* **200**, 177–179.

Savage, R., and Phillips, C. R. (1989). Signals from the dorsal blastopore lip region during gastrulation bias the ectoderm toward a nonepidermal pathway of differentiation in *Xenopus laevis*. *Dev. Biol.* **133**, 157–168.

Scharf, S. R., and Gerhart, J. C. (1980). Determination of the dorso-ventral axis in eggs of *Xenopus laevis*: Complete rescue of UV-impaired eggs by oblique orientation before first cleavage. *Dev. Biol.* **79**, 181–198.

Scharf, S. R., Rowning, B., Wu, M., and Gerhart, J. C. (1989). Hyperdorsoanterior embryos from *Xenopus* eggs treated with D_2O. *Dev. Biol.* **134**, 175–188.

Schneider, S., Herrenknecht, K., Butz, S., Kemler, R., and Hausen, P. (1993). Catenins in *Xenopus* embryogenesis and their relation to the cadherin-mediated cell-cell adhesion system. *Development* **118**, 629–640.

Schroeder, M. M., and Gard, D. L. (1992). Organization and regulation of cortical microtubules during the first cell cycle of *Xenopus* eggs. *Development* **114**, 699–709.

Schwartz, S. P., Aisenthal, L., Elisha, Z., Oberman, F., and Yisraeli, J. K. (1992). A 69-kDa RNA-binding protein from *Xenopus* oocytes recognizes a common motif in two vegetally localized maternal mRNAs. *Proc. Natl. Acad. Sci. USA* **89**, 11895–11899.

Shi, D.-L., Delarue, M., Darribère, T., Riou, J.-F., and Boucaut, J.-C. (1987). Experimental analysis of the extension of the dorsal marginal zone in *Pleurodeles waltl* gastrulae. *Development* **100**, 147–161.

Shih, J., and Keller, R. (1992). The epithelium of the dorsal marginal zone of *Xenopus* has organizer properties. *Development* **116**, 887–899.

Sive, H. L. (1993). The frog prince-ss: A molecular formula for dorsoventral patterning in *Xenopus*. *Genes Dev.* **7**, 1–12.

Slack, J. M. W. (1992). The nature of the mesoderm-inducing signal in *Xenopus*: A transfilter induction study. *Development*. **113**, 661–669.

Slack, J. M. W. (1993). Embryonic induction. *Mech. Dev.* **41**, 91–107.

Slack, J. M. W., and Forman, D. (1980). An interaction between dorsal and ventral regions of the marginal zone in early amphibian embryos. *J. Embryol. Exp. Morphol.* **56**, 283–299.

Slack, J. M. W., Darlington, B. G., Heath, J. K., and Godsave, S. F. (1987). Mesoderm induction in early *Xenopus* embryos by heparin-binding growth factors. *Nature* **326**, 197–200.

Slack, J. M. W., Isaacs, H. V., and Darlington, B. G. (1988). Inductive effects of fibroblast growth factor and lithium ion on *Xenopus* blastula ectoderm. *Development* **103**, 581–590.

Smith, J. C. (1993). Mesoderm-inducing factors in early vertebrate development. *EMBO J.* **12**, 4463–4470.

Smith, J. C., and Slack, J. M. W. (1983). Dorsalization and neural induction: Properties of the organizer in *Xenopus laevis*. *J. Embryol. Exp. Morphol.* **78**, 299–317.

Smith, J. C., Price, B. M. J., Van Nimmen, K., and Huylebroeck, D. (1990). Identification of a potent *Xenopus* mesoderm-inducing factor as a homolog of activin A. *Nature* **345**, 729–731.

Smith, W. C., and Harland, R. M. (1991). Injected Xwnt-8 RNA acts early in *Xenopus* embryos to promote formation of a vegetal dorsalizing center. *Cell* **67**, 753–765.

Smith, W. C., and Harland, R. M. (1992). Expression cloning of noggin, a new dorsalizing factor localized to the Spemann organizer in *Xenopus* embryos. *Cell* **70**, 829–840.

Smith, W. C., Knecht, A. K., Wu, M., and Harland, R. M. (1993). Secreted noggin protein mimics the Spemann organizer in dorsalizing *Xenopus* mesoderm. *Nature* **361**, 547–549.

Sokol, S. Y. (1993). Mesoderm formation in *Xenopus* ectodermal explants overexpressing Xwnt8: Evidence for a cooperating signal reaching the animal pole by gastrulation. *Development* **118**, 1335–1342.

Sokol, S. Y., and Melton, D. A. (1992). Interaction of Wnt and activin in dorsal mesoderm induction in *Xenopus*. *Dev. Biol.* **154**, 348–355.

Sokol, S., Christian, J. L., Moon, R. T., and Melton, D. A. (1991). Injected Wnt RNA induces a complete body axis in *Xenopus* embryos. *Cell* **67**, 741–752.

8. Dorsoanterior Axis in Frogs

Spemann, H. (1938). "Embryonic Development and Induction." Yale University Press, New Haven.
Spemann, H., and Mangold, H. (1924). Über Induktion von Embryonalanlagen durch Implantation artfremder Organisatoren. *Arch. mikr. Anat. Entwicklungsmech. Org.* **100**, 599–638.
Steinbesser, H., De Robertis, E. M., Ku, M., Kessler, D. S., and Melton, D. A. (1993). *Xenopus* axis formation: Induction of *goosecoid* by injected *Xwnt-8* and activin RNAs. *Development* **118**, 499–507.
Stewart, R. M., and Gerhart, J. C. (1990). The anterior extent of dorsal development of the *Xenopus* embryonic axis depends on the quantity of organizer in the late blastula. *Development* **109**, 363–372.
Stewart, R. M., and Gerhart, J. C. (1991). Induction of notochord by the organizer in *Xenopus*. *Roux's Arch. Dev. Biol.* **199**, 341–348.
Sudarwati, S., and Nieuwkoop, P. D. (1971). Mesoderm formation in the anuran *Xenopus laevis* Daudin. *Roux' Arch. Entwicklungsmech. Org.* **166**, 189–204.
Taira, M., Jamrich, M., Good, P. J., and Dawid, I. B. (1992). The LIM domain-containing homeobox gene *Xlim-1* is expressed specifically in the organizer region of *Xenopus* gastrula embryos. *Genes Dev.* **6**, 356–366.
Takasaki, H., and Konishi, H. (1989). Dorsal blastomeres in the equatorial region of the 32-cell *Xenopus* embryo autonomously produce progeny committed to the organizer. *Dev. Growth Diff.* **31**, 147–156.
Thomsen, G. H., and Melton, D. A. (1993). Processed Vg1 protein is an axial mesoderm inducer in *Xenopus*. *Cell* **74**, 433–441.
Thomsen, G., Woolf, T., Whitman, M., Sokol, S., Vaughan, J., Vale, W., and Melton, D. A. (1990). Activins are expressed early in *Xenopus* embryogenesis and can induce axial mesoderm and anterior structures. *Cell* **63**, 485–493.
van Dassow, G., Schmidt, J. E., and Kimelman, D. (1993). Induction of the *Xenopus* organizer. Expression and regulation of *Xnot*, a novel FGF and activin-regulated homeobox gene. *Genes Dev.* **7**, 355–366.
van den Eijnden-Van Raaij, A. J. M., van Zoelent, E. J. J., van Nimmen, K., Koster, C. H., Snoek, G. T., Durston, A. J., and Huylebroeck, D. (1990). Activin-like factor from a *Xenopus laevis* cell line responsible for mesoderm induction. *Nature* **345**, 732–734.
Vincent, J.-P., and Gerhart, J. C. (1987). Subcortical rotation in *Xenopus* eggs: An early step in embryonic axis specification. *Dev. Biol.* **123**, 526–539.
Vincent, J.-P., Oster, G. F., and Gerhart, J. C. (1986). Kinematics of gray crescent formation in *Xenopus* eggs: The displacement of subcortical cytoplasm relative to the egg surface. *Dev. Biol.* **113**, 484–500.
Vincent, J.-P., Scharf, S. R., and Gerhart, J. C. (1987). Subcortical rotation in *Xenopus* eggs: A preliminary study of its mechanochemical basis. *Cell Motil. Cytoskel.* **8**, 143–154.
Vogt, W. (1929). Gestaltungsanalyse am Amphibienkeim mit örtlicher Vitalfärbung. II. Gastrulation und Mesodermbildung bei Urodelen und Anuren. *Roux Arch. Entwicklungsmech. Org.* **120**, 385–786.
Weeks, D. L., and Melton, D. A. (1987). A maternal mRNA localized to the vegetal hemisphere in *Xenopus* eggs codes for a growth factor related to TGF-β. *Cell* **51**, 861–868.
Whitman, M., and Melton, D. A. (1992). Involvement of p21^{ras} in Xenopus mesoderm induction. *Nature* **357**, 252–254.
Yuge, M., Kobayakawa, Y., Fujisue, M., and Yamana, K. (1990). A cytoplasmic determinant for dorsal axis formation in an early embryo of *Xenopus laevis*. *Development* **110**, 1051–1056.
Zimmerman, K., Shih, J., Bars, J., Collazo, A., and Anderson, D. J. (1993). *XASH-3*, a novel *Xenopus achaete-scute* homolog, provides an early marker of planar neural induction and position along the mediolateral axis of the neural plate. *Development* **119**, 221–232.
Zisckind, N., and Elinson, R. P. (1990). Gravity and microtubules in dorsoventral polarization of the *Xenopus* egg. *Dev. Growth Diff.* **32**, 575–581.

Index

A

A23187, 65
Ablation, 203–204
Acetylcholine
  effect in egg activation, 40
  treatment of egg, 40–41
Acetylcholine receptors, muscarinic, m1, human, 40
Acrosomal region, 49
Acrosome, and calcium release, 85–86
Acrosome reaction, 10, 12, 23
  description, 33
  inositol 1,4,5-triphosphate effect, 31
  phorbol diester treatment, 33
  species specificity, 11
Actin, 222–223
  in axis formation, 231
  filamentous
    in cell organization, 231, 242, 244
    in oocyte polarity, 222–223, 226
Activin
  in axis induction, 269–271, 270
  in ectopic axis assay, 262
  mesoderm induction, 260, 269–271, 270
  mRNA, 270
  notochord induction, 271
  in organizer formation, 274
  receptor, 262
Adhesion, cellular, 15
Adhesion plaque protein, 223
Aequorin, 26, 27, 65–66, 70
All-fish gene cassette, 194–195
Amphibian, as axis formation model, 215–218
Amplification, extrachromosomal, 188, 202
Androgenetic embryo, 167–168
Animal cap, 259–260
  dorsal, 271
  ventral, 271
Animal cell, 268, 270
Animal cortex, 241–242
Animal hemisphere, 218, 224

Animal–vegetal axis
  cortex, 220–222
  formation, 215–217, 236, 244, 246
  gravity influence, 263
  microtubule, 224, 238, 240
  mitochondrial mass, 232–233
  mRNA distribution, 218–219
  in oocyte maturation, 235–242, 257–258
  polarization, 218–225
  specification, 232–233
  in *Xenopus* egg, 262
Antibody
  antisperm, 48–49
  monoclonal, *see* Monoclonal antibody
Antifreeze protein, 191
Antifreeze protein gene
  ocean pout, 193, 194–196
  regulatory region, 201
  species comparison, 195
Antisense RNA, 204
Antisperm antibodies, 48–49
Aquaculture, 192–196, 199, 206
Aromatase
  activity, 113, 116
  induction mechanism, 112–113
  medaka, 116
Aromatase cytochrome P-450, *see* Cytochrome P-450
Aster, sperm, 263–264
Asynchronous ovary, 106
Axis
  animal–vegetal, *see* Animal–vegetal axis
  anterior–posterior, 255
  cytoplasmic, 232–233
  dorsal, 254
  dorsal–ventral, 255
  dorsoanterior, 261, 263, 268, 269–273
  ectopic, 260–262, 262
  nuclear/cytoplasmic, 226, 228–231, 229–230
Axis-inducing activity, 262, 265–266, 267, 269–273

287

B

Balbioni body, 232–233, 235
BAPTA, 27–28
Bindin, 24, 52, 84, 85
Blastomere, 267–268, 269
Blastula, 259
Bone morphogenetic protein, 260, 269
Botulinum ADP-ribosyltransferase C3, 41
Bystander effect, 84

C

Caffeine, 75–77, 78
Calcium
  in cytoplasm, 69
  detection methods, 65–67
  effect on cortical granule exocytosis, 35–37
  in egg, unfertilized, 66–67
  in egg activation, 24, 27–28, 85–87, 164–166
  extracellular, 27
  extrusion, 71
  location, 30, 69
  oscillations during oocyte maturation, 34–35
  peak, 68
  protein regulation, 29
  regulation, 29, 64–65, 88
  release, 30, 32, 70, 75–77, 78–81
  release mechanism, 71–73
    G protein, 75–78
    inositol 1,4,5-triphosphate receptor dependent, 73–75
    phosphorylation, 83
    ruthenium red, 76
    ryanodine receptor dependent, 32, 75–78, 88
    species comparison, 72–73
  and sperm, 46, 85–87
  in steroidogenesis regulation, 122
  transient, 28–29, 30
    characteristics, 67–71
    and inositol 1,4,5-triphosphate, 31, 74–75
    origination, 26–27, 77–78, 80, 164–165
    species comparison, 27
Calcium green, 66, 68–69
Calmodulin-dependent protein kinase II, 29
Carbohydrate, *see* Oligosaccharide
Carbonyl reductase, 124
β-Catenin, 262, 272–273
C-Cadherin, 272
CD59, 49
cdc2 kinase
  activation, 133–135
  cDNA clone, 130
  maturation-promoting factor, 130, 132
  in oocyte maturation, 132–134, 137–138
cDNA, 186, 200
Cell cycle, 29, 154, 157, 161–166, 169
Cell fusion, 148, 153
Cell lineage ablation, 203–204
Cellular adhesion, 15
Centrosome, *see also* Microtubule organizing center
  in animal–vegetal axis formation, 228–229, 231
  centrosomal component, 237–238
  inactivation, 229, 232
Cholera toxin, 82
Cholesterol side-chain cleavage cytochrome P-450, 114–115, 124
Chorion, 183, 184
Chromatin, 163–164
Chromosome
  enucleation effect, 153
  extrachromosomal amplification, 188
  lampbrush, 223
  in nuclear transplantation, 149, 150, 162, 163–164
  premature condensation, 161
  uniparental disomy, 167–168
Competence modifier, 260, 269
Complement
  C3, 49
Complement receptors, 48–49
Cortex
  axis-inducing activity, 267–268
  development, 262
  and γ-tubulin distribution, 237–241
  polarization, 240–241
  RNA localization, 266–267
  rotation, 257, 263–264, 267, 273
    gravity effect, 263–264
    ultraviolet radiation, 264
Cortical flash, 69–70
Cortical granule exocytosis, 22, 24, 34–37, 41
  and calcium, 28, 71
  protein kinase C effect, 34
δ-Crystallin gene, 201
Cyanoketone, 118–119

Cyclic adenosine diphosphate-ribose
assay, 80–81
function, 32–33, 78–81, 80, 86
regulation, 81
synthesis, 78, 81
Cyclic AMP, 122, 123–124
Cyclic GMP-dependent protein kinase, 81, 86–87
cascade, 125
Cyclin A, 130–131, 133
Cyclin B
cDNA clone, 131
degradation, 136
and maturation-promoting factor, 131, 132
in oocyte maturation, 132–133, 137–138
synthesis, 135
Cytochalasin, 222
Cytochrome P-450, 115–116, 124; see also specific type
cholesterol side-chain cleavage, 114–115, 124
Cytokeratin, 220–222, 267
Cytoplasm
in axis formation, 264, 265–266
germinal, 218
licensing factor, 162
in nuclear transplantation, 155, 159–161, 169
polarity, species comparison, 235
Cytoplasmic cap, 228, 232
Cytoskeleton, 219–241; see also specific components
in animal–vegetal axis formation, 217, 219–241, 231, 244, 246
cytoplasmic organization, 233, 235
in polarization, 236–238, 240–241

D

Decay accelerating factor, 49
Developmental biology, transgenic fish as model, 199–205
Developmental capacity, 150–151, 159, 170
Developmental frequency, 158, 163, 167, 169
Developmental stage, 156–159
Dextran-coupled fluorescence, 66
sn-1,2-Diacylglycerol, 32–33
Dicarbocyanine dye, 72
17α,20β-dihydroxy-4-pregnen-3-one
cDNA clone, 124
and cyclin B, 135

oocyte maturation, 119–124, 125, 137, 138
production, 120–124
Diphtheria toxin A chain, 203
Disintegrin, 52
DNA
complementary, see cDNA
methylation, 168
in transgenic system, 185–189
Dorsal factor, 264–267
Dorsalization, 274–275
Dorsal lip, 257
Dorsoanterior development, 255–257, 261, 267–269
Dorsoanterior index, 255, 261
17α,20β-DP, see 17α,20β-Dihydroxy-4-pregnen-3-one
Drosophila, 254

E

Ecological issues, 196, 199
Ectoderm, 275–276
Ectopic axis assay, 269
Ectopic spindle, see Spindle
Effector, 22–23, 37
Effector–effector interaction, 44
Egg, see also Oocyte; Zona pellucida; Sea urchin egg
botulinum ADP-ribosyltransferase C3 effect, 41
calcium release, 71
dorsal information localization, 267–269
fish, characteristics, 180, 183
integrin present, 50–51
membrane, 23–25, 48, 110
second messenger, 26–34
sperm–egg interaction, 23–25, 48, 84–87
sperm receptor, 64, 85
ultraviolet radiation treatment, 264
Egg activation, see also Maturation-promoting factor; Sperm
calcium role, 277
description, 21–25, 37, 262–263
fusion-mediated, 37, 44–47, 84–87
integrin role, 50–51
ionophore, 27, 28–29
latent period, 45, 68
and messenger, 26–34, 86–87
in nuclear transplantation, 153, 164–166, 169–170
receptor-mediated, 37–44, 82–84

signal transduction, 37 47
species specificity, 1–3
sperm–egg interaction, 23–25, 37, 50–52, 68
sperm receptor, 47–51, 64, 83
Elastase I promotor, 203
Electrofusion, 153
Electroporation, 183–185
Electrostimulation, 164–165, 169–170
Embryo, 157, 267–269
  androgenetic, 167–168
  developmental stages, 167–168. see also Oocyte
  gynogenetic, 167–168
  urodele, 276
Embryonic stem cell, 258–159
Endocytosis, 109
Endoplasmic reticulum, 88
Enhancer trap, 205
Enucleation, 153
Epi 1, 268
Epidermal marker, 268
Epigenetic modification, 167–168
Epithelial layer, 269
Estradiol-17β
  cDNA clone, 114–116, 124
  function, 108, 110, 137
  regulation, 113, 114–116
  species comparison, 111
  two-cell type model, 110–113

F

F-actin, see Actin, filamentous
Fertilin, 51
Fertilization, see Egg activation
Fibroblast growth factor, 260, 269, 271
Fibroblast growth factor receptor, 271–272
Filamentous actin, see Actin, filamentous
Fish, see also Transgenic fish
  egg, 180, 183
  spawning, 180
  as transgenic model, 202
Fluo-3, 66, 68–69, 76
Fluorescence, ratiometric, 66
Follicle, 113, 120–124, 270
Follicle-stimulating hormone, 104, 111–112
Follistatin, 271
Forskolin, 122
Fura-2, 66, 76–77
Fusion, 164

Fusion hypothesis, in egg activation, 37, 44–47, 84–87

G

G_2 phase, 161–162
Galactosyltransferase, 10
Gamete adhesion, 9–11, 13
Gap junction, 271
Gastrulation, 257, 268–269
GDPβS, 83
Gene, see also specific gene
  bombardment, 185
  disruption, 205
  expression, 155, 200–202
  function study, 203–204
  tagging, 200
  targeting, 204
Gene-cassette, all-fish, 194–195
Germ cell, primordial, 158–159
Germinal vesicle
  actin component, 222–223
  in nuclear transplantation, 155
  in oocyte maturation, 117–118
Germinal vesicle breakdown, 118, 119–120, 125–128, 132
Germ-line transformation, 195
Germplasm, 218
Gonadotropin, see also GTH-I; GTH-II
  17α,20β-DP production, 120–122
  effect on vitellogenin uptake, 110, 121–124
  estradiol-17β stimulation, 110–113
  and follicle response, 120–122, 123–124
  20β-HSD production, 121
  17α-hydroxyprogesterone production, 121–122
  and maturation-inducing hormone receptor, 126–127
  in oocyte maturation, 118–119, 137
  pregnant mare serum gonadotropin, 113
  receptor, 121, 123–124
  regulation, 121–123
  testosterone production stimulation, 112
Goosecoid, 256–257, 262
G protein, see Guanine nucleotide-binding regulatory protein
Granule vesicle
  and meiotic spindle, 244
  in oocyte, 235, 236
Granulosa cell
  estradiol-17β production, 110–113

Index

gonadotropin stimulation, 121, 123–124
  in oocyte, 107
  two-cell type model, 137
Gravity, 242, 263–264
Group synchronous ovary, 104, 106
Growth hormone
  antisense RNA, 204
  gene transfer, 192–195, 194–195
  in transgenic fish, 191, 202
Growth hormone-releasing hormone, 196
GTH-I, 104, 110, 121
GTH-II, 104, 118–119, 121
Guanine nucleotide-binding regulatory protein
  egg activation, 82–84
  inhibitor, 38–40, 41
  pertussis toxin effect, 40
  receptor model, 38–42
Guanylate cyclase, 87
Gynogenetic embryo, 167–168

H

Halocynthia roretzi, 42
Heat shock promotor, 203
Heparin, 72, 75, 76–77
Herpes simplex virus type 1, thymidine kinase, 203
Histone H1 kinase, 129
Homeobox genes, 202
Hox gene, 202
20β-HSD, 121, 123, 124
17α-Hydroxylase/17,20 lyase cytochrome P-450, 115–116
17α-Hydroxyprogesterone, 121–124
3β-Hydroxysteroid dehydrogenase, 107, 124–125
3β-Hydroxysteroid dehydrogenase-isomerase, 114, 115

I

Immune response, bystander effect, 84
Inner cell mass cell, 148, 157
Inositol 1,4,5-triphosphate, 30
  cADPR effect, 80
  calcium regulation, 64, 73–75, 88
  effect on cortical granule exocytosis, 35–37
  egg activation, 29–32, 30–31, 86
  measurement, 31–32
  and mesoderm induction, 272
  monoclonal antibody, 30

production activation, 82
receptor, 71–73
and sperm, 46, 166
Insulin-like growth factor I, 196
Integrin, 50–51, 52
Intermediate filament, 220–222
Ionophore, 27, 28–29
Ion-sensitive microelectrode, 67

K

Kinase
  cdc2
    activation, 133–135
    cDNA clone, 130
    maturation-promoting factor, 130, 132
    in oocyte maturation, 132–134, 137 138
  Cdk2, 130
  histone H1, 129
Kinesin, 263

L

Lamina, nuclear, 160–161
Lamin A, 160 161
Lamin B, 160–161
Lamin C, 160–161
Lampbrush chromosome, 223
Lanthanum, 86
Laser microbeam, 186
Latent period, 68
Licensing factor, 162
Ligand, 22, 50–51
Lithium, 262, 271, 272
Locus activation region, 191–192
Luteinizing hormone, 104, 111–112
Lymphocyte homing, 15
Lysin, 85

M

Maturational competence, 126–127
Maturation-inducing hormone
  cAMP cascade, 125–126
  and oocyte maturation, 119–120, 126–127
  receptor, 125–126, 137
Maturation-promoting factor
  activation mechanism, 133–137
  calcium regulation, 29
  characterization, 129, 131–132

effect on cytokeratin filament, 221
in nuclear transplantation, 155, 160–161
oocyte maturation, 127–127
species comparison, 130–132, 135, 137–138
Maturation spot, 241–242
MCP/CD46, 49
Medaka, 200–201, 204
Meiotic spindle, 241–244
Membrane attack complex inhibitory protein, 49
Membrane cofactor protein, 49
Mesoderm
  inducer, 260, 270
  induction, 257, 259–260, 268, 270, 274
    comparison to dorsalization, 274–275
Mesodermal belt, 257–258
Mesoderm induction assay, 259–260, 269
Metallothionein gene, 193
Metaphase, 29
Metaphase II, 163–164
Metaphase-promoting factor, see Maturation-promoting factor
Microelectrode, ion-sensitive, 67
Microinjection, 180, 183, 185, 190
Microsporocyte, pachytene, 128
Microsurgery, 148, 153
Microtubule
  array, transient, 242, 244
  associated protein, 236
  cold-stable, 225
  in cortical rotation, 263
  distribution, 230
  network, 117–118
  in oocyte polarity, 223–226
  organization in oogenesis, 117–118, 226–228, 238, 240–241
  of sperm aster, 263
Microtubule organizing center, 223–224, 228–229, 236–237, 242–244
Mitochondrial cloud, 232–233, 235
Mitochondrial mass, 232–233, 235
Monoclonal antibody, 30, 31, 49
Mosaicism, 183, 189–191, 195
M phase, 161–162
mRNA
  distribution, 218–219, 221–222, 236
  in nuclear transplantation, 156, 199, 204
Muscarinic receptor, m1, human, 40
Mutagenesis, 205
Myelin basic protein, 204

N

Nanos gene, 219
Neural cell adhesion molecule, 276
Neural induction, 275–277
Neuralization, 275–277
Nieuwkoop center, 257, 268
Noggin, 256–257, 262
  as competence modifier, 269
  as dorsalizing protein, 275
  as mesoderm inducer, 260
  in neuralization, 277
  in organizer formation, 273–274
Norse boat syndrome, 272
Notochord, 271
Nuclear equivalence theory, 149
Nuclear injection, 183
Nuclear transplantation, 199
  cell cycle, 161–166, 169
  chromosome damage, 163–164
  cytoplasm, 159–161, 169
  description, 147–149, 168–170, 178
  development, 178
  developmental capacity, 150–151, 159, 170
  developmental frequency, 158, 163, 167, 169
  donor cell, 151, 156–159, 167, 169
  egg activation, 164–166, 169–170
  embryo development, 156–159
  gene activity regulation, 155–156
  genomic totipotency, 167–168
  importance, 170, 199
  microsurgery, 148, 153
  and nucleus, 154–155, 160
  procedure, 151–153
  recipient cell, 153, 159–161, 169
  reprogramming, 155
  RNA synthesis, 151
  serial, 166–167
  stem cell, embryonic, 158–159
  and virus, 148
Nucleus, 150–151, 154–155, 160, 168–169

O

Ocean pout antifreeze protein gene, 193, 194–196
Oligosaccharide, 8–12, 24
Oncogene, 202–203
Oocyte, *see also* Animal–vegetal axis; Egg; Sea urchin Egg

cell cycle, 117–118
complement receptor, 49
cortex, 238, 240–241
cytoskeleton organization, 219–241
development, 103–104, 106–108, 117–137, 157, 167–168, 267–269
maturation
  gonadotropin influence, 123–124
  hormonal regulation, 137–138
  maturation-promoting factor, 127–127
  steroid influence, 125–126
maturational competence, 126–127
membrane formation, 110
mitochondrial mass formation, 232–233
nuclear organization, 108, 117–118, 226, 228
pigmentation, 218
polarity, 226
size, 108
spindle assembly, 241–244
stage, 229, 232–233, 235–232
ultraviolet radiation, 264
Xenopus laevis, 218–244
Organizer
  arc, 256, 268
  formation, 262–274, 273–274
  function, 254, 274–277
  gene, 262
  localization, 256, 257
Ovary, 104, 106–108

P

Pachytene microsporocyte, 128
Perivitelline space, 24
Pertussis toxin, 40, 41, 82
PH-30, 51–52
4β-Phorbol diesters, 33–34
Phosphatidylinositol, 64–65
Phosphatidylinositol 4,5-bisphosphate, 73, 84
Phosphoinositide, 272
Phospholipase C
  PLCβ, 82
  PLCτ, 82–83
Phosphorylation, 84
Pigmentation, 218, 235
Pintallavis, 256–257
Planar induction, 275–276
Pluripotency, 168
p40^{MO15}, 134
Polarity, species comparison, 235

Polarization, see Axis
Polymerase chain reaction, 187
Polyspermy block, 24–25, 41
P21^{ras}, 270
Pregnant mare serum gonadotropin, 113
Primary neural induction, 275–277
Primordial germ cell, 158–159
Procaine, 76
Promoter trap, 205
Pronucleus, 41, 42
Proteasome, 26S, 136–137, 138
Protein kinase, cGMP-dependent, 81, 86–87
Protein kinase C, 29, 33, 34, 277
Protein kinase II, calmodulin-dependent, 29

R

Raf-1 kinase, 272
Rana pipiens, 225
ras, 41, 270
Ratiometric fluorescence, 66
Receptor, see also specific type
  definition, 22
  and effector activation, 37
Receptor hypothesis, in egg activation, 37–44
Retrovirus, 189
RGD sequence, 50
rho, 41
Ring canal, 231
RNA
  antisense, 204
  in ectopic axis assay, 260–262
  localization, 266–267
  messenger, see mRNA
  synthesis, 151
RSV promotor, 201
Ruthenium red, 72, 76, 79–80
Ryanodine, 64, 71–73, 76–77
Ryanodine receptor, 30, 71–73, 79–80, 81
  in calcium regulation, 32–33, 75–78, 88

S

Salmon ovary, 107–108
Sea urchin egg
  activation, 42–43
  calcium changes during fertilization, 32–33, 65–71
  inositol 1,4,5-triphosphate receptor, 71–73
  ryanodine receptor, 71–73
  sperm–egg interaction, 24, 51–52, 88–89

Second messenger, 38, 41, 45–46
Second messenger system, 125
Sendai virus, 148, 153
Serial nuclear transplantation, 166–167
Signal transducer, 22
sp56, 10
Spawning, 180
Spemann's organizer, see Organizer
Sperm, see also Egg activation
  acrosome reaction, 33
  activation of egg receptor, 37–38
  activator, 85–87
  antisperm antibodies, 48–49
  aster, 263–264
  attachment to egg, 1–2, 4, 9–10, 47–51, 51–53
  and calcium induction, 26–27, 85–87
  complement system interaction, 48–49
  effects of fertilization, 25
  egg activation mechanism, 42–44, 83–84, 166
  egg interaction, 23–25, 24, 37, 50–52, 68
  egg membrane interaction, 23–25
  electroporation, 184–185
  fusion, 44–47, 51–52
  galactosyltransferase, 10
  and inositol 1,4,5-triphosphate, 74
  ligand, 50–51
  lysin, 85
  plasma membrane, 10
  polyspermy block, 24–25, 41
  protein, 46–47, 85
  receptor, 9, 13–14, 47–51, 64, 83, 166
  sp56, 10
  surface ligand, 51–53
  ZP3, 4, 10, 13–14
Sperm-derived factor, 44–47
S phase, 161–163, 162, 163
Spindle, 241–244, 242
src, 44
Stem cell, 204
  embryonic, 158–159
Sterilization, transgenic fish, 196, 199
Steroid, C_{212}, 119–120
SV40 promoter, 201
Synchronous ovary, 104

T

Talin, 223
T-cell antigen receptor, 84
Teratocarcinoma cell, 159

Testosterone, 111
Thapsigargin, 70–71
Thecal cell
  cAMP production, 122
  gonadotropin stimulation, 121–123, 123–124
  hormone production, 110–113
  in oocyte, 106–107
  two-cell type model, 137
Thimerosal, 87
Thymidine kinase, herpes simplex virus, 203
Totipotency, 157–158, 158, 168, 170
  nuclear transplantation, 168
Transcription, 155–156, 202
Transfection, 180, 183–186
Transfilter recombinant, 271
Transformation, 178
  germ-line, 195
Transforming growth factor β, 260
Transgene
  expression, 190–192, 200–203
  inheritance, 186–187, 189–190
  mosaicism, 189–190
Transgene marker, 205
Transgenic animal technology, see also Transgenic fish
  application, 194, 205–206
  description, 177–179, 203
Transgenic fish
  advantages, 179–180
  application, 192–196, 199–206
  cell lineage ablation, 203–204
  description, 178–179
  difficulties, 200
  DNA, 186–189, 200–202
  ecological issues, 196, 199
  gene targeting, 204, 205
  germ-line transformation, 195
  growth hormone, 192–195
  methodology, 180, 183–187
  problems, 193–194
  stability, 194
  sterilization, 196, 199
  transgene expression, 186–187, 189–192, 200–203, 205
  zebrafish as model, 199–201
Transplantation, nuclear, see Nuclear transplantation
Trichloroacetic acid, 220
Tubulin, 224, 229
  α-tubulin, 224
  γ-tubulin, 237–241, 241–242

Index

Tungsten particles, 185
Tyrosine, 42, 43
Tyrosine kinase, 42–44, 83–84
Tyrosine kinase-dependent pathway, 82–84

U

Ubiquitin, 136
Ultraviolet radiation, 264
Uniparental disomy, 167–168
Urodele embryo, 276

V

Vegetal cell, 257, 259, 268, 270
Vegetal cortical domain, 273
Vegetal hemisphere, 218
Vertical induction, 275–276
Vg1, 219
  in axis induction, 267, 269–270, 273
  in ectopic axis assay, 262
  as mesoderm inducer, 260, 269–270
  mRNA, 235–236, 241
  in organizer formation, 273
Vimentin, 220–222
Vinculin, 223
Vitellogenin, 108–110, 185–186
VYTHE peptide, 134

X

XCAT-2, 219
Xenopus laevis
  advantages for study, 258
  dorsoanterior axis specification, 257
  egg
    calcium release, 72
    inositol 1,4,5-triphosphate and calcium, 73–74

oocyte, 218–244
sperm-egg interaction, 257
Xwnt, 273
Xwnt-8, 262
  expression, 256–257, 271
  function, 260, 262, 269
  mesoderm induction, 260, 271–272

Y

Yolk, 109, 218, 235
Yolk nucleus, 232–233, 235

Z

Zebrafish, 199–200, 200–201, 204, 206
Zinc finger gene, 202
Zona pellucida, 23, 24
  characteristics, 4
  and fertilization, 1–3, 24–25, 52–53
  filament, 8, 13–14
  glycoprotein, 4
ZP2
  effect of inositol 1,4,5-triphosphate, 30
  sperm interaction, 23
  $ZP2_f$, 25, 30, 40, 41
ZP3
  characteristics, 2–3
  and fertilization, 4, 9, 10, 11–13, 23
  gene, 4–6
  glycopeptide, 9–10
  immunization with, 12
  inositol 1,4,5-triphosphate effect, 31
  and oligosaccharide, 8–11, 12
  protein kinase C effect, 33–34
  species comparison, 4–6, 6–8, 10–13, 14–15
  structure, 6–8, 11–13, 14–15
$ZP3_f$, 24–25, 33